Progress in Physics
Volume 12

A. Jaffe
D. Ruelle
series editors

Olivier Piguet Klaus Sibold

Renormalized Supersymmetry

The Perturbation Theory of $N = 1$ Supersymmetric Theories in Flat Space–Time

1986

Birkhäuser
Boston · Basel · Stuttgart

Olivier Piguet
Département de Physique Théorique
Université de Genève
1211 Genève
Switzerland

Klaus Sibold
Max-Planck-Institut für Physik
und Astrophysik
Werner-Heisenberg-Institut
für Physik
8000 München
Federal Republic of Germany

Library of Congress Cataloging in Publication Data
Piguet, O. (Olivier)
Renormalized supersymmetry.
(Progress in physics; vol. 12)
Bibliography: p.
Includes index.
1. Supersymmetry. 2. Renormalization (Physics)
3. Perturbation (Mathematics) I. Sibold, K, (Klaus)
II. Title. III. Series: Progress in physics
(Boston, Mass.); v. 12.
QC174.17.S9P54 1986 530.1'42 86-17118

CIP-Kurztitelaufnahme der Deutschen Bibliothek
Piguet, Olivier:
Renormalized supersymmetry : the perturbation
theory of N = 1 supersymmetr. theories in flat
space-time / O. Piguet ; K. Sibold. – Boston ;
Basel ; Stuttgart : Birkhäuser, 1986.
(Progress in physics ; Vol. 12)
ISBN 3-7643-3346-4
NE: Sibold, Klaus:; GT

Softcover reprint of the hardcover 1st edition 1986

ISBN-13: 978-1-4684-7328-5 e-ISBN-13: 978-1-4684-7326-1
DOI: 10.1007/978-1-4684-7326-1

Dedicated to our wives, Françoise and Elke

PREFACE

The present book grew out of lecture notes prepared for a "Cours du troisieme cycle de la Suisse Romande", 1983 in Lausanne. The original notes are considerably extended and brought up to date. In fact the book offers at many instances completely new derivations. Half-way between textbook and research monograph we believe it to be useful for students in elementary particle physics as well as for research workers in the realm of supersymmetry.

In writing the book we looked back not only on ten years of supersymmetry but also on ten years of our own life and work. We realize how deeply we are indebted to many friends and colleagues. Some shared our efforts, some helped and encouraged us, some provided the facilities to work. Their list comprises at least C. Becchi, S. Bedding, P. Breitenlohner, T.E. Clark, S. Ferrara, R. Gatto, M. Jacob, W. Lang, J.H. Lowenstein, D. Maison, H. Nicolai, J. Prentki, A. Rouet, H. Ruegg, M. Schweda, R. Stora, J. Wess, W. Zimmermann, B. Zumino. During the last ten years we had the privilege to work at CERN (Geneva), Département de Physique Théorique (University of Geneva), Institut für Theoretische Physik (University of Karlsruhe) and at the Max-Planck-Institut für Physik und Astrophysik (Munich) for which we are most grateful. Gratefully acknowledged is also the support we received by "the Swiss National Science Foundation" (O.P.), the "Deutsche Forschungsgemeinschaft" (Heisenberg-Fellowship; K.S.).

Last but not least we express our gratitude to Mrs. R. Jurgeleit for her heroic efforts in typing: she transformed a difficult manuscript into a pleasantly looking book.

Olivier Piguet
Département de Physique Théorique
Université de Genève
Geneva

Klaus Sibold
Max-Planck-Institut für Physik
und Astrophysik, Werner-Heisenberg-
Institut für Physik
Munich

February 1986

TABLE OF CONTENTS

INTRODUCTION

CHAPTER I

THE SUPERSYMMETRY ALGEBRA AND ITS REPRESENTATION OF FIELDS 1

Sect. 1		The supersymmetry algebra, superspace, superfields	1
Sect. 2		The superconformal algebra	9

CHAPTER II

SPECIFIC MODELS IN THE TREE APPROXIMATION 14

Sect. 3		Chiral models	15
	3.1	The free chiral field	15
	3.2	Interacting chiral fields	18
	3.3	The O'Raifeartaigh model	20
Sect. 4		Abelian gauge models	23
	4.1	Free abelian gauge fields	23
	4.2	SQED	33
	4.3	S'QED and S'QED'	37
Sect. 5		Non-abelian gauge models	41
	5.1	Non-abelian gauge transformations	41
	5.2	BRS-invariance	53
	5.3	General solution of the Slavnov identity	60
	5.4	Interpretation of the parameters a_k	68
	5.5	Gauge independence	72
	5.6	Summary of section 5	76
Sect. 6		Supercurrents	77
	6.1	Generalities	77
	6.2	Chiral models	80
	6.2.1	Massless Wess-Zumino model	81
	6.2.2	Massive Wess-Zumino model	84
	6.2.3	General chiral model	85
	6.3	Abelian gauge theory	87
	6.4	Non-abelian gauge theory	93
	6.5	Identification of component currents	104
	6.6	Superfield form of internal symmetry currents	109

CHAPTER III

PERTURBATION THEORY IN SUPERSPACE 112

Sect. 7		A simple example	112
Sect. 8		Feynman rules and power counting	119
Sect. 9		The subtraction scheme	128
Sect. 10		Normal products	143
Sect. 11		The action principle	152
Sect. 12		Symmetric operators	161

CHAPTER IV

RENORMALIZATION: HARD ANOMALIES 167

Sect.	13	Rigid symmetries	168
	13.1	Consistency conditions, the algebraic technique	168
	13.2	Symmetry breaking	170
	13.3	Supersymmetry	171
	13.3.1	Wess-Zumino model	174
	13.3.2	O'Raifeartaigh model	175
	13.3.3	SQED	177
	13.3.4	S'QED	178
	13.3.5	SYM	179
	13.4	Rigid gauge invariance	180
	13.5	R-invariance	181
	13.5.1	Wess-Zumino model	182
	13.5.2	O'Raifeartaigh model	184
	13.5.3	SQED	185
	13.5.4	S'QED	185
	13.5.5	SYM	185
Sect.	14	Abelian gauge invariance	186
	14.1	SQED	186
	14.2	S'QED	191
Sect.	15	Non-Abelian gauge invariance	193
	15.1	Statement of the problem	193
	15.2	The consistency condition	197
	15.3	Solution of the consistency condition: the anomaly	200
	15.4	The anomaly in the Slavnov-identity	207
Sect.	16	Renormalized Supercurrents	212
	16.1	The Wess-Zumino model	213
	16.1.1	The massless case	215
	16.1.2	The massive case	216
	16.1.3	Summary	222
	16.2	Supersymmetric QED	223
	16.2.1	Massless vector field	225
	16.2.2	Massive vector field	231
	16.2.3	Massive vector field, massless matter fields	234
	16.2.4	The gauge invariance of the supercurrent	236
	16.2.5	The non-renormalization of the axial anomaly	243
	16.3	Supersymmetric Yang-Mills theory	249
	16.3.1	General preparation	249
	16.3.2	The BRS-invariance of current and breaking	251
	16.3.3	Renormalized supercurrent and Callan-Symanzik equation	259
	16.3.4	The "conserved" supercurrent	263

CHAPTER V

RENORMALIZATION: SOFT ANOMALIES 267

Sect.	17	Mass generation - the O'Raifeartaigh model	267
Sect.	18	The off-shell infrared problem in SYM	278
	18.1	Statement of the problem. Tree approximation	278
	18.2	Higher orders: Solution of the cohomology	286

18.3	Higher orders: The absence of infrared anomalies and the Callan-Symanzik equation	293
18.4	Discussion of the result. Open questions	302
Appendix A	Notations, conventions and useful formulae	304
Appendix B	Generating functionals	312
Appendix C	$\mathcal{b}_0$-cohomology	317
Appendix D	Symmetric insertions and differential operators	329
Appendix E	Solution of some superfield constraints	332
References		337
Subject index		343

INTRODUCTION

After a decade of supersymmetry a thorough exposition and critical assessment of the results obtained so far does not seem to be inappropriate. The present book attempts to give such an account in the very limited and yet astonishingly rich area of renormalized perturbation theory of models which have only one supersymmetry and are defined in flat four-dimensional space-time. Hence, we exclude from our considerations supergravity and extended supersymmetries and even for simple supersymmetry we shall not dwell with the efforts of phenomenological applications. Since at the moment quite a few other reviews appeared in print (s. list of reviews) this restriction is at least economical. Our aim is the careful construction of higher orders in perturbation theory which permits a systematic search for anomalies. The algebraic technique which we use for establishing Ward identities avoids reference to any specific scheme of regularization and renormalization and is necessary for theories which do not admit an invariant regularization due to the occurrence of anomalies. In fact it provides an intrinsic formulation of the theories in question based on symmetry principles and yields an intrinsic definition of the anomalies which may break some of the postulated symmetries. Moreover this algebraic characterization has a meaning even beyond perturbation theory. It becomes clear this way that anomalies are genuine elements of the algebraic structure of the theory which simply happen not to be relevant in its classical approximation.

Let us now give an outline of the book. Chapters I, II provide a self-contained introduction into N = 1, rigid supersymmetry. Whereas chapter I deals with the superconformal algebra, its representation on superfields and with superspace, chapter II presents the tree approximation of all the models to be dealt with later on. In particular, an extensive discussion of the nonabelian gauge transformations in its most general form is presented, since this is mandatory for higher orders. Supercurrents are constructed, thus all symmetry currents of the models in supersymmetric form derived. This part of the book is ordered according to models. In the subsequent, the renormalization part, the ordering is according to symmetries, a fact which underlines our point of view that the symmetries ought to define the theory in question. The way to this is prepared by chapter III which develops perturbation theory in superspace and provides the concrete basis for all higher order calculations. Whereas the detailed (and cumbersome) description of a specific subtraction scheme is not essential for the main part of the subsequent chapters it is crucial for the actual proof of what everything afterwards is based upon: the action principle. The reader who is willing to accept the action principle as a fundamental theorem valid a priori or has its own proof need not rely on section 9. The study of anomalies presented in chapter IV and V is based on the action principle. In chapter IV we search for ultraviolet anomalies, indicating symmetry breaking by hard terms and establish as our two main results that there is none for supersymmetry and exactly the supersymmetric extension of the known one for the gauge invariance. The technique we apply, namely solving the algebraic consistency conditions, proves to be straightforward in the former and rather involved in the latter case. But, up to the present time, there is no other uniqueness proof available. The condition for absence of the anomaly in one loop (and then to all orders) is the usual one: restricting the representation of the matter fields. The existence of supercurrents to all orders for the models considered is proved in section 16. Its subsections follow again the order according to models and thus serve also as a convenient recapitulation of the relevant effective actions. The main result is that for all (massless) models two superconformal structures exist: one in which the R-current is not conserved, R-weights and dimensions are the anomalous ones and the superconformal anomalies form a chiral multiplet. In the other one the R-current is conserved (between physical states), R-weights and dimensions stay naive and the superconformal anomalies lie in a real vectorsuperfield. Chapter V deals with soft or infrared anomalies of which one type indicates mass generation and can be removed by a suitable redefinition of the perturbation series (expansion in $\sqrt{\hbar}\ \ln \hbar$). It is not specific for supersymmetric models but occurs also in others. The other type is a genuine supersymmetry problem arising when a local gauge invariance is present and requires the introduction of fields of canonical dimension zero. In the abelian case this problem can be circumvented and is therefore treated already in chapter IV, but in the nonabelian theory it has to be solved and we present a solution in chapter V.

Five appendices are devoted to additonal technical information.

A word has still to be said to the references. The wealth of literature on the subject rendered hopeless any attempt to be complete. We have either tried to find the earliest references on each topic or else have chosen those with which we are most familiar. This choice is, of course, a personal one and we apologize for all omissions.

The subject index is supposed to be complementary to the table of contents, hence both should be consulted when a specific topic is being sought.

CHAPTER I.

THE SUPERSYMMETRY ALGEBRA AND ITS REPRESENTATION ON FIELDS

1. The supersymmetry algebra, superspace, superfields

The characteristic ingredient which distinguishes supersymmetric quantum field theories from ordinary ones is the occurrence of spinoral charges $Q_\alpha^{(i)}$ amongst the symmetry generators, which embrace at least those from Poincaré invariance ($M_{\mu\nu}$, P_μ). The smallest number of them is thus one (range of i) and, in four-dimensional space-time, to which we restrict ourselves in the following, Q_α can furthermore be required to be a Weyl spinor ($\alpha = 1,2$):

$$[M_{\mu\nu}, Q_\alpha] = -\frac{1}{2}\sigma_{\mu\nu\alpha}{}^{\beta} Q_\beta \tag{1.1}$$

$$[M_{\mu\nu}, \bar{Q}^{\dot\alpha}] = \frac{1}{2}\bar{\sigma}_{\mu\nu}{}^{\dot\alpha}{}_{\dot\beta}\bar{Q}^{\dot\beta} \tag{1.2}$$

$$\bar{Q}_{\dot\alpha} = (Q_\alpha)^+ \tag{1.3}$$

(cf. App. A for conventions).

In addition we impose

$$[Q_\alpha , P_\mu] = [\bar{Q}_{\dot\alpha} , P_\mu] = 0 \tag{1.4}$$

and the key relation

$$\{Q_\alpha , \bar{Q}_{\dot\alpha}\} = 2\,\sigma^\mu_{\alpha\dot\alpha} P_\mu \tag{1.5}$$

together with

$$\{Q_\alpha , Q_\beta\} = \{Q_{\dot\alpha} , Q_{\dot\beta}\} = 0. \tag{1.6}$$

The Poincaré-algebra

$$[M_{\mu\nu} , P_\rho] = i\,(g_{\mu\rho} P_\nu - g_{\nu\rho} P_\mu) \tag{1.7}$$

$$[M_{\mu\nu} , M_{\rho\sigma}] = -i\,(g_{\mu\rho}M_{\nu\sigma} - g_{\mu\sigma}M_{\nu\rho} + g_{\nu\sigma}M_{\mu\rho} - g_{\nu\rho}M_{\mu\sigma}) \tag{1.8}$$

and (1.1) - (1.6) form the smallest algebra containing one spinorial generator : it is the N = 1, rigid (flat 4-dimensional space-time) supersymmetry algebra [I.1,2,3].

We now wish to represent this algebra on fields Φ i.e. we look for transformations $\delta_X\Phi$ such that

$$i\,[X,\Phi\,] = \delta_X\Phi \qquad X\in\{M,P,Q,\bar{Q}\} \tag{1.9}$$

and the application of any second generator does not result in a transformation outside of the set of given transformations i.e.

$$i\,[X_2,i[X_1,\Phi]] = i\,[X_2, \delta_{X_1}\Phi] \tag{1.10}$$

implies via Jacobi-identity

$$-i\,[\,i\,[X_1,X_2],\Phi] = \delta_{X_1}\,\delta_{X_2}\,\Phi - \delta_{X_2}\,\delta_{X_1}\,\Phi \tag{1.11}$$

and for

$$[X_1,X_2] = i\,X_3 \tag{1.12}$$

$$\delta_{X_3}\,\Phi = [\delta_{X_1}, \delta_{X_2}]\,\Phi \tag{1.13}$$

hence

$$\delta_{X_3} = [\delta_{X_1}, \delta_{X_2}] \tag{1.14}$$

independent of the field Φ.*

The most concise way of deriving the transformation law is to use the "group" which is obtained by exponentiating the generators $P, Q, \bar{Q}$ [I.4,5]. For this purpose one introduces in addition to commuting parameters a_μ characterising a translation in space-time, anti-commuting spinorial parameters ξ_α, $\bar{\xi}_{\dot{\alpha}}$ labelling supersymmetry transformations. (Rules for manipulating such objects are found in App. A.). A group element may then be written as

$$G(a,\xi,\bar{\xi}) = e^{i(a^\mu P_\mu + \xi^\alpha Q_\alpha + \bar{\xi}_{\dot{\alpha}} \bar{Q}^{\dot{\alpha}})} \tag{1.15}$$

and the action from the left of $G(a,\xi,\bar{\xi})$ on $G(x,\theta,\bar{\theta})$ can be worked out with the simplified Hausdorff formula

$$e^A\, e^B = e^{A+B+\frac{1}{2}[A,B]} \tag{1.16}$$

and the algebra (1.4) (1.5) (1.6):

$$G(a,\xi,\bar{\xi})G(X,\theta,\bar{\theta}) = G(x + a + i\xi\sigma\bar{\theta} - i\theta\sigma\bar{\xi},\ \theta+\xi,\ \bar{\theta}+\bar{\xi}). \tag{1.17}$$

Group element multiplication has caused a motion in parameter space:

$$(X,\theta,\bar{\theta}) \rightarrow (x + a + i\xi\sigma\bar{\theta} - i\theta\sigma\bar{\xi},\ \theta+\xi,\ \bar{\theta}+\bar{\xi}) \tag{1.18}$$

which we may reproduce by differential operators acting on functions Φ defined on the parameter space:

Let $\Phi = \Phi(x,\theta,\bar{\theta})$

---

* One should note that not any arbitrary set of fields will do the job: for closure of the algebra on a specific set one might be forced to use equations of motion obeyed by the fields. This happens in fact in extended supersymmetries.

then $P_\mu \Phi = - i \partial_\mu \Phi$

$$Q_\alpha \Phi = - i \left(\frac{\partial}{\partial\theta^\alpha} + i \sigma_{\alpha\dot{\alpha}} \bar{\theta}^{\dot{\alpha}} \partial\right) \Phi$$

$$\bar{Q}_{\dot{\alpha}} \Phi = - i \left(- \frac{\partial}{\partial\bar{\theta}^{\dot{\alpha}}} - i \theta^\alpha \sigma_{\alpha\dot{\alpha}} \partial\right) \Phi \tag{1.19}$$

will generate the motion:

$$e^{i(aP+\xi Q+\bar{\xi}\bar{Q})} \Phi(x,\theta,\bar{\theta}) = \Phi(x+a+i\xi\sigma\bar{\theta} - i\theta\sigma\bar{\xi},\ \theta+\xi,\ \bar{\theta}+\bar{\xi}) \tag{1.20}$$

and the differential operators

$$\delta^P_\mu \Phi = \partial_\mu \Phi$$

$$\delta^Q_\alpha \Phi = \left(\frac{\partial}{\partial\theta^\alpha} + i \sigma_{\alpha\dot{\alpha}} \bar{\theta}^{\dot{\alpha}} \partial\right) \Phi \tag{1.21}$$

$$\delta^{\bar{Q}}_{\dot{\alpha}} \Phi = \left(- \frac{\partial}{\partial\bar{\theta}^{\dot{\alpha}}} - i \theta^\alpha \sigma_{\alpha\dot{\alpha}} \partial\right) \Phi$$

are what we looked for: a representation of the algebra on fields.

$$\{\delta^Q_\alpha , \delta^{\bar{Q}}_{\dot{\alpha}}\} = - 2 i \sigma^\mu_{\alpha\dot{\alpha}} \delta^P_\mu \tag{1.22}$$

$$\{\delta^Q_\alpha , \delta^Q_\beta\} = \{\delta^{\bar{Q}}_{\dot{\alpha}} , \delta^{\bar{Q}}_{\dot{\beta}}\} = \left[\delta^Q_\alpha , \delta^P_\mu\right] = \left[\delta^{\bar{Q}}_{\dot{\alpha}} , \delta^P_\mu\right] = 0 \tag{1.23}$$

Analogously one finds

$$\delta^M_{\mu\nu} = (x_\mu \partial_\nu - x_\nu \partial_\mu - \frac{i}{2} \theta\sigma_{\mu\nu}\partial_\theta + \frac{i}{2} \bar{\theta}\bar{\sigma}_{\mu\nu}\partial_{\bar{\theta}})\Phi \tag{1.24}$$

These differential operators are in accord with the operator interpretation (1.9).

Identifying now the parameter space of translations with usual space-time we see that we may interpret in the same vein the fermionic parameters with additional fermionic coordinates, i.e $(x_\mu, \theta_\alpha, \theta_{\dot{\alpha}})$ are the coordinates of a point in an enlarged space: the superspace

[I.4]. Any function $\Phi(x,\theta,\bar{\theta})$ transforming according to (1.9) (1.20) is called a superfield. Since the θ's anticommute, any superfield can be expanded in θ and yields a finite series

$$\Phi(x,\theta,\bar{\theta}) = \Phi^{(0.0)}(x) + \theta^{\alpha}\Phi_{\alpha}^{(1,0)}(x) + \ldots + \theta^2\bar{\theta}^2\Phi^{(2,2)}(x)$$

$$\equiv \sum_{(\omega)=0}^{\Omega} (\theta)^{(\omega)}\Phi^{(\omega)}(x) \tag{1.25}$$

where the $\Phi^{(\omega)}(x)$ are ordinary fields - the "component"-fields. The equation

$$i\,[Q_{\alpha},\Phi] = \delta^{Q}_{\alpha}\,\Phi \tag{1.26}$$

provides also the supersymmetry transformations for the components on equating equal θ-powers on both sides of the equations. The spin-statistics relations will be maintained by adopting the rule that the θ's anticommute with every spinor (including Q_{α}, $\bar{Q}_{\dot{\alpha}}$) and commute with every bosonic variable. Before going into more detail of this subject we have to note that the general superfield $\Phi(x,\theta,\theta)$ is not irreducible under the supersymmetry transformations. Let us look back at (1.15) and (1.17) from which we eventually abstracted the transformation law. Action from the left yielded differential operators δ^{Q}_{α} , $\delta^{\bar{Q}}_{\dot{\alpha}}$ representing field variations, action from the right should lead to differential operators which can be expected to have good covariance as well [I.5]. So, starting from

$$G(x,\theta,\bar{\theta})\;G(a,\xi,\bar{\xi}) = G(x+a-i\xi\sigma\bar{\theta}+i\theta\sigma\bar{\xi},\;\theta+\xi,\;\bar{\theta}+\bar{\xi}) \tag{1.27}$$

we define

$$D_{\alpha} \equiv \frac{\partial}{\partial\theta^{\alpha}} - i\,\sigma_{\alpha\dot{\alpha}}\bar{\theta}^{\dot{\alpha}}\partial$$

$$\bar{D}_{\dot{\alpha}} \equiv -\frac{\partial}{\theta\bar{\partial}^{\dot{\alpha}}} + i\theta^{\alpha}\sigma_{\alpha\dot{\alpha}}\,\partial \tag{1.28}$$

and see that

$$\{D_\alpha, \bar{D}_{\dot\alpha}\} = 2i\sigma^\mu \partial_\mu \tag{1.29}$$

$$\{D_\alpha, D_\beta\} = \{\bar{D}_{\dot\alpha}, \bar{D}_{\dot\beta}\} = 0 \tag{1.30}$$

(hence $D_\alpha D_\beta D_\gamma \equiv 0$).

But we also find

$$\{D_\alpha, \delta^{\bar Q}_{\dot\alpha}\} = \{D_\alpha, \delta^{Q}_{\beta}\} = 0 \tag{1.31}$$

(same for $\bar{D}_{\dot\alpha}$) hence D_α, $\bar{D}_{\dot\alpha}$ are covariant under supersymmetry transformations: if Φ was a superfield $D_\alpha \Phi$ also is one (actually one with a spinor index). This property permits us to constrain superfields covariantly in superspace*:

$D_\alpha \Phi = 0$ $\qquad$ Φ is called anti-chiral $\qquad$ (1.32)

$DD\,\Phi = 0$ $\qquad$ Φ is called linear $\qquad$ (1.33)

$\bar{D}_{\dot\alpha} \Phi = 0$ $\qquad$ Φ is called chiral $\qquad$ (1.34)

$\overline{DD}\,\Phi = 0$ $\qquad$ Φ is called linear $\qquad$ (1.35)

Also, since $\xi\delta^Q + \bar\xi\delta^{\bar Q}$ is real, one may impose

$\Phi(x,\theta,\bar\theta) = \bar\Phi(x,\theta,\bar\theta)$ $\qquad$ Φ is a real superfield. $\qquad$ (1.36)

The actual solution of the constraints and the derivation of component transformations is now greatly simplified [I.5] by observing that a general group element can be represented by not only (1.15) but also by

---

* Note that constraints must not yield differential equations in x-space. Those given here do not.

$$G_1(x,\theta,\bar\theta) = e^{i(xP + \theta Q)}\, e^{i\bar\theta\bar Q} \tag{1.37}$$

$$G_2(x,\theta,\bar\theta) = e^{i(xP + \bar\theta\bar Q)}\, e^{i\theta Q} \tag{1.38}$$

Left action by $G(a,\xi,\bar\xi)$ yields

$$G(a,\xi,\bar\xi)G_1(x,\theta,\bar\theta) = G_1(x+a-2i\theta\sigma\bar\xi,\ \theta+\xi,\ \bar\theta+\bar\xi) \tag{1.39}$$

$$G(a,\xi,\bar\xi)G_2(x,\theta,\bar\theta) = G_2(x+a+2i\xi\sigma\bar\theta,\ \theta+\xi,\ \bar\theta+\bar\xi) \tag{1.40}$$

Abstracting again for superfields we see that in addition to (1.19) we have two new "types" of superfield, related to each other by

$$\Phi(x,\theta,\bar\theta) = \Phi_1(x-i\theta\sigma\bar\theta,\theta,\bar\theta) = \Phi_2(x+i\theta\sigma\bar\theta,\theta,\bar\theta) \tag{1.41}$$

an operation called "shift".

Accordingly supersymmetry transformations and covariant derivatives change:

$$(\delta^Q_\alpha\,\Phi)_1 = \frac{\partial}{\partial\theta^\alpha}\Phi_1 \qquad (D_\alpha\Phi)_1 = \left(\frac{\partial}{\partial\theta^\alpha} - 2i\sigma_{\alpha\dot\alpha}\bar\theta^{\dot\alpha}\partial\right)\Phi_1$$

$$(\delta^{\bar Q}_{\dot\alpha}\,\Phi)_1 = \left(-\frac{\partial}{\partial\bar\theta^{\dot\alpha}} - 2i\theta^\alpha\sigma_{\alpha\dot\alpha}\partial\right)\Phi_1 \qquad (\bar D_{\dot\alpha}\Phi)_1 = -\frac{\partial}{\partial\bar\theta^{\dot\alpha}}\Phi_1 \tag{1.42}$$

$$(\delta^Q_\alpha\,\Phi)_2 = \left(\frac{\partial}{\partial\theta^\alpha} + 2i\sigma_{\alpha\dot\alpha}\bar\theta^{\dot\alpha}\partial\right)\Phi_2 \qquad (D_\alpha\Phi)_2 = -\frac{\partial}{\partial\theta^\alpha}\Phi_2$$

$$(\delta^{\bar Q}_{\dot\alpha}\,\Phi)_2 = -\frac{\partial}{\partial\bar\theta^{\dot\alpha}}\Phi_2 \qquad (\bar D_{\dot\alpha}\Phi)_2 = \left(-\frac{\partial}{\partial\bar\theta^{\dot\alpha}} + 2i\theta^\alpha\sigma_{\alpha\dot\alpha}\partial\right)\Phi_2$$

(1.43)

Written in the 1 basis (1.33) therefore reads

$$-\frac{\partial}{\partial\bar\theta^{\dot\alpha}}\Phi_1 = 0 \tag{1.44}$$

i.e. Φ_1 has to be independent of $\bar{\theta}$. The general solution - the chiral field - we write as*

$$A_1 = A + \theta^{\alpha}\psi_{\alpha} + \theta^{\alpha}\theta_{\alpha} F \tag{1.45}$$

and $(\delta_{\alpha}^{Q}\Phi)_1$ immediately says for the components:

$$\begin{aligned}
\delta_{\alpha}^{Q} A &= \psi_{\alpha} & \delta_{\dot\alpha}^{\bar{Q}} A &= 0 \\
\delta_{\alpha}^{Q} \psi_{\beta} &= -2\varepsilon_{\alpha\beta} F & \delta_{\dot\alpha}^{\bar{Q}} \psi_{\alpha} &= 2i\sigma^{\mu}_{\alpha\dot\alpha}\partial_{\mu} A \\
\delta_{\alpha}^{Q} F &= 0 & \delta_{\dot\alpha}^{\bar{Q}} F &= i\,\partial_{\mu}\psi^{\alpha}\sigma^{\mu}_{\alpha\dot\alpha}
\end{aligned} \tag{1.46}$$

Analogously the transformation laws of the anti-chiral field and of the real superfield can be found [I.3,5], they are listed in App. A.

Superfields also facilitate greatly the construction of composite objects transforming covariantly: The sum and the product of superfields of one and the same type is a superfield (of this type). If one wants to multiply superfields of different types one has first to shift the fields to one common type. Let us as an example consider a chiral field A and its complex conjugate the anti-chiral field $\bar{A}$. Powers A^n, $\bar{A}^n$ are chiral, resp. anti-chiral fields. A product of A with $\bar{A}$ can e.g. be formed in the real basis (type Φ) by

$$A(x,\theta,\bar{\theta}) = A_1(x-i\theta\sigma\bar{\theta},\theta,\bar{\theta}) = e^{-i\theta\sigma\bar{\theta}\partial} A_1(x,\theta) \tag{1.47}$$

$$\bar{A}(x,\theta,\bar{\theta}) = \bar{A}_2(x+i\theta\sigma\bar{\theta},\theta,\bar{\theta}) = e^{+i\theta\sigma\bar{\theta}\partial} \bar{A}_2(x,\bar{\theta}) \tag{1.48}$$

i.e.

$$\Phi = e^{-i\theta\sigma\bar{\theta}\partial} A_1(x,\theta)\; e^{+i\theta\sigma\bar{\theta}\partial} \bar{A}_2(x,\bar{\theta}) \tag{1.49}$$

transforms as a superfield of type Φ.

---

* Unless confusion arises we shall denote by the same letter A chiral superfields and their lowest component

2. The superconformal algebra

Just as the Poincaré algebra (1.7) (1.8) is generalized to the conformal algebra by taking into account the generators of dilatations, D, and those of special conformal transformations, K_μ, one may ask what generalization of the supersymmetry algebra (1.1)-(1.8) is obtained by doing so. The answer, in the form of variations on component fields, has been given in [I.3]. It turns out that not only a second class of spinorial generators is needed, which one might have expected, but in order to obtain closure of the algebra one has to introduce a chiral transformation in addition. This result may be understood as "grading" the algebra U(2,2) = SU(2,2) x U(1) i.e. replacing some of the commutators by anti-commutators. Let us now write down in the conventions of [I.6] the algebra.

$$\begin{aligned}
&[M_{\mu\nu}, M_{\rho\sigma}] = -i(g_{\mu\rho}M_{\nu\sigma} - g_{\mu\sigma}M_{\nu\rho} + g_{\nu\sigma}M_{\mu\rho} - g_{\nu\rho}M_{\mu\sigma})\\
&[M_{\mu\nu}, P_\lambda] = i(P_\mu g_{\nu\lambda} - P_\nu g_{\mu\lambda})\\
&[M_{\mu\nu}, K_\lambda] = i(K_\mu g_{\nu\lambda} - K_\nu g_{\mu\lambda})\\
&[D, P_\mu] = -iP_\mu \qquad [D, K_\mu] = +iK_\mu \qquad [P_\mu, K_\nu] = 2i(g_{\mu\nu}D - M_{\mu\nu})\\
&[P_\mu, P_\nu] = [K_\mu, K_\nu] = [D, M_{\mu\nu}] = 0
\end{aligned} \tag{2.1}$$

(2.1) comprises the subalgebra of the conformal generators. M,P,K,D denote Lorentz transformations, translations, special conformal transformations, dilatations respectively. Another subalgebra of primary importance below is formed by the above mentioned chiral transformations, called R, and the supersymmetry algebra:

$$\begin{aligned}
&\{Q_\alpha, \bar{Q}_{\dot\alpha}\} = 2\,\sigma^\mu_{\alpha\dot\alpha}\,P_\mu \qquad &&\{Q_\alpha, Q_\beta\} = \{\bar{Q}_{\dot\alpha}, \bar{Q}_{\dot\beta}\} = 0\\
&[M^{\mu\nu}, Q_\alpha] = -\tfrac{1}{2}\,\sigma^{\mu\nu}_\alpha{}^\beta Q_\beta \qquad &&[M^{\mu\nu}, \bar{Q}_{\dot\alpha}] = -\tfrac{1}{2}\,\bar\sigma^{\mu\nu}_{\dot\alpha\dot\beta}\,\bar{Q}^{\dot\beta}\\
&[P^\mu, Q_\alpha] = 0 \qquad &&[P^\mu, \bar{Q}_{\dot\alpha}] = 0\\
&[R, M_{\mu\nu}] = [R, R_\mu] &&= [R, R] = 0\\
&[R, Q_\alpha] = Q_\alpha \qquad &&[R, \bar{Q}_{\dot\alpha}] = -\bar{Q}_{\dot\alpha}
\end{aligned} \tag{2.2}$$

For the full superconformal algebra another spinorial generator S "special supersymmetry transformations" is added to the above system:

$$[M^{\mu\nu}, S_\alpha] = -\frac{1}{2}\sigma^{\mu\nu\beta}_{\alpha} S_\beta \qquad [M^{\mu\nu}, \bar{S}_{\dot\alpha}] = -\frac{1}{2}\bar{\sigma}^{\mu\nu}_{\dot\alpha\dot\beta}\bar{S}^{\dot\beta}$$

$$\{S_\alpha, \bar{S}_{\dot\alpha}\} = 2\sigma^{\mu}_{\alpha\dot\alpha} K_\mu \qquad \{S_\alpha, S_\beta\} = \{\bar{S}_{\dot\alpha}, \bar{S}_{\dot\beta}\} = 0$$

$$[D, Q_\alpha] = -\frac{i}{2} Q_\alpha \qquad [D, \bar{Q}_{\dot\alpha}] = -\frac{i}{2}\bar{Q}_{\dot\alpha}$$

$$[D, S_\alpha] = +\frac{i}{2} S_\alpha \qquad [D, \bar{S}_{\dot\alpha}] = +\frac{i}{2}\bar{S}_{\dot\alpha}$$

$$[R, S_\alpha] = -S_\alpha \qquad [R, \bar{S}_{\dot\alpha}] = +\bar{S}_{\dot\alpha}$$

$$[R, D] = [R, K_\mu] = 0 \tag{2.3}$$

$$[K^\mu, Q_\alpha] = -i\,\sigma^{\mu}_{\alpha\dot\alpha}\bar{S}^{\dot\alpha} \qquad [K^\mu, \bar{Q}_{\dot\alpha}] = -i\,S^\alpha \sigma^{\mu}_{\alpha\dot\alpha}$$

$$[K^\mu, S_\alpha] = [K^\mu, \bar{S}_{\dot\alpha}] = 0$$

$$[P^\mu, S_\alpha] = -i\,\sigma^{\mu}_{\alpha\dot\alpha}\bar{Q}^{\dot\alpha} \qquad [P^\mu, \bar{S}_{\dot\alpha}] = -i\,Q^\alpha\sigma^{\mu}_{\alpha\dot\alpha}$$

$$\{Q_\alpha, \bar{S}_{\dot\alpha}\} = \{\bar{Q}_{\dot\alpha}, S_\alpha\} = 0$$

$$\{Q_\alpha, S_\beta\} = +i\,(\sigma^{\mu}_{\alpha\beta} M_{\mu\nu} + 2i\,\epsilon_{\alpha\beta} D - 3\,\epsilon_{\alpha\beta} R)$$

$$\{\bar{Q}_{\dot\alpha}, \bar{S}_{\dot\beta}\} = -i\,(\bar{\sigma}^{\mu\nu}_{\dot\alpha\dot\beta} M_{\mu\nu} - 2i\,\epsilon_{\dot\alpha\dot\beta} D - 3\,\epsilon_{\dot\alpha\dot\beta} R)$$

It is the anti-commutator of Q_α with S_β which enforces the introduction of the R-transformations. For the subalgebra (2.2) it plays only the role of an automorphism, which will be permitted for many systems as soon as they are supersymmetric.

Extending the method of section 1, one can represent the generators of the superconformal algebra as differential operators on superfields. The result of the corresponding analysis is displayed in the appendix A (A.28), (A.29). It is again noteworthy that for chiral fields their dilatational weight (naive dimension) has to be related to their R-weight if the superconformal algebra is required to close on such fields.

In analogy to the behaviour under Lorentz transformations one may ask now whether not the system of charges might have a kind of covariance under supersymmetry. After all, there are many non-vanishing (anti-) commutators of Q_α with the other generators. E.g. we are looking for functions $\hat{Q} = \hat{Q}(Q^i, \theta, \theta)$ which transform as (x-independent) superfields

$$i[Q_\alpha, \hat{Q}] = \frac{\partial}{\partial\theta^\alpha}\hat{Q} \qquad i[\bar{Q}_{\dot\alpha}, \hat{Q}] = -\frac{\partial}{\partial\bar{\theta}^{\dot\alpha}}\hat{Q} \tag{2.4}$$

Indeed,

$$\hat{R} = R - i\,\theta^\alpha Q_\alpha + i\,\bar{\theta}_{\dot\alpha}\bar{Q}^{\dot\alpha} - 2\,\theta\sigma^\mu\bar{\theta}P_\mu \tag{2.5}$$

is a superfield in this sense. A somewhat weaker requirement has been introduced in [I.6]: we shall call a function $\hat{Q}$ a <u>quasi-superfield</u> if it satisfies

$$i[\theta^\alpha Q_\alpha + \bar{\theta}_{\dot\alpha}\bar{Q}^{\dot\alpha}, \hat{Q}] = (\theta^\alpha\frac{\partial}{\partial\theta^\alpha} - \bar{\theta}_{\dot\alpha}\frac{\partial}{\partial\bar{\theta}_{\dot\alpha}})\,\hat{Q} \tag{2.6}$$

i.e. we perform a combination of two very specific supersymmetry transformations, namely those with parameters $\theta,\bar{\theta}$. But now one can easily convince oneself that the solution of (2.6) is given by

$$\hat{Q} = e^{i(\theta^\alpha Q_\alpha + \bar{\theta}_{\dot\alpha}\bar{Q}^{\dot\alpha})}\;Q\;e^{-i(\theta^\alpha Q_\alpha + \bar{\theta}_{\dot\alpha}\bar{Q}^{\dot\alpha})} \tag{2.7}$$

which means that any generator Q gives rise to a quasi-superfield $\hat{Q}$ by "boosting" it with a supersymmetry transformation of special parameters θ and $\bar{\theta}$. As a consequence we find that the algebra of the $\hat{Q}$'s is the same as that of the Q's but its members have now a definite transformation law under supersymmetry. If we have established for a system that it is supersymmetric we have immediately relations amongst the superconformal generators which follow from (2.7). In perturbation theory, for instance, this will lead to relations amongst anomalies which beset some of the superconformal transformations.

Let us give an explicit list of all the $\hat{Q}$'s. Since below the charges Q will be represented as Ward-identity operators we shall use the letter W with sub- and superscripts to denote the generators and their Lorentz-indices.

$$i[\theta^{\alpha} W^{Q}_{\alpha} + \bar{\theta}_{\dot{\alpha}}\bar{W}^{\dot{\alpha}}_{Q}, \hat{W}^{A}] = (\theta \frac{\partial}{\partial\theta} - \bar{\theta}\frac{\partial}{\partial\bar{\theta}})\hat{W}^{A} \tag{2.8}$$

for $A \in \{M,P,K,D,Q,\bar{Q},R,S,\bar{S}\}$

$$\begin{aligned}
\hat{W}^{P}_{\mu} &= W^{P}_{\mu} \\
\hat{W}^{Q}_{\alpha} &= W^{Q}_{\alpha} - 2i(\sigma^{\mu}\bar{\theta})_{\alpha} W^{P}_{\mu} \\
\hat{\bar{W}}^{Q}_{\dot{\alpha}} &= \bar{W}^{Q}_{\dot{\alpha}} + 2i(\theta\sigma^{\mu})_{\dot{\alpha}} W^{P}_{\mu} \\
\hat{W}^{R} &= W^{R} - i\theta W^{Q} + i\bar{\theta}\bar{W}^{Q} - 2\,\theta\sigma^{\mu}\bar{\theta}\, W^{P}_{\mu} \\
\hat{W}^{M}_{\mu\nu} &= W^{M}_{\mu\nu} + \frac{i}{2}\theta\sigma_{\mu\nu}W^{Q} + \frac{i}{2}\bar{\theta}\bar{\sigma}_{\mu\nu}\bar{W}^{Q} - \frac{1}{2}\epsilon_{\mu\nu\rho\lambda}\theta\sigma^{\rho}\bar{\theta}\, W^{P\lambda} \\
\hat{W}^{D} &= W^{D} - \frac{1}{2}\theta W^{Q} - \frac{1}{2}\bar{\theta}\bar{W}^{Q}
\end{aligned} \tag{2.9}$$

$$\hat{W}^K_\mu = W^K_\mu - \theta\sigma_\mu\bar{W}^S - W^S\sigma_\mu\bar{\theta} + \theta\sigma^\nu\bar{\theta}(3g_{\mu\nu}W^R + \epsilon_{\mu\nu\rho\sigma}\, W^{M\rho\sigma})$$

$$- \;\; i\bar{\theta}^2\theta\sigma_\mu\,\bar{W}^Q + i\, W^Q\,\sigma_\mu\bar{\theta}\theta^2 - \theta^2\bar{\theta}^2\, W^P_\mu$$

$$\hat{W}^S_\alpha = W^S_\alpha - \theta^\beta(\sigma^\mu_{\alpha\beta}\, W^M_{\mu\nu} + 3\,\epsilon_{\alpha\beta}\, W^R - 2i\,\epsilon_{\alpha\beta}\, W^D) \qquad (2.10)$$

$$- 2i\;\theta^2\, W^Q_\alpha - i\;\theta\sigma^\mu\bar{\theta}(\sigma_\mu\bar{W}^Q)_\alpha - 2\theta^2(\sigma^\mu\bar{\theta})_\alpha\, W^P_\mu$$

$$\hat{\bar{W}}^S_{\dot{\alpha}} = \bar{W}^S_{\dot{\alpha}} - \bar{\theta}^{\dot{\beta}}(\bar{\sigma}^\mu_{\dot{\alpha}\dot{\beta}}\, W^M_{\mu\nu} + 3\,\epsilon_{\dot{\alpha}\dot{\beta}}\, W^R + 2i\,\epsilon_{\dot{\alpha}\dot{\beta}}\, W^D)$$

$$+ 2i\;\bar{\theta}^2\,\bar{W}^Q_{\dot{\alpha}} + i\;\theta\sigma^\mu\bar{\theta}(W^Q\sigma_\mu)_{\dot{\alpha}} - 2\bar{\theta}^2(\theta\sigma^\mu)_{\dot{\alpha}}\, W^P_\mu$$

CHAPTER II.

SPECIFIC MODELS IN THE TREE APPROXIMATION

In this chapter we introduce the models which will be discussed systematically in the sequel. A short account of the free theories is followed by the study of the tree approximation. Since our aim is eventually a treatment to all orders in perturbation theory we expose the material in this lowest order in a way which is best suited for the recursive extension to higher orders.

The aim is to construct a Lagrangian field theory which is supersymmetric and may or may not have additional symmetries. Looking into the supersymmetry transformation laws (1.26) for any superfield

$$
\begin{aligned}
i\,[Q_\alpha, \Phi] &= \delta^Q_\alpha \Phi \equiv \left(\frac{\partial}{\partial\theta^\alpha} + i\,\sigma_{\alpha\dot\alpha}\bar\theta^{\dot\alpha}\partial\right)\Phi \\
i\,[\bar Q_{\dot\alpha}, \Phi] &= \delta^{\bar Q}_{\dot\alpha}\,\Phi \equiv \left(-\frac{\partial}{\partial\bar\theta^{\dot\alpha}} - i\,\theta^\alpha\sigma_{\alpha\dot\alpha}\partial\right)\Phi
\end{aligned}
\qquad (II.1)
$$

$$
i\,[Q_\alpha, A_1] = \frac{\partial}{\partial\theta^\alpha} A_1 \qquad\qquad i\,[\bar Q_{\dot\alpha}, A_1] = (-\,2i\,\theta^\alpha\sigma_{\alpha\dot\alpha}\,\partial)\,A_1
$$

it is clear that the highest θ-component of any superfield transforms into a total derivative under supersymmetry transformation hence may give rise to an invariant action. For a chiral field A, the highest θ-component is θ^2 (in the chiral basis (1.44)), hence double differentiation with respect to θ picks out its component field. Since the difference between $\partial/\partial\theta^\alpha$ and D_α is a total x-divergence we can project covariantly under the space-time integral:

$$
\int dS\, A_1 \equiv \int d^4x\, D^\alpha D_\alpha\, A_1 = \int d^4x\, \frac{\partial}{\partial\theta_\alpha}\frac{\partial}{\partial\theta^\alpha} A_1 = \int dS\, A \qquad (II.2)^*
$$

---

* (Anti-) chiral fields are sometimes called scalar fields, real superfields are called vector (super) fields, hence dS scalar measure and dV vector measure.

(The last equality is due to the shift.)

Following Berezin [II.1] we may also call this integration over the variables $\boldsymbol{\theta}$.

Analogously we have for an anti-chiral field $\bar{A}_2$ (highest component $\bar{\theta}^2$)

$$\int d\bar{S}\ \bar{A}_2 \equiv \int d^4x\ \bar{D}_{\dot{\alpha}}\bar{D}^{\dot{\alpha}}\ \bar{A}_2 = \int d^4x \frac{\partial}{\delta\bar{\theta}^{\dot{\alpha}}}\frac{\partial}{\delta\bar{\theta}_{\dot{\alpha}}}\bar{A}_2 = \int d\bar{S}\ \bar{A} \qquad (II.3)^*$$

and for a real superfield Φ (highest component $\theta^2\bar{\theta}^2$):

$$\int dV\ \Phi \equiv \int d^4x\ DD\ \overline{DD}\ \Phi \qquad (II.4)^*$$

At our disposal for the integrands A, $\bar{A}$, Φ we have elementary fields to be multiplied in a manner which represents the type (cf section 1) and the covariant derivatives D, $\bar{D}$. Out of these ingredients we will have to form the supersymmetric actions which will (or sometimes will not) have other desired properties.

3. Chiral models

3.1 The free chiral field

The power A^n of an elementary chiral field**

$$A = A + \theta\psi + \theta^2\ F \qquad (3.1)$$

(A complex scalar, ψ Weyl spinor, F complex scalar) will never contain derivatives, hence $\int dS\ A^n$ will at most produce mass or interaction terms. But the correctly built product of A with its complex conjugate antichiral field $\bar{A}$ (1.49)

---

* cf preceding footnote

** The subindices 1,2 denoting the type will now be dropped, if no confusion can arise.

$$\Phi = \bar{A}(x,\theta,\bar{\theta})\, A(x,\theta,\bar{\theta}) = e^{i\theta\sigma\bar{\theta}\partial}\, \bar{A}_2(x,\bar{\theta})\, e^{-i\theta\sigma\bar{\theta}\partial}\, A_1(x,\theta) \tag{3.2}$$

does contain derivatives due to the shift necessary for having multiplied fields of the same type. Indeed, one finds [I.3,5]:

$$\Gamma^{(o)}_{kin} = \frac{1}{16}\int dV\, \bar{A}\, A = \int dx\, (\partial\bar{A}\, \partial A + \frac{1}{2}\bar{\psi}\bar{\sigma}\partial\psi + \bar{F}F) \tag{3.3}$$

i.e. a good kinetic action for a massless scalar field A (canonical dimension 1), a massless Weyl-spinor ψ (dim. 3/2) and a complex scalar F (dim. 2). Since there are no time-derivatives on F it does not propagate and is an auxiliary field.

$$A^2 = --- + \theta^2\, (2AF - \frac{1}{2}\psi\psi), \tag{3.4}$$

$$\begin{aligned} \Gamma^{(o)}_m &= \frac{1}{8} m\, (\int dS\, A^2 + \int d\bar{S}\, \bar{A}^2) \\ &= -\frac{m}{2}\int dx\, (2(AF + \bar{A}\bar{F}) - \frac{1}{2}(\psi\psi + \bar{\psi}\bar{\psi})) \end{aligned} \tag{3.5}$$

yields obviously a mass term for the spinor ψ, whereas $AF + \bar{A}\bar{F}$ can be seen to make massive A only after elimination of $F, \bar{F}$ by their equation of motion arising from $\Gamma_{kin} + \Gamma_m$:

$$F = m\,\bar{A} \qquad \bar{F} = m\, A \tag{3.6}$$

$$(\Gamma^{(o)}_{kin} + \Gamma^{(o)}_m)\Big|_{(3.6)} = \int dx (\partial\bar{A}\partial A - m^2\bar{A}A + \frac{i}{2}\bar{\psi}\bar{\sigma}\partial\psi + \frac{m}{4}(\psi\psi + \bar{\psi}\bar{\psi})) \tag{3.7}$$

Canonical quantization of this action will yield a Hilbert space of states for a massive complex scalar field A and a massive Weyl spinor ψ (and its conjugate $\bar{\psi}$), but no trace of F will be left: F is simply another interpolating field for A, given by (3.6). It is for closure of the supersymmetry algebra off-shell that the auxiliary field F is needed. Due to supersymmetry the scalar and the spinor have the same mass m.

Having convinced ourselves from the component formulation that the action

$$\Gamma^{(0)} = \Gamma^{(0)}_{kin} + \Gamma^{(0)}_{m} \tag{3.8}$$

describes a decent free theory we shall continue now the discussion in the superfield language. The next aim is the derivation of propagators. As already indicated by the notations (and explained in App. B) we understand the action $\Gamma^{(0)}$ as the lowest approximation to the vertex functional, hence we can go over to the Green's functional by Legendre transformation

$$Z_c[J] = \Gamma[A] + \int dS\ J\ A + \int d\bar{S}\ \bar{J}\ \bar{A} \tag{3.9}$$

with

$$J = -\frac{\delta\Gamma}{\delta A} \qquad \bar{J} = -\frac{\delta\Gamma}{\delta\bar{A}} \tag{3.10}$$

the inverse being

$$A = \frac{\delta Z_c}{\delta J} \qquad \bar{A} = \frac{\delta Z_c}{\delta\bar{J}} \quad . \tag{3.11}$$

Here we have used functional derivatives with respect to superfields and corresponding δ-functions [II.2] (cf. rules in App. A). The propagators are then given by

$$< T\ A(1)\ A(2) > = \frac{\delta}{i\delta J(1)}\ A\ (J,\bar{J})\ (2) \tag{3.12}$$

$$< T\ A(1)\ \bar{A}(2) > = \frac{\delta}{i\delta J(1)}\ \bar{A}\ (J,\bar{J})\ (2) \tag{3.13}$$

where the arguments (1), (2) refer to superspace points. In order to calculate explicitly the propagators we derive first (3.10) and solve these equations for $A = A(J,\bar{J})$.

$$-J = \frac{1}{16}\ \overline{DD}\ \bar{A} + \frac{m}{4}\ A \tag{3.14}$$

$$-\bar{J} = \frac{1}{16}\ DD\ A + \frac{m}{4}\ \bar{A} \tag{3.15}$$

Multiplying (3.14) by DD, (3.15) by - 4 m and adding we find

$$- (\Box + m^2)\ \bar{A} = - DD\ J + 4\ m\ \bar{J} \tag{3.16}$$

$$- (\Box + m^2)\ A = - \overline{DD}\ \bar{J} + 4\ m\ J \tag{3.17}$$

(using (A.12)).

Hence

$$< T\ A(1)\ A(2) > = + i\ \frac{4m\ \delta_S(1,2)}{\Box + m^2} \tag{3.18}$$

$$< T\ A(1)\ \bar{A}(2) > = - i\ \frac{DD_2\ \delta_S(1,2)}{\Box + m^2} \tag{3.19}$$

With $\delta_S(1,2) = \overline{DD}_1\delta_V(1,2) = \overline{DD}_2\ \delta_V(1,2)$ one can actually check that the propagators have the correct chirality properties.

3.2 Interacting chiral fields

As already noted above terms of the type $\int dS\ A^n$ $n = 1,2,\ldots$ are supersymmetrically invariant and represent self-interactions of a chiral field. If we restrict ourselves to power-counting renormalizable couplings i.e. to Lagrangian vertices of dimension less than or equal to three for chiral, less than or equal to two for general ones the most general supersymmetric action for a set of N chiral fields A_i ($i = 1, \ldots, N$) reads

$$\Gamma^{(0)} = \frac{1}{16} \int dV\ \bar{A}_i A_i + \int dS\ (\lambda_i A_i + m_{ij}\ A_i\ A_j + g_{ijk}\ A_i\ A_j\ A_k) \tag{3.20}$$
$$+ \int d\bar{S}\ (\bar{\lambda}_i \bar{A}_i + \bar{m}_{ij}\ \bar{A}_i\ \bar{A}_j + \bar{g}_{ijk}\ \bar{A}_i\ \bar{A}_j\ \bar{A}_k)$$

(sum over repeated indices; m, g symmetric in their indices; kinetic terms already diagonalized and normalized). If we wish to maintain parity (cf. App. A) [II.3] we have

$$\lambda_i = \bar{\lambda}_i \qquad m_{ij} = \bar{m}_{ij} \qquad g_{ijk} = \bar{g}_{ijk} \tag{3.21}$$

For special values of λ, m, g the action (3.20) may have besides supersymmetry other invariances e.g. R-invariance or rigid gauge invariance:

$$\delta_\omega A_k = - i \omega^i T^i_{k\ell} A_\ell$$
$$\delta_\omega \bar{A}_k = i \omega^i T'^i_{k\ell} \bar{A}_\ell \tag{3.22}$$

where the T^i generate a representation of an internal symmetry group G.

An example of the latter is provided by G = SU(n) $n \geq 3$,

$$T^i_{k\ell} = T'^i_{\ell k} = i f^i_{k\ell} \qquad \text{(adjoint representation)}$$
$$\lambda = \bar{\lambda} = 0, \quad m_{ij} = \bar{m}_{ij} = m \delta_{ij}, \quad g_{ijk} = \bar{g}_{ijk} = g d_{ijk} \tag{3.23}$$

with $d_{ijk} = T_r (\tau_i \tau_j \tau_k)$ (τ: generator of the fundamental representation). An example of the former: $\Gamma^{(o)}$ is R-invariant with R-weights $n(A_i) = \frac{2}{3}$ for $\lambda = 0$, $m = 0$ (i.e. for dimensionless couplings).

These invariances of the action can be expressed on the functional level with the help of corresponding functional differential operators:

$$W_x \equiv - i \left(\int dS \, \delta_x A_i \frac{\delta}{\delta A_i} + \int d\bar{S} \, \delta_x \bar{A}_i \frac{\delta}{\delta \bar{A}_i} \right), x \in \{\alpha, \dot{\alpha}, R, G\} \tag{3.24}$$

with $\delta_x A_i$ being the field variation of the i-th field under the symmetry x, simply as

$$W_x \Gamma^{(o)} = 0 \tag{3.25}$$

This equation is called a Ward-identity and it is under this form that the invariance property of the theory can be extended to higher orders of perturbation theory. In our recursive approach theories are in fact <u>defined</u> by the Ward-identities, i.e. for a prescribed set of fields and Ward-identities the action should be uniquely determined - up to the coefficients of the independent invariants. Those are then to be fixed by normalization conditions. In the course of this one has to guarantee also vanishing vacuum expectation values of the fields and stability of the chosen minimum of the effective potential.

Let us look at the simplest example based on one field only: the Wess-Zumino model [II.4], which maintains parity

$$\Gamma^{(o)} = \frac{1}{16}\int dV\bar{A}A + \frac{1}{4}\int dS\,(\lambda A + \frac{m}{2}A^2 + \frac{g}{12}A^3) + \frac{1}{4}\int d\bar{S}\,(\lambda\bar{A} + \frac{m}{2}\bar{A}^2 + \frac{g}{12}\bar{A}^3) \tag{3.26}$$

(λ, m, g real)

The effective potential is given by

$$V = \bar{F}F = (\lambda + mA + \frac{g}{4}A^2)(\lambda + m\bar{A} + \frac{g}{4}\bar{A}^2) \geq 0 \tag{3.27}$$

The absolute minimum for it can be reached for

$$A = -\frac{2m}{g} \pm \frac{2}{|g|}\sqrt{-g\lambda + m^2} \tag{3.28}$$

But a shift in the scalar component A cannot break supersymmetry, as seen from the transformation law (A.30). Therefore at best a redefinition of parity is required, namely, when the shift (3.28) is complex [II.4,5]. We may thus without loss of generality impose as normalization conditions (in momentum space):

$$\begin{aligned} &\langle A \rangle = 0 \\ &\Gamma^{(o)}_{F\bar{F}} = 1 \\ &\Gamma^{(o)}_{AF} = -m \\ &\Gamma^{(o)}_{AAF} = -\frac{g}{2} \end{aligned} \tag{3.29}$$

Requiring the supersymmetry Ward-identity ((3.25) with $x = \alpha, \dot{\alpha}$) and these normalization conditions to hold fixes $\Gamma^{(o)}$ uniquely to be (3.26) with $\lambda = 0$.

3.3 The O'Raifeartaigh model

As another example of the above considerations we discuss the simplest model showing spontaneous breakdown of supersymmetry: the O'Raifeartaigh model [II.3]. It is constructed with three chiral multiplets A_0, A_1, A_2 and defined by parity (cf. App. A)

$$P: A_k(z) \to \bar{A}_k(z^P) \tag{3.30}$$

an internal discrete symmetry

$$I: A_0 \to A_0 \qquad A_{1,2} \to - A_{1,2} \tag{3.31}$$

R-invariance

$$i\,[R,A] = \delta_R\, A \qquad \delta_R\, A = i\,(n + \theta \frac{\partial}{\partial\theta}\, A) \tag{3.32}$$

with

$$n\,(A_0) = n\,(A_2) = -2 \qquad\qquad n\,(A_1) = 0 \tag{3.33}$$

and, of course, supersymmetry. The classical action which emerges as solution of these postulates is

$$\begin{aligned}\Gamma^{(o)} = \frac{1}{16} \int dV \bar{A}_k A_k &+ \int dS\, (\frac{\lambda}{4}\, A_0 + \frac{m}{4}\, A_1 A_2 + \frac{g}{32}\, A_0 A_1^2) \\ &+ \int d\bar{S}\, (\frac{\lambda}{4}\, \bar{A}_0 + \frac{m}{4}\, \bar{A}_1 \bar{A}_2 + \frac{g}{32}\, \bar{A}_0 \bar{A}_1^2\end{aligned} \tag{3.34}$$

(λ, m, g real). We now have to look for the minimum of the effective potential

$$V = \bar{F}_k F_k = |\lambda + \frac{g}{8}\, A_1^2|^2 + |m\, A_2 + \frac{g}{4}\, A_0 A_1|^2 + m^2\, \bar{A}_1 A_1 \geq 0\,. \tag{3.35}$$

$$A_2 = - \frac{g}{4m}\, A_0\, A_1 \tag{3.36}$$

minimizes the second term,

$$A_1 = 0 \tag{3.37}$$

minimizes the first + the last term if

$$m^2 > |\tfrac{1}{8}\lambda g| .$$

A_o is not determined. The choice

$$\langle A_o \rangle = 0 \tag{3.38}$$

maintains R-invariance and is therefore made for convenience. Since

$$V \geq |\lambda|^2 \tag{3.39}$$

the shift $F_o \to F_o + \lambda$ is dictated. Looking into the supersymmetry transformations (A.30) we find that ψ_o transforms inhomogeneously, performing the shift in the action (3.34) additional bilinear terms arise:

$$- \frac{g\lambda}{8} \int dx \, (A_1^2 + \bar{A}_1^2) \tag{3.40}$$

which cause mass splitting in the multiplet A_1 : $m^2 \pm \frac{1}{4}\lambda g$ for Re A_1, Im A_1 respectively (m is the mass of the spinor ψ_1).

Hence supersymmetry is spontaneously broken and ψ_o is the corresponding Goldstone particle.

The precise formulation of our symmetry requirements in functional form is therefore

$$W_\alpha \Gamma \equiv -i \int dx \left(\psi_{k\alpha} \frac{\delta}{\delta A_k} + 2F_k \frac{\delta}{\delta \psi_k^\alpha} - 2i\sigma_{\alpha\dot{\beta}} \partial \bar{A}_k \frac{\delta}{\delta \bar{\psi}_{k\dot{\beta}}} + i\sigma_{\alpha\dot{\beta}} \partial \bar{\psi}_k^{\dot{\beta}} \frac{\delta}{\delta \bar{F}_k} + 2f \frac{\delta}{\delta \psi_o^\alpha} \right) \Gamma = 0 \tag{3.41}$$

for spontaneously broken supersymmetry,

$$W_R \Gamma \equiv -i \left(\int dS \, \delta_R A_k \frac{\delta}{\delta A_k} + \int d\bar{S} \, \delta_R \bar{A}_k \frac{\delta}{\delta \bar{A}_k} \right) \Gamma = 0 \tag{3.42}$$

for R-invariance,

$$\Gamma(A) = \Gamma(A^P) \tag{3.43}$$

$$\Gamma(A) = \Gamma(A^I) \tag{3.44}$$

for the discrete symmetries.

As normalization conditions we may prescribe (in momentum space)

$$\Gamma_{F_k \bar{F}_k} = 1 \qquad k = 0,1,2 \tag{3.45}$$

$$\Gamma_{A_1 F_2} = - m \tag{3.46}$$

$$\Gamma_{A_1 A_1} = - \xi \tag{3.47}$$

$$\Gamma_{F_0 A_1 A_1} = - \frac{g}{4} \tag{3.48}$$

$$\Gamma_{F_0} = 0 \tag{3.49}$$

These postulates uniquely fix the action in the tree approximation. Indeed (3.45) fixes the normalization of the kinetic terms; (3.46) the ψ_1-mass; (3.47) the mass splitting in the A_1-multiplet; (3.48) the coupling constant and (3.49) expresses the parameter f occurring in the Ward-identity (3.41) in terms of the mass splitting ξ and the coupling g. Everything else is now determined by the symmetries (3.41) - (3.44). In particular the masses in the A_0-multiplet are fixed: in the tree approximation they are still degenerate, namely both zero, but only that of the Goldstone fermion ψ_0 is expected to stay zero to all orders. Indeed, describing that and how in higher orders a mass is generated for A_0 will be one of the main problems of renormalization.

4. Abelian gauge models

4.1 Free abelian gauge fields

Superfields of type Φ ((1.20), (1.25)) can be constrained to be real and contain then a real vector field v_μ:

$$\Phi = \bar{\Phi} \tag{4.1}$$

$$\Phi = C + \theta\chi + \bar{\theta}\bar{\chi} + \frac{1}{2}\theta^2 M + \frac{1}{2}\bar{\theta}^2\bar{M} + \theta\sigma^\mu\bar{\theta}v_\mu$$
$$+ \frac{1}{2}\bar{\theta}^2\theta\lambda + \frac{1}{2}\theta^2\bar{\theta}\bar{\lambda} + \frac{1}{4}\theta^2\bar{\theta}^2 D \quad (4.2)$$

Hence they are natural candidates for gauge fields in a supersymmetric gauge theory. If v_μ is to be identified with an ordinary gauge vector field, Φ has dimension zero and the hope is that the fields with sub-canonical dimensions correspond to longitudinal parts which represent pure gauge degrees of freedom. To characterize them supersymmetrically means finding projection operators which decompose Φ (non-locally) into transverse and longitudinal parts. The ordinary longitudinal projector $\frac{\partial_\mu\partial_\nu}{\Box}$ (i.e. the denominator $\Box$) suggests a combination of 4 D's divided by $\Box$, the fact that Φ carries no external Lorentz index enforces combinations of $D, D, \bar{D}, \bar{D}$ with spinor indices saturated. But the sequences $D\bar{D}D\bar{D}$, $\bar{D}D\bar{D}D$ reduce by (A.11) to $D\bar{D}\bar{D}D$ and $\bar{D}DD\bar{D}$, i.e. at most these and $DD\overline{DD}$, $\overline{DD}DD$ are independent. But $D\bar{D}\bar{D}D = \bar{D}DD\bar{D}$ (proof by (A.9), (A.11)) thus one ends up with

$$P_{L_1} = -\frac{DD\overline{DD}}{16\Box} \qquad P_{L_2} = -\frac{\overline{DD}DD}{16\Box} \quad (4.3)$$

$$P_T = \frac{D\overline{DD}D}{8\Box} \quad (4.4)$$

It is easily checked via (A.12) that with these coefficients the P's are projectors, i.e. they are

idempotent: $$P^2 = P \quad (4.5)$$

orthogonal: $$P_{L_1}P_{L_2} = P_{L_2}P_{L_1} = P_{L_1}P_T = P_{L_2}P_T = 0 \quad (4.6)$$

and sum up to unity: $$1 = P_T + P_{L_1} + P_{L_2} \quad (4.7)$$

P_{L_2} projects to a chiral field, P_{L_1} to an anti-chiral, P_T to a real one. The explicit computation yields

$$D\bar{D}\bar{D}D\Phi = 4D' - 4i\theta\sigma^\mu\partial_\mu\bar{\lambda}' + 4i\partial_\mu\lambda'\sigma^\mu\bar{\theta} + 8\theta\sigma^\nu\theta\bar{\partial}^\mu f_{\mu\nu}$$
$$+ 2\theta^2\bar{\theta}\Box\bar{\lambda}' + 2\bar{\theta}^2\theta\Box\lambda' + \theta^2\bar{\theta}^2\Box D' \quad (4.8)$$

$$D D\bar{D}\bar{D}\Phi = 4(D'' + 2i\partial v) + 8i\partial\lambda''\sigma\bar{\theta} - 8\bar{\theta}^2 \Box \bar{M}$$
$$- 4i\theta\sigma\bar{\theta}\partial(D'' + 2i\partial v) - 4\bar{\theta}^2\theta\lambda'' - \theta^2\bar{\theta}^2 \Box (D'' + 2i\partial v) \qquad (4.9)$$

$$\bar{D}\bar{D}DD = 4(D'' - 2i\partial v) - 8i\theta\sigma\partial\bar{\lambda}'' - 8\theta^2 \Box M$$
$$- 4i\theta\sigma\bar{\theta}\partial (D'' - 2i\partial v) - 4\theta^2\bar{\theta}\bar{\lambda}'' - \theta^2\bar{\theta}^2\Box(D'' - 2i\partial v) \qquad (4.10)$$

$$D' = D + \Box C \qquad \lambda' = \lambda + i\sigma^\mu\partial_\mu\bar{\chi} \qquad f_{\mu\nu} = \partial_\mu v_\nu - \partial_\nu v_\mu$$
$$D'' = D - \Box C \qquad \lambda'' = \lambda - i\sigma^\mu\partial_\mu\bar{\chi} \qquad (4.11)$$

Due to the occurrence of $f_{\mu\nu}$ in (4.8) one has to identify $P_T\Phi$ with the transverse part of Φ and hence $P_{L_1}\Phi$, $P_{L_2}\Phi$ with the longitudinal parts. The gauge transformations of Φ involve therefore necessarily a chiral and an antichiral field and are given by:

$$\delta\Phi = i\ (\Lambda - \bar{\Lambda})$$
$$= (A+\theta\psi+\theta^2F - i\theta\sigma^\mu\bar{\theta}\partial_\mu A - \frac{i}{2}\theta^2\bar{\theta}\bar{\sigma}^\mu\partial_\mu\psi - \frac{1}{4}\theta^2\bar{\theta}^2\Box A$$
$$- \bar{A} - \bar{\theta}\bar{\psi} - \bar{\theta}^2\bar{F} - i\theta\sigma^\mu\bar{\theta}\partial_\mu\bar{A} + \frac{i}{2}\bar{\theta}^2\theta\sigma^\mu\partial_\mu\bar{\psi} + \frac{1}{4}\theta^2\bar{\theta}^2\Box\bar{A} \qquad (4.12)$$

(Here we shifted $\Lambda, \bar{\Lambda}$ to a common - the real - basis in order to compare with Φ.)

We immediately read off the transformation laws for the components, in particular the usual one for v_μ:

$$\delta v_\mu = \partial_\mu\ (A + \bar{A}) \qquad (4.12a)$$

and realize that the combinations D', λ', λ' (4.11) are gauge invariant. So, indeed, $D\bar{D}\bar{D}C\Phi$ is gauge invariant, as seen from the comonents, and of course also immediately in terms of superfields:

$$\delta D\bar{D}\bar{D}D\Phi = iD\bar{D}\bar{D}D(\Lambda-\bar{\Lambda}) = iD\bar{D}\bar{D}D\Lambda = iD[\bar{D}\bar{D},D]\Lambda \sim D\sigma\partial\bar{D}\Lambda = 0.$$

Having found the transformation law and already an invariant it is obvious how to construct an invariant action: $D\bar{D}\bar{D}D\Phi$ contains as $\theta\sigma^\nu\bar{\theta}$-

- component the term $\partial^\mu f_{\mu\nu}$, so multiplication by the field Φ and integration over the entire superspace will produce the term $\int dx\, v^\nu \partial^\mu f_{\mu\nu}$ the ordinary gauge invariant term for a vector field:

$$\Gamma_{kin} = \frac{1}{128} \int dV \Phi D\overline{DD}D\Phi = \int dx\, (- \frac{1}{4} f^{\mu\nu} f_{\mu\nu} + \frac{i}{4} \lambda' \not\partial \bar{\lambda}' + \frac{1}{8} D'^2) \tag{4.13}$$

I.e. on including also a gauge invariant Weyl spinor field along with the gauge vector field one has a free action which is gauge invariant and supersymmetric (on shell, where D' = 0) [II.6].

We now wish to quantize the theory, in particular to calculate the propagator for Φ. In order to do so we have to break gauge invariance, i.e. to add to the invariant action a gauge fixing term. Since the construction of the physical state space depends on the gauge chosen we always give also the equation of motion satisfied by the ghosts.

α - gauge

Following the above line of reasoning the most obvious choice of gauge is to add back to the field Φ its longitudial parts:

$$\Gamma^{(o)} = \frac{1}{128} \int dV\, (\Phi D\overline{DD}D\Phi - \frac{1}{2\alpha} \Phi\{DD,\overline{DD}\}\Phi + 8M^2\Phi^2) \tag{4.14}$$

At the component level this corresponds to complementing the transverse projector in $v_\nu\, (\partial^\nu\partial^\lambda - g^{\nu\lambda}\Box)\, v_\lambda$ by the longitudinal one:

$\frac{1}{\alpha} v_\nu\, (- \partial^\nu\partial^\lambda)\, v_\lambda$. α is the normal gauge parameter with $\alpha = 1$ corresponding to Feynmann, $\alpha = 0$ to Landau gauge. We have also added a mass term (which is not gauge invariant).

The broken gauge invariance we may express by a Ward-identity namely

$$W_g\, \Gamma^{(o)} \equiv - i \int dVi\, (\Lambda-\bar{\Lambda}) \frac{\delta}{\delta\Phi} \Gamma^{(o)} \tag{4.15}$$

$$W_g\, \Gamma^{(o)} = \frac{1}{64} \int dV \left\{ - \frac{1}{2\alpha} (\Lambda-\bar{\Lambda})\, \{DD,\overline{DD}\}\Phi + 8M^2\, (\Lambda-\bar{\Lambda})\Phi \right\} \tag{4.16}$$

which is decomposable into two local Ward-identities:

$$w_\Lambda \equiv \frac{\delta}{\delta\Lambda} W_g \qquad w_{\bar\Lambda} \equiv - \frac{\delta}{\delta\bar\Lambda} W_g \tag{4.17}$$

$$\begin{aligned} w_\Lambda \Gamma^{(0)} &= \frac{1}{8\alpha} (\Box + \alpha M^2)\, \overline{DD}\Phi \\ w_{\bar\Lambda} \Gamma^{(0)} &= \frac{1}{8\alpha} (\Box + \alpha M^2)\, DD\Phi \end{aligned} \tag{4.18}$$

Let us now perform the Legendre transformation

$$Z_c = \Gamma + \int dV\, J\Phi \tag{4.19}$$

and rewrite the Ward-identities for Z_c with $\frac{\delta\Gamma}{\delta\Phi} = - J$ (4.20)

$$\begin{aligned} - \overline{DD}\, J &= \frac{1}{8\alpha} (\Box + \alpha M^2)\, \overline{DD}\, \frac{\delta Z_c}{\delta J} \\ - DD\, J &= \frac{1}{8\alpha} (\Box + \alpha M^2)\, DD\, \frac{\delta Z_c}{\delta J} \end{aligned} \tag{4.21}$$

On the functional $Z = e^{iZ_c}$ for general Green's functions we thus finally have

$$\begin{aligned} - i\, \overline{DD}\, J \cdot Z &= \frac{1}{8\alpha} (\Box + \alpha M^2)\, \overline{DD}\, \frac{\delta Z}{\delta J} \\ - i\, DD\, J \cdot Z &= \frac{1}{8\alpha} (\Box + \alpha M^2)\, DD\, \frac{\delta Z}{\delta J} \end{aligned} \tag{4.22}$$

On the physical mass shell, $J = 0$, these equations therefore give us

$$\begin{aligned} 0 &= (\Box_1 + \alpha M^2) < \overline{DD}_1\, \Phi(1)\, X > \\ 0 &= (\Box_1 + \alpha M^2) < DD_1\, \Phi(1)\, X > \end{aligned} \tag{4.23}$$

(here X denotes an arbitrary sequence of fields Φ with derivatives possibly acting on them). I.e. the fields $\overline{DD}\Phi$ and $DD\Phi$ are free fields with respect to a Klein-Gordon wave operator of mass^2 αM^2. If one is able to prove a similar Ward-identity for the interacting theory freedom of the longitudinal fields implies unitarity (in the state space constructed appropriate to their nature).

We shall calculate now the propagator for Φ [II.7]. As in section 3 (and App. B) we derive first

$$\frac{\delta\Gamma^{(0)}}{\delta\Phi} = \frac{1}{64}\left(\bar{D}DD\bar{D}\Phi - \frac{1}{2\alpha}\{DD,\bar{D}\bar{D}\}\Phi + 8M^2\Phi\right) \tag{4.24}$$

require then (4.20) and solve for $\Phi = \Phi(J)$.

$$\bar{D}DD\bar{D}\Phi - \frac{1}{2\alpha}\{DD,\bar{D}\bar{D}\}\Phi + 8M^2\Phi = -64\,J \tag{4.25}$$

Applying the projectors (4.3) (4.4) with their rules (4.5) (4.6) we find

$$(8\Box + 8M^2)\,\bar{D}DD\bar{D}\Phi = -64\,\bar{D}DD\bar{D}\,J \tag{4.26}$$

$$\left(\frac{8}{\alpha}\Box + 8M^2\right)\{DD,\bar{D}\bar{D}\}\Phi = -64\,\{DD,\bar{D}\bar{D}\}J \tag{4.27}$$

inserting into (4.25) one has

$$\Phi = \frac{16}{M^2}\left(-J + \frac{1}{8}\frac{\bar{D}DD\bar{D}J}{\Box + M^2} - \frac{1}{16\alpha}\frac{\{DD,\bar{D}\bar{D}\}J}{\frac{1}{\alpha}\Box + \alpha M^2}\right) \tag{4.28}$$

hence for the propagator

$$\langle T\Phi(1)\Phi(2)\rangle = \frac{\delta}{i\delta J(1)}\,J(\Phi)(2) = \frac{16i}{M^2}\left(1 - \frac{1}{8}\frac{\bar{D}DD\bar{D}}{\Box + M^2} - \frac{1}{16}\,\frac{\{DD,\bar{D}\bar{D}\}}{\Box + \alpha M^2}\right)\delta_V(1,2) \tag{4.29}$$

at $\alpha = 1$

$$\langle T\Phi(1)\Phi(2) = \frac{8i}{\Box + M^2}\,\delta_V(1,2) \tag{4.30}$$

For M = 0, one can immediately combine (4.26) (4.27) to obtain

$$\langle T\Phi(1)\Phi(2)\rangle = \frac{i}{\Box^2}\left(\bar{D}DD\bar{D} - \frac{\alpha}{2}\{DD,\bar{D}\bar{D}\}\right)\delta_V(1,2) \tag{4.31}$$

at $\alpha = 1$

$$\langle T\Phi(1)\Phi(2)\rangle = \frac{8i}{\Box}\,\delta_V(1,2) \tag{4.32}$$

<u>α, B - gauge</u>

The above gauge-fixing (4.14)

$$\Gamma^{(o)}_{g.f.} = \frac{1}{128} \int dV \left(- \frac{1}{2\alpha} \Phi\{DD,\overline{DD}\}\Phi\right) \tag{4.33}$$

can also be imposed via a Lagrange-multiplier field B. Let us require for a chiral field B of dimension 1

$$\frac{\delta\Gamma}{\delta B} = \alpha\overline{DD}\ \overline{B} + \frac{1}{8}\ \overline{DD}DD\Phi$$
$$\frac{\delta\Gamma}{\delta\overline{B}} = \alpha DD\ B + \frac{1}{8}\ DD\overline{DD}\Phi \tag{4.34}$$

Then the general solution is

$$\Gamma = \alpha\int dV\ B\overline{B} + \frac{1}{8} \int dV(BDD\Phi + \overline{B}\ \overline{DD}\Phi) + \overline{\Gamma} \tag{4.35}$$

where $\overline{\Gamma}$ is independent of B and upon using (4.34) on shell one sees that

$$\Gamma = \frac{1}{128} \int dV \left(- \frac{1}{2\alpha}\right)\Phi\{DD,\overline{DD}\}\Phi + \overline{\Gamma} \tag{4.36}$$

i.e. one obtains back the gauge fixing term (4.33).

Proceeding parallel to the discussion of the α-gauge we consider now (4.35) with

$$\overline{\Gamma} = \Gamma^{(o)}_{kin} + \frac{1}{128} \int dV\ 8M^2\Phi^2 \tag{4.37}$$

and call the sum $\Gamma^{(o)}$. Choosing B to be invariant under gauge transformations we derive as Ward-identities

$$w_\Lambda\ \Gamma^{(o)} = \frac{1}{8}\ (\overline{DD}DD\ B + M^2\ \overline{DD}\ \Phi)$$
$$w_{\overline{\Lambda}}\ \Gamma^{(o)} = \frac{1}{8}\ (DD\overline{DD}\ \overline{B} + M^2\ DD\ \Phi) \tag{4.38}$$

Using the equation of motion for the B field this reads

$$(w_\Lambda - \frac{1}{8\alpha}\,\overline{DD}\,\frac{\delta}{\delta\bar{B}})\;\Gamma^{(0)} = \frac{1}{8\alpha}\,(\Box + \alpha M^2)\;\overline{DD}\;\Phi$$
$$(w_{\bar{\Lambda}} - \frac{1}{8\alpha}\,DD\,\frac{\delta}{\delta B})\;\Gamma^{(0)} = \frac{1}{8\alpha}\,(\Box + \alpha M^2)\;DD\;\Phi \tag{4.39}$$

Without going through all the steps as we did before it is already clear from (4.39) that on shell again $DD\Phi$, $\overline{DD}\Phi$ are Klein-Gordon free fields of mass^2 αM^2. The B-derivatives on the left hand side of (4.39) may be interpreted as inhomogeneous gauge transformations for the B fields.

On shell we have for the B's from (4.38)

$$\Box B - \frac{1}{8} M^2\,\overline{DD}\;\Phi = 0$$
$$\Box \bar{B} - \frac{1}{8} M^2\,DD\;\Phi = 0 \tag{4.40}$$

The above formulation involving the field B is useful if one wants to study α-dependence in the theory. The special value $\alpha = 0$ (Landau gauge) requires nevertheless a separate treatment, which we present now. At $\alpha = 0$ we may still require (4.34) and obtain (4.35) at this value - and in this sense the Landau gauge is here embedded in the α-family - but, of course, (4.36) cannot be derived. For the Ward-identities we have again (4.38), hence on shell

$$\overline{DD}\;DD\;\Phi = 0 \tag{4.41}$$

$$\Box B - \frac{1}{8} M^2\,\overline{DD}\;\Phi = 0 \tag{4.42}$$

i.e. now the ghost field $DD\Phi$ satisfies the free field equation appropriate to a massless antichiral field.

Parameter - free gauge

We require

$$\frac{\delta\Gamma}{\delta B} = \frac{-1}{128}\,\overline{DD}\;\Phi$$
$$\frac{\delta\Gamma}{\delta\bar{B}} = \frac{-1}{128}\,DD\;\Phi \tag{4.43}$$

to be maintained for a chiral field B of dimension 2. The general solution is

$$\Gamma = \frac{-1}{128} \int dV\ (B + \bar{B})\Phi + \bar{\Gamma}. \qquad (4.44)$$

with $\bar{\Gamma}$ independent of B. Due to the dimension of B in fact no terms other than those occurring in (4.44) are possible in any action containing B (and being supersymmetric). In this sense "parameter-free" is understood; in the case above, dim B = 1, Landau gauge still belonged to the α - family - here there is none. In fact, (4.43) taken on shell restricts the real vector field to be a linear one:

$$DD\ \Phi = \overline{DD}\ \Phi = 0 \qquad (4.45)$$

implies * (cf. (4.11))

$$D'' = 0 \qquad \lambda'' = 0 \qquad \partial v = 0 \qquad (4.45a)$$

i.e. we are in a Landau-gauge. The Ward-identities read:

$$\begin{aligned} w_{\bar{\Lambda}}\ \Gamma^{(o)} &= -\frac{1}{128}\ \overline{DD}\ \bar{B} + \frac{1}{8} M^2\ \overline{DD}\ \Phi \\ w_{\Lambda}\ \Gamma^{(o)} &= -\frac{1}{128}\ DD\ B + \frac{1}{8} M^2\ DD\ \Phi \end{aligned} \qquad (4.46)$$

for the action

$$\Gamma^{(o)} = \frac{1}{128} \int dV(\Phi D\bar{D}\bar{D}D\Phi - (B+\bar{B})\Phi + 8M^2\Phi^2). \qquad (4.47)$$

Using the B-equation of motion they can be rewritten as

$$\begin{aligned} (w_{\bar{\Lambda}} + 16M^2 \frac{\delta}{\delta B})\ \Gamma^{(o)} &= -\frac{1}{128}\ \overline{DD}\ \bar{B} \\ (w_{\Lambda} + 16M^2 \frac{\delta}{\delta \bar{B}})\ \Gamma^{(o)} &= -\frac{1}{128}\ DD\ B \end{aligned} \qquad (4.48)$$

---

* This differential equation for v_μ shows why the linear multiplet is not a good one to start with as a genuine gauge vector multiplet. Here the constraint appears as a gauge condition which is permitted.

Hence on shell B satisfies the equation of motion of a massless chiral field.

Let us also calculate the propagators.

We obtain in the usual manner

$$D\overline{DD}D\Phi - \frac{1}{2}(B+\overline{B}) + 8M^2\Phi = -64 J$$

$$-\frac{1}{2}DD\Phi \qquad\qquad = -64 J_{\overline{B}} \qquad\qquad (4.49)$$

$$-\frac{1}{2}\overline{DD}\Phi \qquad\qquad = -64 J_B$$

hence

$$8(\Box+M^2)\ D\overline{DDD}\Phi \qquad\qquad = -64\ D\overline{DDD}J$$

$$-\frac{1}{2}\overline{DDDD}\Phi \qquad\qquad = -64\ \overline{DD}J_{\overline{B}} \qquad\qquad (4.50)$$

$$-\frac{1}{2}DD\overline{DD}\Phi \qquad\qquad = -64\ DDJ_B$$

and therefore $\Phi = \dfrac{1}{8\Box}\left(\dfrac{-8D\overline{DDD}J}{\Box+M^2} - 64\ \overline{DD}\ \overline{J}_B - 64\ DDJ_B\right)$

$$< T\Phi(1)\Phi(2) > = \frac{8i}{\Box+M^2}\ \frac{D\overline{DDD}}{8\Box}\ \delta_V(1,2)$$

$$< T\Phi(1)B(2) > = \frac{8i}{\Box}\ DD_1\ \delta_S(1,2) \qquad\qquad (4.51)$$

$$< T\Phi(1)\overline{B}(2) > = \frac{8i}{\Box}\ \overline{DD}_1\ \delta_{\overline{S}}(1,2)$$

For M = 0 we have furthermore

$$< TB(1)B(2) > = < TB(1)\overline{B}(2) > = 0 . \qquad\qquad (4.52)$$

Wess-Zumino gauge

In this gauge one eliminates the ghost fields $C,\chi,\chi,M,\overline{M}$ by a non-supersymmetric gauge transformation:

$$\delta_g C = i(A-\bar{A}) \qquad \delta_g \chi = \psi \qquad \delta_g M = 2F \tag{4.53}$$

and combines thereafter any supersymmetry transformation with a gauge-re-establishing gauge transformation, since the gauge $C = \chi = \bar{\chi} = M = \bar{M} = 0$ is, of course, not stable under supersymmetry. As an example consider

$$(\delta_s + \delta_g)\chi_\alpha = \xi_\alpha M + \sigma^\mu_{\alpha\dot{\alpha}} \bar{\xi}^{\dot{\alpha}} (v_\mu - i\partial_\mu C) + i\psi_\alpha \tag{4.54}$$

$\overset{!}{=}$ 0 implies

$$\psi_\alpha = -i\ (\xi_\alpha M + \sigma^\mu_{\alpha\dot{\alpha}} \bar{\xi}^{\dot{\alpha}} (v_\mu - i\partial_\mu C)) \tag{4.55}$$

$$\psi_\alpha \Big|_{\text{WZ-gauge}} = -\, i\, \sigma^\mu_{\alpha\dot{\alpha}} \bar{\xi}^{\dot{\alpha}} v_\mu \tag{4.56}$$

i.e. the supersymmetry transformation $\xi^\alpha\delta_\alpha + \bar{\xi}_{\dot{\alpha}}\bar{\delta}^{\dot{\alpha}}$ on χ can be compensated by the v-dependent gauge transformation δ_g with parameter ψ_α [II.6,8]. Since we shall not use this gauge in higher orders we will not pursue this subject any further.

4.2 SQED

The supersymmetric extension of quantum electrodynamics [II.6] has to comprise the abelian gauge invariant interaction of the photon and the electron with additional fields in order to achieve supersymmetry. Since the partner of the v_μ in an abelian gauge superfield - the Weyl spinor λ - is not charged it cannot be identified with even a part of the electron. Hence other spinors are needed and since the smallest multiplets are chiral ones we use those to provide the spinors. Supersymmetry will then combine them with scalars as partners. The number of multiplets is fixed once we have decomposed a Dirac spinor into Weyl spinors. A Dirac spinor is the direct sum of a $(\frac{1}{2}, 0)$ and a $(0, \frac{1}{2})$ representation of SL (2, C) in which the terms are interchanged by parity:

$$\psi_D = \frac{1}{\sqrt{2}}\begin{pmatrix}\psi_{+\alpha} \\ \psi_-^{\dot\alpha}\end{pmatrix} \qquad P:\ \psi_{+\alpha} \to \psi_-^{\dot\alpha} \tag{4.57}$$

Hence we need two chiral multiplets A_+, A_- (and their conjugates) and they are transformed under parity as $A_+ \to \bar{A}_-$.

Under gauge transformation a Dirac spinor changes according to:

$$\begin{aligned} \delta_g \psi_D &= -\, i\, g\, \omega\, \psi_D \\ \delta_g \bar{\psi}_D &= \, i\, g\, \omega\, \bar{\psi}_D \end{aligned} \tag{4.58}$$

$$\begin{aligned} \text{i.e. } \delta_g \psi_+ &= -\, i\, g\, \omega\, \psi_+ \qquad & \delta_g \bar{\psi}_+ &= i\, g\, \omega\, \bar{\psi}_+ \\ \delta_g \psi_- &= +\, i\, g\, \omega\, \psi_- \qquad & \delta_g \bar{\psi}_- &= -i\, g\, \omega\, \bar{\psi}_- \end{aligned} \tag{4.59}$$

The gauge invariant Dirac spinor mass term yields

$$m\,\bar{\psi}_D\,\psi_D = \frac{m}{2}\,(\bar{\psi}_+,\psi_-)\begin{pmatrix}0 & \sigma^0 \\ \bar{\sigma}^0 & 0\end{pmatrix}\begin{pmatrix}\psi_+ \\ \bar{\psi}_-\end{pmatrix} = \frac{m}{2}\,(\psi_+\psi_- + \bar{\psi}_+\bar{\psi}_-) \tag{4.60}$$

$$\text{i.e.} \qquad m \int dS\, A_+A_- + m \int d\bar{S}\, \bar{A}_+\bar{A}_-$$

as supersymmetric extension. But we still have to supersymmetrize the gauge transformations. $\omega(x)$ has to be read off from (4.12)

$$\delta_g v_\mu = \partial_\mu (A + \bar{A})\ , \tag{4.12a}$$

i.e. it contains a contribution from a chiral field (Λ) as well as from an antichiral one ($\bar{\Lambda}$). Therefore the only transformation law which maintains chirality and matches (4.59) is

$$\begin{aligned} \delta_\Lambda A_\pm &= \mp\, i\, g\, \Lambda\, A_\pm \qquad & \delta_\Lambda \bar{A}_\pm &= 0 \\ \delta_{\bar{\Lambda}} A_\pm &= 0 \qquad & \delta_{\bar{\Lambda}} \bar{A}_\pm &= \pm\, i\, g\, \bar{\Lambda}\, \bar{A}_\pm \end{aligned} \tag{4.61}$$

Together with

$$\delta\Phi = i\ (\Lambda-\bar{\Lambda}) \tag{4.62}$$

these will be the gauge transformations of SQED.

The only remaining problem is now the gauge invariant generalization of the matter kinetic terms

$$\int dV\ (\bar{A}_+A_+ + \bar{A}_-A_-)$$

That the kinetic terms are diagonal in the charge assignments +,- follows from

$$\bar{\psi}_D \not{\partial} \psi_D = (\bar{\psi}_+,\psi_-) \begin{pmatrix} 0 & \sigma^0 \\ \bar{\sigma}^0 & 0 \end{pmatrix} \begin{pmatrix} 0 & \sigma\partial \\ \bar{\sigma}\partial & 0 \end{pmatrix} \begin{pmatrix} \psi_+ \\ \psi_- \end{pmatrix} = \psi_-\sigma\partial\bar{\psi}_- + \bar{\psi}_+\bar{\sigma}\partial\psi_+ \tag{4.63}$$

For this we integrate up the infinitesimal transformation on A_+ to a finite one:

$$\begin{aligned} A_+ &\longrightarrow e^{-ig\Lambda}\ A_+ \\ \bar{A}_+ &\longrightarrow e^{+ig\bar{\Lambda}}\ \bar{A}_+ \end{aligned} \tag{4.64}$$

$$\text{i.e. } \bar{A}_+A_+ \longrightarrow \bar{A}_+ e^{ig(\bar{\Lambda}-\Lambda)}\ A_+ \tag{4.65}$$

and clearly $\bar{A}_+\ e^{g\Phi}A_+$ is the invariant sought. Under parity:

$\Phi \rightarrow -\Phi$, hence $\bar{A}_-\ e^{-g\Phi}A_-$ is the corresponding invariant made up from $\bar{A}_-\ A_-$. Collecting all results together:

$$\begin{aligned} \Gamma^{(o)}_{inv} = {} & \frac{1}{128} \int dV \Phi D\bar{D}\bar{D}D\Phi + \frac{1}{16} \int dV\ (\bar{A}_+ e^{g\Phi} A_+ + \bar{A}_- e^{-g\Phi} A_-) \\ & - \frac{1}{4} m \int dS\ A_+\ A_- - \frac{1}{4} m \int d\bar{S}\ \bar{A}_+\ \bar{A}_- \end{aligned} \tag{4.66}$$

is the minimal supersymmetric extension of QED. It is invariant under the transformations (4.61) (4.62) and preserves parity (A.22) (A.23) [II.6]. The exponential is understood to mean its series expansion and the preceding statements hold order by order in the number of fields.

For quantization we have like in the free theory to fix the gauge and must therefore characterize the theory by a gauge Ward-identity. Since the complete action $\Gamma^{(o)}$ is given by

$$\Gamma^{(o)} = \Gamma^{(o)}_{inv} + \Gamma_{g.f.} \tag{4.67}$$

(with $\Gamma_{g.f.}$ any one of the gauge fixing terms of sect. 4.1) we have added to the previous free action only a few more <u>invariants</u>, thus all of our old Ward-identities derived above change only on the left hand side: in the transformations participate now also the matter fields. I.e.

$$\begin{aligned} w_\Lambda &\equiv DD\frac{\delta}{\delta\Phi} - gA_+\frac{\delta}{\delta A_+} + gA_-\frac{\delta}{\delta A_-} \\ w_{\bar{\Lambda}} &\equiv \bar{D}\bar{D}\frac{\delta}{\delta\Phi} - g\bar{A}_+\frac{\delta}{\delta\bar{A}_+} + g\bar{A}_-\frac{\delta}{\delta\bar{A}_-} \end{aligned} \tag{4.68}$$

For the interacting theory in the tree approximation (4.67) we have therefore again

$$w_\Lambda\,\Gamma^{(o)} = \frac{1}{8\alpha}(\Box + \alpha M^2)\,\bar{D}\bar{D}\,\Phi \tag{4.69}$$

in the α-gauge,

$$w_\Lambda\,\Gamma^{(o)} = \frac{1}{8}(\bar{D}\bar{D}DD\,B + M^2\,\bar{D}\bar{D}\,\Phi) \tag{4.70}$$

in the α, B-gauge,

$$w_\Lambda\,\Gamma^{(o)} = -\frac{1}{128}\,\bar{D}\bar{D}\,\bar{B} + \frac{1}{8}M^2\,\bar{D}\bar{D}\,\Phi \tag{4.71}$$

in the parameter-free gauge. (Analogously for the conjugates. We have permitted a photon mass term which is optional.)

Due to the specific matter-bilinear coupling $\Gamma^{(o)}$ is strictly invariant under R-transformation with $n(\Phi) = 0$, $n(A_+) = n(A_-) = -1$ (and suitable weights for B).

$$W_R^{(-1)}\,\Gamma^{(o)} = 0 \tag{4.72}$$

For $n(A_+) = n(A_-) \equiv n$, the R-invariance is only softly broken by the matter mass terms:

$$W_R^{(n)}\Gamma = -\frac{m}{2}(n+1)(\int dSA_+A_- - \int d\bar{S}\bar{A}_+\bar{A}_-) \quad (4.73)$$

These observations become particularly important at the level of the corresponding currents.

4.3 S'QED and S'(QED)' *

For any phenomenological application of supersymmetry its breaking is mandatory. Supersymmetry implies mass equalities and such degeneracies are very rare. But equally rare are workable mechanisms of supersymmetry breaking. For instance for the desirable spontaneous breakdown up to now only two ways are known. One corresponds to the breaking type presented in chiral models (section 3.3) and is characterized by the fact that the F-component of a chiral field acquires a vacuum expectation value, hence the spinor ψ in its multiplet transforms inhomogeneously. Obviously the other possibility is that the D-component of a vector superfield acquires a vacuum expectation value and its partner λ transforms as a Goldstone spinor. This effect is realized in a model [II.9] where one adds to $\Gamma^{(o)}_{inv}$ (4.66) a linear term D. Parity is now broken explicitly, but by a dimension 2 term only. Let us study the consequences first in the Wess-Zumino gauge (4.53).

The effective potential is (the negative of the Lagrangian for constant fields)

$$\begin{aligned} V = &-\frac{1}{8}D^2 + \frac{v}{4}D - \frac{g}{4}D(\bar{A}_+A_+ - \bar{A}_-A_-) - \bar{F}_+F_+ - \bar{F}_-F_- \\ &- m(A_+F_- + A_-F_+ + \bar{A}_+\bar{F}_- + \bar{A}_-\bar{A}_+) \end{aligned} \quad (4.74)$$

hence with the equations of motion

$$D = v - g(\bar{A}_+A_+ - \bar{A}_-A_-) \quad (4.75)$$

---

* The prime indicates spontaneous breaking of the respective symmetry.

$$F_\pm = - m\bar{A}_\mp \qquad \bar{F}_\pm = - mA_\mp \tag{4.76}$$

$$V = \frac{1}{8}(v - g(\bar{A}_+A_+ - \bar{A}_-A_-))^2 + m^2(\bar{A}_+A_+ + \bar{A}_-A_-) \geq 0 \tag{4.77}$$

$$V = \frac{v^2}{8} + (m^2 - \frac{vg}{4})\bar{A}_+A_+ + (m^2 + \frac{vg}{4})\bar{A}_-A_- + \frac{g^2}{8}(\bar{A}_+A_+ - \bar{A}_-A_-)^2 \tag{4.78}$$

One has to distinguish two cases ($vg > 0$ without loss of generality)

(1) $$m^2 - \frac{vg}{4} \geq 0 \tag{4.79}$$

Then $$V \geq \frac{v^2}{8} \tag{4.80}$$

and this (absolute) minimum is reached for

$$A_+ = A_- = 0 \tag{4.81}$$

i.e. $$D = v \qquad F_\pm = \bar{F}_\pm = 0 \tag{4.82}$$

Therefore supersymmetry is spontaneously broken with the Goldstone spinor λ and the masses^2 for $A_\pm$ split by

$$\pm\xi = \mp\frac{vg}{4} \tag{4.83}$$

to be compared with the $\psi_\pm$ masses which are still m. Gauge invariance is unbroken, the vector v_μ is massless.

(2) $$m^2 + \xi < 0 \tag{4.84}$$

then the absolute minimum of V is reached for $A_- = 0$ and

$$A_+A_+ = \frac{-(m^2 + \xi)}{g^2/4} \equiv a^2 \tag{4.85}$$

By a gauge transformation one may reduce this to a real shift in A_+ and the minimum will be at

$$D = -\frac{4m^2}{g} \qquad A_+ = \frac{2}{|g|}\sqrt{|m^2+\xi|} \qquad F_+ = -\frac{2m}{|g|}\sqrt{|m^2+\xi|} \tag{4.86}$$

Performing all the shifts and diagonalizing spinor mass terms one ends up with a massive vector v_μ, i.e. spontaneously broken gauge symmetry, a massless spinor with inhomogeneous transformation law, i.e spontaneously broken supersymmetry. the mass spectrum is: one complex scalar of mass $\sqrt{2m^2}$, one vector and one real scalar of mass $\sqrt{\frac{1}{2}g^2a^2}$, two spinors of mass $\sqrt{m^2+\frac{1}{2}g^2a^2}$ and the massless Goldstone spinor. This case will not be discussed any further.

Returning to case (1) we first wish to re-establish the general gauge. From (4.78) it is clear that the general gauge form of the potential is

$$\begin{aligned} V = \frac{v^2}{8} &+ (m^2+\xi)\,\bar{A}_+ e^{gC} A_+ + (m^2-\xi)\,\bar{A}_- e^{-gC} A_- \\ &+ \frac{g^2}{8}\left(\bar{A}_+ e^{gC} A_+ - \bar{A}_- e^{-gC} A_-\right)^2 \end{aligned} \tag{4.87}$$

hence none of the conclusions reached before is affected. The vacuum expectation values of C and of M - which does not show up in (4.87) - are not determined which is allright since these fields are subjected to inhomogeneous gauge transformations. But this means that we can go over from SQED to S'QED (the spontaneously broken phase) just by replacing $D \to D+v$ i.e. we require the spontaneously broken supersymmetry Ward-identity to hold:

$$\begin{aligned} W_\alpha \Gamma &\equiv -i\left(\int dV\, \delta_\alpha \hat{\Phi}\frac{\delta}{\delta\hat{\Phi}} - i\int dS\, \delta_\alpha A_\pm \frac{\delta}{\delta A_\pm} - i\int d\bar{S}\, \delta_\alpha \bar{A}_\pm \frac{\delta}{\delta \bar{A}_\pm}\right)\Gamma = 0 \\ \hat{\Phi} &= \Phi + \frac{1}{4}\theta^2\bar{\theta}^2 v \end{aligned} \tag{4.88}$$

as well as the strict gauge Ward-identity:

$$W_\Lambda \Gamma \equiv \left(\overline{DD}\frac{\delta}{\delta\hat{\Phi}} \mp g A_\pm \frac{\delta}{\delta A_\pm}\right)\Gamma = 0 \tag{4.89}$$

(analogously for $\bar{W}_\alpha$, $W_{\bar{\Lambda}}$).

The invariant action in the tree approximation will then be completely fixed by the normalization conditions (at p = o in momentum space)

$$\Gamma_{DD} = \frac{1}{4} \qquad \Gamma_D = 0$$

$$\Gamma_{A_+F_-} = m \qquad \Gamma_{A_+\bar{A}_+}(p=0) = -\frac{1}{4}\xi \tag{4.90}$$

$$\Gamma_{F_+\bar{F}_+} = 1$$

$$\Gamma_{F_-\bar{F}_-} = 1 \qquad \Gamma_{A_+\bar{A}_+D} = \frac{1}{4}g$$

Adding the gauge fixing term (in α-gauge form, for instance) will, of course, modify the gauge Ward-identity to be prescribed

$$\begin{aligned} w_\Lambda \Gamma &= \frac{1}{8\alpha} \; \Box \, \overline{DD} \, \Phi \\ w_{\bar{\Lambda}} \Gamma &= \frac{1}{8\alpha} \; \Box \, DD \, \Phi \, , \end{aligned} \tag{4.91}$$

and also the normalization condition for Γ_{DD}:

$$\Gamma_{DD} = \frac{1}{4}\left(1 - \frac{1}{\alpha}\right) \tag{4.92}$$

The gauge parameter α itself is fixed by its appearance in the gauge Ward-identity. The peculiar normalization condition (4.92) depending on α is chosen for reasons which become clear in higher orders. As in the symmetric theory the ghosts are free as a consequence of the gauge Ward-identity (4.91).

Section 5: Non-abelian gauge models *

5.1 Non-abelian gauge transformations

A heuristic and rather suggestive derivation of the non-abelian gauge transformations compatible with supersymmetry starts from rigid gauge transformations on a multiplet of chiral fields (3.22):

$$\begin{aligned} \delta_\omega A_k &= -\, i\, \omega^i\, \tau^i_{k\ell}\, A_\ell \\ \delta_\omega \bar{A}_k &= \quad i\, \bar{A}_\ell\, \tau^i_{\ell k}\, \omega^i\ . \end{aligned} \tag{5.1}$$

Here we have assumed parity to hold $T' = T \equiv \tau$; τ_i generate the fundamental representation of a group G; the ω^i are real parameters. The corresponding transformations in SQED suggest that we generalize ω^i to chiral superfields Λ_i and assume that for finite transformations:

$$\begin{aligned} A_k &\rightarrow (e^{-i\Lambda})_{k\ell}\, A_\ell \qquad\qquad \Lambda \equiv \Lambda^i \tau^i \\ \bar{A}_k &\rightarrow \bar{A}_\ell\, (e^{i\bar{\Lambda}})_{\ell k} \end{aligned} \tag{5.2}$$

The gauge invariant interaction term of SQED

$$\bar{A}\, e^{\Phi}\, A \tag{5.3}$$

then tells one to put the gauge vector fields associated with the group G also into a matrix:

$$\Phi \equiv \Phi^i \tau^i \tag{5.4}$$

and to try as transformation law [II.10,11]

$$e^{\Phi} \rightarrow e^{-i\bar{\Lambda}}\, e^{\Phi}\, e^{i\Lambda} = e^{\Phi'} \tag{5.5}$$

For infinitesimal $\Lambda, \bar{\Lambda}$ the first few terms read

---

* The method used in this section to find the general form of the non-abelian gauge transformations is adapted from the treatment in chapter IV.

$$\Phi' = \Phi + \delta\Phi$$

$$\delta\Phi = i\,(\Lambda - \bar{\Lambda}) + i\,\frac{1}{2}\,[\Phi, \Lambda+\bar{\Lambda}] + i\,\frac{1}{12}\,[\Phi[\Phi, \Lambda-\bar{\Lambda}]] + O(\Phi^3) \tag{5.6}$$

$$\equiv i\, Q_s(\Phi, \Lambda)$$

(The index s stands for special: we shall see later that more general gauge transformations have to be considered.) Together with the occurence of an exponential of Φ in the interaction there arises another effect: an infinite power series in the field Φ for its transformation law. Clearly all of these multiple commutators are in the algebra, so nothing is wrong with that.

Projecting out the $\theta\bar{\theta}$ - component in (5.6) we see

$$\delta v_\mu = \partial_\mu(A+\bar{A}) + \frac{i}{2}\,[v_\mu, A+\bar{A}\,] + \frac{i}{4}\,[\chi, \bar{\psi}]\,\sigma_\mu - \frac{i}{4}\,[\psi, \bar{\chi}]\,\sigma_\mu$$

$$+ \frac{1}{2}\,[C, \partial_\mu(A-\bar{A})\,] + \frac{1}{12}\,[C[C, \partial(A+\bar{A})]] + \ldots. \tag{5.7}$$

i.e. in the Wess-Zumino gauge ($C = \chi = \bar{\chi} = 0$ (4.53)) we just have the ordinary Yang-Mills transformation of the vector multiplet v_μ; in any supersymmetric gauge we have contributions of the fields $C, \chi, \bar{\chi}$ with subcanonical dimensions 0, $\frac{1}{2}$, $\frac{1}{2}$ resp., which make up in particular the higher commutators.

Our next task is to generalize the free kinetic term coming from SQED (4.13)

$$\int dV\, \Phi\, \overline{DD}DD\Phi = -\int dS\, \overline{DD}D\Phi\, \overline{DD}D\Phi \tag{5.8}$$

to the non-abelian case [II.10,11]. One observes that*

---

* (5.5) implies

$$e^{-\Phi} \to e^{-i\Lambda}\, e^{-\Phi}\, e^{i\bar{\Lambda}} \qquad \text{(check with } e^{-\Phi}\, e^{\Phi} = 1\text{)}$$

$$e^{-\Phi} D_\alpha e^{\Phi} \to e^{-i\Lambda} e^{-\Phi} e^{i\bar{\Lambda}} D_\alpha (e^{-i\bar{\Lambda}} e^{\Phi} e^{i\Lambda})$$

$$= e^{-i\Lambda} e^{-\Phi} D_\alpha (e^{\Phi} e^{i\Lambda})$$

$$= e^{-i\Lambda} e^{-\Phi} D_\alpha e^{\Phi} e^{i\Lambda} + e^{-i\Lambda} D_\alpha e^{i\Lambda} \tag{5.9}$$

The inhomogeneous term contains only chiral fields, the kinetic term (5.8) $\overline{DD}$, hence try

$$\overline{DD} (e^{-\Phi} D_\alpha e^{\Phi}) \to \overline{DD} (e^{-i\Lambda} e^{-\Phi} D_\alpha e^{\Phi} e^{i\Lambda}) + \overline{DD} (e^{-i\Lambda} D_\alpha e^{i\Lambda})$$

$$= e^{-i\Lambda} \overline{DD} (e^{-\Phi} D_\alpha e^{\Phi}) e^{i\Lambda} + e^{-i\Lambda} \overline{DD} D_\alpha e^{i\Lambda}$$

$$= e^{-i\Lambda} \overline{DD} (e^{-\Phi} D_\alpha e^{\Phi}) e^{i\Lambda} \tag{5.10}$$

(due to (A.10)).

Therefore

$$\mathrm{Tr}\, \overline{DD} (e^{-\Phi} D^\alpha e^{\Phi})\, \overline{DD} (e^{-\Phi} D_\alpha e^{\Phi}) \equiv \mathrm{Tr}\, F^\alpha F_\alpha \tag{5.11}$$

is invariant under the gauge transformation (5.5). Since it is a chiral field, $\mathrm{Tr} \int dS\, F^\alpha F_\alpha$ is supersymmetrically invariant and will be the desired kinetic term for the vector multiplet. In order to convince oneself of this, one calculates the tri- and quadrilinear contributions:

$$\left(\int dS\, F^\alpha F_\alpha\right)_3 = i\, f^{ijk} \int dV\, D\Phi_i\, \Phi_j\, \overline{DDD}\, \Phi_k \tag{5.12}$$

$$\left(\int dS\, F^\alpha F_\alpha\right)_4 = -\, f^{ijn} f^{nk\ell} \int dV \Big(\frac{1}{3} D\Phi^i\, \Phi^j\, \Phi^k\, \overline{DD}\, \Phi^\ell + \frac{1}{4} D\Phi^i\, \Phi^j\, \overline{DD}\, (D\Phi^k \Phi^\ell) \Big) \tag{5.13}$$

and checks in the Wess-Zumino gauge (4.53):

$$-\frac{1}{128} \mathrm{Tr} \int dS\, F^\alpha F_\alpha = \mathrm{Tr} \int dx \left(-\frac{1}{4} F^{\mu\nu} F_{\mu\nu} - \frac{i}{4} \lambda \not{\partial} \bar{\lambda} + \frac{1}{8} D^2\right). \tag{5.14}$$

Here $F_{\mu\nu}$, $\not{D}\bar{\lambda}$ are given by

$$F_{\mu\nu} = \partial_\mu v_\nu - \partial_\nu v_\mu + \frac{i}{2}[v_\mu, v_\nu], \tag{5.15}$$

$$\not{D}\bar{\lambda} = \partial_\mu \bar{\lambda} + i\,[v_\mu, \bar{\lambda}], \tag{5.16}$$

i.e. they are curl of the vector field and derivative of $\bar{\lambda}$, covariant with respect to

$$\delta_\omega\, v_\mu = \partial_\mu \omega - \frac{i}{2}[\omega, v_\mu] \tag{5.17}$$

$$\delta_\omega\, \bar{\lambda} = \frac{i}{2}[\bar{\lambda}, \omega], \tag{5.18}$$

the transformation law in the Wess-Zumino gauge. We may summarize the result: a massless Weyl-spinor multiplet in the adjoint representation together with an ordinary Yang-Mills field forms a supersymmetric system (on shell, where D = 0).

Scalar self-interaction and mass terms will be locally invariant and therefore permitted in the action once they are rigidly invariant. If T^i are the generators of any unitary representation of our group G, the matter fields transform under rigid transformations according to

$$\delta_{rig}\, A = -\, i\, \tilde{\omega}\, A \tag{5.19}$$

with $\tilde{\omega} \equiv \omega^i\, T^i$, ω^i = const, and vector notation for A_a. For the special case that the A transform under the adjoint representation one may rewrite (5.19)

$$\delta_{rig}\, A = i\,[A, \omega]$$

$$\text{with} \quad A \equiv A^i\, \tau^i,\ \omega \equiv \omega^i \tau^i\ , \quad \omega^i = \text{const.} \tag{5.19a}$$

In any case invariant interaction and mass terms will be of the usual form

$$\Gamma_m = m_{ab} \int dS\, A_a A_b + \bar{m}_{ab} \int d\bar{S}\, \bar{A}_a \bar{A}_b$$

$$\Gamma_h = h_{abc} \int dS\, A_a A_b A_c + \bar{h}_{abc} \int d\bar{S}\, \bar{A}_a \bar{A}_b \bar{A}_c \tag{3.20}$$

A linear term is only permitted for a singlet under the gauge group. The total invariant action reads therefore

$$
\begin{aligned}
\Gamma_{inv} &= \Gamma_{YM} + \frac{1}{16} \int dV\, \bar{A}\, e^{\tilde{\Phi}}\, A + \Gamma_m + \Gamma_h \\
\Gamma_{YM} &= \frac{-1}{128} \frac{1}{g^2} \operatorname{Tr} \int dS\, F^{\alpha} F_{\alpha} \qquad (5.20) \\
\tilde{\Phi} &\equiv \Phi^i\, T^i
\end{aligned}
$$

On this action is also based the discussion of symmetry breaking of which we shall mention only few results. Spontaneous breaking of the internal symmetry leaving supersymmetry intact is easily possible and leads then to massive vector supermultiplets. Spontaneous breaking of supersymmetry occurs rather rarely and requires either singlets of matter fields (F - type breaking) or the presence of an abelian factor in the gauge group (D - type breaking). The first situation is exemplified by the O'Raifeartaigh model (section 3.3) (although it is possible that the breaking is operative only after gauging the group), the second by S'QED (section 4.3). These remarks conclude our somewhat sketchy presentation of SYM in its standard form.

What we still have to do is to study the uniqueness of the group transformation law (5.5). We have not shown, for instance, that (5.5) is the unique extension of (5.17), the latter being the postulated link of SYM to YM. Let us therefore modify (5.6) to

$$
\begin{aligned}
\delta\Phi = i\, (\Lambda - \bar{\Lambda}) + \frac{i}{2}\, x\, [\Phi, \Lambda + \bar{\Lambda}] \\
+ a_2\, i\, \{\Phi, \Lambda - \bar{\Lambda}\} + o(\Phi^2, \Lambda)
\end{aligned}
$$

and check whether there exist terms bilinear in Φ, linear in $\Lambda(\bar{\Lambda})$ such that we have a composition law:

$$
[\delta_2, \delta_1]\, \Phi = \delta_3\, \Phi \qquad (5.21)
$$

up to that order in the fields.

The trial is motivated by the desire to keep the abelian approximation $(i(\Lambda-\Lambda))$ and to stay in the algebra: $\{\Phi,\Lambda-\bar{\Lambda}\}$ is therefore more properly to be understood as $\{\Phi,\Lambda-\bar{\Lambda}\}$ - Tr $\{\Phi,\Lambda-\bar{\Lambda}\}$ we shall denote this for the time being by $\{...\}'$. The first - very simple, but non-trivial - test is, of course, the order 1, already written, in Φ:

$$\delta_2\delta_1\Phi = \frac{i}{2} x\, [i(\Lambda_2-\bar{\Lambda}_2),\ \Lambda_1+\bar{\Lambda}_1] + a_2\, i\, \{i(\Lambda_2-\bar{\Lambda}_2),\ \Lambda_1-\bar{\Lambda}_1\}'$$

$$[\delta_2,\delta_1]\Phi = -\frac{x}{2}[\Lambda_2-\bar{\Lambda}_2,\ \Lambda_1+\bar{\Lambda}_1] + \frac{x}{2}[\Lambda_1-\bar{\Lambda}_1,\ \Lambda_2+\bar{\Lambda}_2]$$

$$= -x\,[\Lambda_2,\Lambda_1] + x\,[\bar{\Lambda}_2,\bar{\Lambda}_1]$$

$$= i\,x\,(i[\Lambda_2,\Lambda_1] - i\,[\bar{\Lambda}_2,\bar{\Lambda}_1]) = i\,\delta_3\,\Phi \tag{5.22}$$

i.e. the transformation δ_3 has parameter

$$\Lambda_3 = x\,[\Lambda_2,\Lambda_1] \tag{5.23}$$

and the anti-commutator terms have dropped out without restriction on a_2. In the next order the corresponding calculation fixes $x = 1$ and enforces as additional terms in $\delta\Phi$:

$$\delta\Phi = i(\Lambda-\Lambda) + \frac{i}{2}[\Phi,\Lambda+\bar{\Lambda}] + i\,a_2\,\{\Phi,\Lambda-\bar{\Lambda}\}'$$
$$+\, i\left(\frac{1}{12} - \frac{a_2^2}{3}\right)[\Phi[\Phi,\ \Lambda-\bar{\Lambda}]] + O(\Phi^3,\Lambda) \tag{5.24}$$

Hence, at $a_2 = 0$ one obtains the same results as previously (5.6), but $a_2 \neq 0$ is quite allowed! It is therefore clear that at this order in the fields the transformation (5.6) are not the most general ones yielding a closed algebra. We thus have to answer two questions: (1) Can one consistently continue to higher orders in the fields? (2) Are there more free parameters to come? In order to answer these questions we shall reformulate this purely group theoretical problem with the help of the Becchi-Rouet-Stora transformations (BRS-transformations):

One performs the replacement

$$i\,\Lambda \to c_+ \equiv c_+^i \tau^i \qquad i\,\bar{\Lambda} \to \bar{c}_+ \equiv \bar{c}_+^i \tau^i \tag{5.25}$$

in the gauge transformation law, with c_+^i anti-commuting chiral fields (the Faddeev-Popov-fields, $\Phi\pi$- fields), $\bar{c}_+^i$ the conjugate anti-chiral ones. I.e. one defines

$$\begin{aligned} b\Phi &= c_+ - \bar{c}_+ + \tfrac{1}{2}\,[\Phi, c_+ + \bar{c}_+] + a_2\,\{\Phi, c_+ - \bar{c}_+\}' \\ &\quad + \left(\tfrac{1}{12} - \tfrac{a_2^2}{3}\right)[\Phi[\Phi, c_+ - \bar{c}_+]] + O(\Phi^3) \equiv Q \end{aligned} \tag{5.26}$$

for the transformation of Φ, and for the new fields c_+:

$$bc_+ = -\,c_+c_+ \qquad b\bar{c}_+ = -\,\bar{c}_+\bar{c}_+ \tag{5.27}$$

The reason for switching to BRS-transformations is a technical one: First of all the requirements

$$b^2\Phi = 0 \tag{5.28}$$

$$b^2c_+ = 0 \qquad b^2\bar{c}_+ = 0 \tag{5.29}$$

i.e. nilpotency of the transformations on the fields Φ, c_+, embodies the entire algebraic structure of the gauge transformations δ on Φ. Indeed [R.5,1], setting

$$bc_+^i = -\,x^i_{[jk]}\, c_+^j\, c_+^k \tag{5.30}$$

and requiring (5.28) enforces relations on the $x^i_{[jk]}$ which show that they are the structure constants of a Lie group i.e. (5.25) and (5.27) specify only in a very precise sense about which group one is talking, whereas (5.29) dictates "Lie group". Analogously (5.28) determines the actual transformation law for Φ. Secondly, the equations (5.28) (5.29) permit a much easier inductive procedure than (5.20).

Let us define transformations b_k by decomposing

$$b = b_0 + b_1 + b_2 + \ldots \tag{5.31}$$

into transformations which raise the order in the fields as indicated by the index k:

$$b_o\Phi = c_+ - \bar{c}_+ \equiv Q_1$$

$$b_1\Phi = \frac{1}{2}\,[\Phi, c_+ + \bar{c}_+] + a_2\,\{\Phi, c_+ - \bar{c}_+\}' \equiv Q_2 \tag{5.32}$$

$$b_k\Phi \equiv Q_{k+1}$$

$$b_o c_+ = 0 \qquad\qquad b_o\bar{c}_+ = 0$$

$$b_1 c_+ = -\,c_+c_+ \qquad\qquad b_1\bar{c}_+ = -\,\bar{c}_+\bar{c}_+ \tag{5.33}$$

$$b_k c_+ = 0 \quad k \geq 2 \qquad b_k\bar{c}_+ = 0 \quad k \geq 2$$

We now proceed by induction. Let us suppose that we have satisfied (5.28) up to and including the order n in the fields, i.e.

$$(b^2\Phi)_n = (bQ)_n = 0 \qquad n \geq 2 \tag{5.34}$$

Then we have to solve

$$0 \stackrel{!}{=} (bQ)_{n+1} = b_n Q_1 + b_{n-1}Q_2 + \ldots\, b_1 Q_n + b_o Q_{n+1} \tag{5.35}$$

Abbreviating

$$H_{n+1} \equiv b_n Q_1 + b_{n-1}Q_2 + \ldots + b_1 Q_n \tag{5.36}$$

it is seen that (5.35) implies a consistency condition for H_{n+1} since $b_o\, b_o \equiv 0$:

$$b_o H_{n+1} = 0 \tag{5.37}$$

This equation, in turn, certainly has the solution

$$H_{n+1} = b_o \hat{H}_{n+1} \quad . \tag{5.38}$$

If it has only this one, then (5.35) reads

$$0 = b_o (\hat{H}_{n+1} + Q_{n+1}) \tag{5.39}$$

hence

$$Q_{n+1} = - \hat{H}_{n+1} + Q'_{n+1} \tag{5.40}$$

with $b_o \, Q'_{n+1} = 0.$ (5.41)

Now (5.41) is again of the type (5.37) (with one c_+ less involved), hence again one may hope that

$$Q'_{n+1} = b_o \hat{Q}_{n+1} \tag{5.42}$$

is the only solution and one has completely determined Q_{n+1}.

It is instructive to go explicitly through the first orders n.

n = 1 $\quad (bQ)_1 = b_oQ_1 = b_o(c_+-\bar{c}_+) = 0$ (5.43)

is correct by definition (5.32) and (5.33).

n = 2

(5.34) reads: $b_1Q_1 + b_oQ_2 = 0$ (5.44)

The consistency (5.37) arises for $b_1Q_1 = H$:

$$b_ob_1Q_1 = -b_1b_oQ_1 = 0 \tag{5.45}$$

Here the second equality uses (5.43) whereas the first

$$(b_ob_1 + b_1b_o)Q_1 = 0 \tag{5.46}$$

is nothing but (5.29) written in terms of the expansion (5.31) up to order 1 in the fields.

$$\text{Now}\quad H = b_1Q_1 = -\,c_+c_+ + \bar{c}_+\bar{c}_+ \tag{5.47}$$

$$= b_0\left(-\tfrac{1}{2}[\Phi,c_+ + \bar{c}_+]\right) \tag{5.48}$$

$$\hat{H} = -\tfrac{1}{2}[\Phi,c_+ + \bar{c}_+] \tag{5.49}$$

$$\text{i.e.}\quad Q_2 = +\tfrac{1}{2}[\Phi,c_+ + \bar{c}_+] + a_2\{\Phi,c_+ - \bar{c}_+\}' \tag{5.50}$$

$$Q_2 = \tfrac{1}{2}[\Phi,c_+ + \bar{c}_+] + a_2\, b_0(\Phi^2)' \tag{5.51}$$

This calculation is therefore exactly the one we did before for the gauge variations. Hence we know also that there is just the freedom of having the a_2-term. The ' in (5.50) and (5.51) indicates removal of the trace and arises from the additional requirement that $b\Phi = Q$ has to be in the adjoint representation under rigid transformations. Looking back at (5.44) it is also clear that if there is any solution for Q_2, such that (5.44) holds, it will be ambiguous by a $b_0(\ldots)$ since $b_0b_0 \equiv 0$.

For the induction we learn that the proof of (5.37) requires relations of the type

$$(b_{k-1}b_0 + b_{k-2}b_1 + \ldots + b_1b_{k-2} + b_0b_{k-1})\Phi = 0 \tag{5.52}$$

coming from $b^2\Phi = 0$, and solving the "cohomology" problem:

$$b_0X = 0 \Longrightarrow X = b_0\hat{X} \tag{5.53}$$

$(X = H_{n+1},\ Q'_{n+1},\quad n \geq 2)$.

For the actual proof of (5.37) we assume now in accordance with the induction hypothesis (5.34)

$$(bQ)_k = 0 \quad \text{for} \quad 1 \leq k \leq n \tag{5.54}$$

i.e. (5.52) for the same range of k. Since $b_n\, Q = 0$ $(n \geq 2)$,

$$b_0\,H_{n+1} = b_0b_{n-1}Q_2 + b_0b_{n-2}Q_3 + \ldots + b_0b_2Q_{n-1} + b_0b_1Q_n. \tag{5.55}$$

Using (5.52) on each term, this becomes

$$= - (b_{n-1}b_0 + b_{n-2}b_1 + \dots + b_2b_{n-3} + b_1b_{n-2})\, Q_2$$

$$- (b_{n-2}b_0 + b_{n-3}b_1 + \dots + b_2b_{n-4} + b_1b_{n-3})\, Q_3$$

$$\cdot$$
$$\cdot$$
$$\cdot$$

$$- (b_2b_0 + b_1b_1)\, Q_{n-1}$$

$$- b_1b_0Q_n \quad . \qquad (5.56)$$

Upon reordering the sum, this is

$$= - b_1\, (b_{n-2}Q_2 + b_{n-3}Q_3 + \dots + b_1Q_{n-1} + b_0Q_n)$$

$$- b_2\, (b_{n-3}Q_2 + b_{n-4}Q_3 + \dots + b_1Q_{n-2} + b_0Q_{n-1})$$

$$\cdot$$
$$\cdot$$
$$\cdot$$

$$- b_{n-2}\, (b_1Q_2 + b_0Q_3)$$

$$- b_{n-1}\, (b_0Q_2) \qquad (5.57)$$

We now show that each line vanishes:

$$b_0Q_2 = b_0b_1\Phi = - b_1b_0\Phi = c_+c_+ - \bar{c}_+\bar{c}_+ \qquad (5.58)$$

i.e. $b_{n-1}b_0Q_2 = 0$ for $n \geq 2$

$$b_1Q_2 + b_0Q_3 = (b_1b_1 + b_0b_2)\Phi = - b_2b_0\Phi = 0 \qquad (5.59)$$

$$\cdot$$
$$\cdot$$
$$\cdot$$

$$b_{n-3}Q_2 + b_{n-4}Q_3 + \dots + b_1 Q_{n-2} + b_0 Q_{n-1}$$

$$= (b_{n-3}b_1 + b_{n-4}b_2 + \dots + b_1 b_{n-3} + b_0 b_{n-2})\,\Phi$$

$$= - b_{n-2}b_0\Phi = 0 \tag{5.60}$$

$$b_{n-2}Q_2 + \dots + b_0 Q_n = (b_{n-2}b_1 + b_{n-3}b_2 + \dots + b_0 b_{n-1})\,\Phi$$

$$= - b_{n-1}b_0\Phi = 0 \tag{5.61}$$

hence $\quad b_0\, H_{n+1} = 0 \quad .$ (5.62)

The solution of the cohomology problem (5.53) is shown in appendix C. This proves (5.38) and (5.42).

As the result we give now the anwsers to the questions posed before: (1) One can consistently continue the transformation law to all orders in the fields. (2) At each order $n \geq 2$ in the fields there appear new free parameters:

$$Q_{n+1} = - \hat{H}_{n+1} + a_n b_0 (\Phi^n)' \tag{5.63}$$

this undetermined part $a_n b_0 (\Phi^n)'$ we may write more precisely in components:

$$\sum_{\omega=1}^{\Omega(n)} a_{\omega n}\, s^{\omega}_{i(i,\dots i_n)}\, (c_+ - \bar{c}_+)_{i_1}\, \Phi_{i_1} \dots \Phi_{i_n} \tag{5.64}$$

Here $s^{\omega}_{i(i_1\dots i_n)}$ are invariant tensors under the group G completely symmetric in $(i_1\dots i_n)$, $\Omega(n)$ is the number of such tensors and $a_{\omega n}$ are arbitrary parameters.* In the example above, n = 2 e.g. for SU(N), there

---

* The expression (5.64) thus transforms as the adjoint representation of the rigid gauge group if c_+ and Φ do so. This will be seen later to follow from the requirement of rigid gauge invariance.

is one such tensor: d_{ijk} and it was formed by taking out the trace from $b_o(\Phi^2)$. The entire transformation Q is determined once one has fixed these terms.

Let us repeat: demanding a closed algebra of gauge transformations (5.20) or equivalently (5.26) with (5.29) we find as solution (5.63) with free parameters $a_{\omega n}$ (5.64). Only the case $a_{\omega n} = 0$ coincides with the solution (5.5) (5.6). Construction of invariants under this general transformations is postponed until section 5.3.

5.2. BRS - invariance

In this section we wish to reformulate gauge invariance as BRS (Becchi-Rouet-Stora)-invariance and in particular to derive the Ward-identity associated with it - the Slavnov-identity. Eventually it will be this latter which defines the theory.

Before doing so let us emphasize again that the BRS-transformations have two aspects which have conceptually nothing to do with each other. The first is that they comprise the algebraic (and geometric) structure of the gauge group - a classical object. This property has been used in the preceding section. The second is their role in quantized gauge field theory - which we shall use presently. Quantization via gauge fixing introduces ghosts into the theory. In the non-abelian case those are no longer free and so they destroy unitarity. In order to compensate this effect additional fields are employed - the Faddeev-Popov fields - and it is their behaviour which is governed by the BRS-transformations. This conversion process from gauge Ward-identity to Slavnov-identity has been described in detail in [R.5], so that here we content ourselves with the derivations of the Slavnov identity and do not motivate this any further. Let us start our presentation with the BRS-transformations (5.26) at $a_k = 0$.

$$s\Phi = c_+ - \bar{c}_+ + \frac{1}{2}[\Phi, c_+ + \bar{c}_+] + \frac{1}{12}[\Phi[\Phi, c_+ - \bar{c}_+]] + \ldots \equiv Q_s(\Phi, c_+) \tag{5.65}$$

$$s\,c_+ = -\,c_+c_+$$

$$s\,A = -\,\tilde{c}_+\,A \qquad \tilde{c}_+ \equiv c^iT^i$$

$$s\,\bar{A} = \bar{A}\,\bar{\tilde{c}}_+ \tag{5.66}$$

Γ_{inv} (5.20) is of course invariant under (5.65) (5.66) since on the fields Φ,A the BRS-transformations are just gauge transformations with fields c_+ as parameters. But the main content of this invariance reveals itself when we consider the naive variation of the gauge fixing:

$$s\,\Gamma^{(\alpha)}_{g.f.} = \frac{1}{128}\int dV(-\frac{1}{\alpha})s(DD\Phi\overline{DD}\bar{\Phi}) = \frac{1}{128}\int dV(-\frac{1}{\alpha})(DDQ_s\overline{DD}\bar{\Phi} + DD\Phi\overline{DD}\bar{Q}_s) \tag{5.67}$$

$$s\,\Gamma^{(\alpha,B)}_{g.f.} = \frac{1}{8}\int dV(B\;DD\;Q_s + \bar{B}\;\overline{DD}\;\bar{Q}_s) \tag{5.68}$$

$$s\,\Gamma^{(B)}_{g.f.} = -\,\frac{1}{128}\int dV\;(B + \bar{B})\;Q_s \tag{5.69}$$

where $\quad s\,B = s\,\bar{B} = 0$ (5.70)

has been assumed for the B-gauges, and observe that they can be compensated by varying corresponding terms $\Gamma_{\Phi\pi}$ containing another set of $\Phi\pi$-fields $c_-,\bar{c}_-$ (chiral, anti-commuting):

$$s\,\Gamma^{(\alpha)}_{\Phi\pi} = s\,\frac{1}{128}\int dV(-\frac{1}{\alpha})\;(DDQ_sc_- - \bar{c}_-\;\overline{DD}\;Q_s) \tag{5.71}$$

with $s\,c_- = \overline{DD}\;\bar{\Phi} \qquad s\bar{c}_- = DD\;\Phi$ (5.72)

$$s\,\Gamma^{(\alpha,B)}_{\Phi\pi} = s\,\frac{1}{8}\int dV(DDQ_sc_- + \overline{DD}\;\bar{Q}_s\bar{c}_-) \tag{5.73}$$

$$s\,\Gamma^{(B)}_{\Phi\pi} = s\;(-\,\frac{1}{128})\int dV\;Q_s(c_- + \bar{c}_-) \tag{5.74}$$

with $s\,c_- = B \qquad s\,\bar{c}_- = \bar{B}$ (5.75)

Compensation occurs indeed if we let the operation s anticommute with Q. The terms bilinear in c_+, c_- resp. define a conserved charge $Q_{\Phi\pi}$, the Faddeev-Popov charge, with the (arbitrary) assignment ± 1 for $c_\pm$

(and $\bar{c}_{\pm}$). One also notes

$$s^2 c_- = 0 \qquad \text{for B-gauges} \tag{5.76}$$

$$s^2 c_- = \overline{DD}\, Q \qquad \text{for } \alpha\text{-gauges.} \tag{5.77}$$

We may summarize the above by stating that the actions

$$\Gamma^{(o)} = \Gamma_{inv} + \Gamma_{g.f.} + \Gamma_{\Phi\pi} \tag{5.78}$$

(for all three gauge fixings) are naively invariant under the BRS-transformations (5.65),(5.66),(5.70),(5.72),(5.75).

Naive BRS-invariance of the action (5.78) can thus be expressed as

$$\begin{aligned} s\,\Gamma^{(o)} \equiv \mathrm{Tr} \int dV \left(Q_s \frac{\delta\Gamma^{(o)}}{\delta\Phi} + \Phi\left(\frac{\delta\Gamma^{(o)}}{\delta c_-} + \frac{\delta\Gamma^{(o)}}{\delta\bar{c}_-}\right) \right) \\ + \int dS \left(- \mathrm{Tr}\; c_+ c_+ \frac{\delta\Gamma^{(o)}}{\delta c_+} - (\tilde{c}_+ A)_a \frac{\delta\Gamma^{(o)}}{\delta A_a} \right) \\ + \int d\bar{S} \left(- \mathrm{Tr}\; \bar{c}_+ \bar{c}_+ \frac{\delta\Gamma^{(o)}}{\delta\bar{c}_+} + (\bar{A}\,\tilde{\bar{c}}_+)_a \frac{\delta\Gamma^{(o)}}{\delta\bar{A}_a} \right) = 0 \end{aligned} \tag{5.79}$$

for the α - gauge,

$$\begin{aligned} s\Gamma^{(o)} \equiv \mathrm{Tr} \int dV\, Q_s \frac{\delta\Gamma^{(o)}}{\delta\Phi} + \mathrm{Tr}\left(\int dS\; B\, \frac{\delta\Gamma^{(o)}}{\delta c_-} + \int d\bar{S}\; \bar{B}\, \frac{\delta\Gamma^{(o)}}{\delta\bar{c}_-} \right) \\ + (c_+, A \text{ - terms}) = 0 \end{aligned} \tag{5.80}$$

for the B-gauges.

In order to deal with the non-linear field transformations in higher orders we introduce external fields coupled to them:

$$\begin{aligned} \Gamma_{ext.field} = \mathrm{Tr} \int dV \rho Q_s - \mathrm{Tr}\left(\int dS\, \sigma\, c_+ c_+ + \int d\bar{S}\, \bar{\sigma}\, \bar{c}_+ \bar{c}_+\right) \\ - dS \int Y\, \tilde{c}_+ A - \int d\bar{S}\, \bar{A}\, \tilde{\bar{c}}_+ \bar{Y} \end{aligned} \tag{5.81}$$

$\rho \equiv \rho^i \tau^i$ ρ^i real vector superfield

$\sigma \equiv \sigma^i \tau^i$ σ^i chiral superfield

$Y \equiv (Y_a)$ Y_a chiral superfield

We may maintain naive BRS-invariance by prescribing that the external fields do not transform, since on the elementary fields transforming non-linearly, s is nilpotent:

$$s^2\Phi = s^2 c_+ = s^2 A = 0 \tag{5.82}$$

With the help of the external fields - this was the reason for their introduction - we may express all composite field variations by differentiation with respect to the corresponding external field:

$$\frac{\delta\Gamma^{(0)}}{\delta\rho} = Q_s \qquad \frac{\delta\Gamma^{(0)}}{\delta\sigma} = - c_+ c_+ \qquad \frac{\delta\Gamma^{(0)}}{\delta Y} = - \tilde{c}_+ A \tag{5.83}$$

for $\Gamma^{(0)} = \Gamma_{inv} + \Gamma_{g.f.} + \Gamma_{\Phi\pi} + \Gamma_{ext.f.}$

Inserting these relations into the naive Slavnov-identity (5.79), (5.80) we obtain:

$$\begin{aligned} \mathcal{s}(\Gamma^{(o)}) \equiv \operatorname{Tr} \int dV \left(\frac{\delta\Gamma^{(0)}}{\delta\rho} \cdot \frac{\delta\Gamma^{(0)}}{\delta\Phi} + \Phi \left(\frac{\delta\Gamma^{(0)}}{\delta c_-} + \frac{\delta\Gamma^{(0)}}{\delta \bar{c}_-} \right) \right) \\ + \int dS \left(\operatorname{Tr} \frac{\delta\Gamma^{(0)}}{\delta\sigma} \cdot \frac{\delta\Gamma^{(0)}}{\delta c_+} + \frac{\delta\Gamma^{(0)}}{\delta Y} \cdot \frac{\delta\Gamma^{(0)}}{\delta A} \right) \\ + \int d\bar{S} \left(\operatorname{Tr} \frac{\delta\Gamma^{(0)}}{\delta\bar{\sigma}} \cdot \frac{\delta\Gamma^{(0)}}{\delta \bar{c}_+} + \frac{\delta\Gamma^{(0)}}{\delta \bar{Y}} \cdot \frac{\delta\Gamma^{(0)}}{\delta \bar{A}} \right) = 0 \end{aligned} \tag{5.84}$$

for the α-gauge,

$$\begin{aligned} \mathcal{s}(\Gamma^{(o)}) \equiv \operatorname{Tr} \int dV \frac{\delta\Gamma^{(0)}}{\delta\rho} \cdot \frac{\delta\Gamma^{(0)}}{\delta\Phi} + \operatorname{Tr} \left(\int dS B \frac{\delta\Gamma^{(0)}}{\delta c_-} + \int d\bar{S}\, \bar{B} \frac{\delta\Gamma^{(0)}}{\delta \bar{c}_-} \right) \\ + (c_+,\ A\text{-terms}) = 0 \end{aligned} \tag{5.85}$$

for the B-gauges.

It is in this Γ-non-linear functional form* that the Slavnov-identity permits the generalization to all orders of perturbation theory.

We now have to explore somewhat this non-linear functional. Let us deal with the B-gauge formulation first. Eventually we shall define the theory by (5.85) and the gauge conditions (4.34) (4.43)

$$(\alpha,B) \qquad \frac{\delta\Gamma}{\delta B} = \alpha\ \overline{DD}\ \bar{B} + \frac{1}{8}\overline{DD}\ DD\ \Phi \tag{5.86}$$

$$(B) \qquad \frac{\delta\Gamma}{\delta B} = -\frac{1}{128}\overline{DD}\ \Phi \tag{5.87}$$

(and their conjugates).

Hence a <u>necessary</u> condition which has to be satisfied by the solution Γ is

$$\frac{\delta}{\delta B}\ \mathcal{s}(\Gamma) = 0 \tag{5.88}$$

which yields by use of (5.86) (5.87)

$$(\alpha,B) \qquad \frac{1}{8}\overline{DD}\ DD\frac{\delta\Gamma}{\delta\rho} + \frac{\delta\Gamma}{\delta c_-} = 0 \tag{5.89}$$

$$(B) \qquad -\frac{1}{128}\overline{DD}\frac{\delta\Gamma}{\delta\rho} + \frac{\delta\Gamma}{\delta c_-} = 0 \tag{5.90}$$

(and their conjugates). But these are just the equations of motion for the ghosts c_-, $\bar{c}_-$. Using them in (5.85) we arrive at

$$\mathcal{s}(\Gamma) = \mathrm{Tr}\int dV\ \frac{\delta\Gamma}{\delta\rho}\left(\frac{\delta\Gamma}{\delta\Phi} - \frac{1}{8}DDB - \frac{1}{8}DD\bar{B}\right) + (c_+\ ,\ A\text{-terms}) \tag{5.91}$$

for (α,B)

$$\mathcal{s}(\Gamma) = \mathrm{Tr}\int dV\ \frac{\delta\Gamma}{\delta\rho}\left(\frac{\delta\Gamma}{\delta\Phi} + \frac{1}{128}(DDB + \overline{DDB})\right) + (c_+\ ,\ A\text{-terms}) \tag{5.92}$$

for (B).

---

* This is underlined by the notation. sΓ (5.79) (5.80) was the linear action of the BRS-variation on Γ.

Recalling now (4.35)

$$\bar{\Gamma} = \Gamma - \alpha \mathrm{Tr} \int dV\, B\bar{B} - \frac{1}{8} \mathrm{Tr} \int dV\, (B\, DD\Phi + \bar{B}\, \overline{DD}\bar{\Phi}) \tag{5.93}$$

for (α, B), and (4.44)

$$\bar{\Gamma} = \Gamma + \frac{1}{128} \mathrm{Tr} \int dV\, (B+\bar{B})\Phi \tag{5.94}$$

for (B),

we may rewrite the bracket multiplying $\frac{\delta\Gamma}{\delta\rho}$ as $\frac{\delta\bar{\Gamma}}{\delta\Phi}$, i.e. performing this transition from Γ to $\bar{\Gamma}$ we obtain

$$s(\Gamma) = \frac{1}{2} B_{\bar{\Gamma}} \bar{\Gamma} \tag{5.95}$$

(appropriate integration and Tr understood) with

$$\begin{aligned} B_{\bar{\Gamma}} &= \frac{\delta\bar{\Gamma}}{\delta\rho} \cdot \frac{\delta}{\delta\Phi} + \frac{\delta\bar{\Gamma}}{\delta\Phi} \cdot \frac{\delta}{\delta\rho} + \frac{\delta\bar{\Gamma}}{\delta\sigma} \cdot \frac{\delta}{\delta c_+} + \frac{\delta\bar{\Gamma}}{\delta c_+} \cdot \frac{\delta}{\delta\sigma} \\ &+ \frac{\delta\bar{\Gamma}}{\delta Y} \cdot \frac{\delta}{\delta A} + \frac{\delta\bar{\Gamma}}{\delta A} \cdot \frac{\delta}{\delta Y} + (\text{conj. for chiral fields}). \end{aligned} \tag{5.96}$$

And $B_{\bar{\Gamma}}$ acts as a <u>linear</u> functional operator on $\bar{\Gamma}$: The non-linear action of s on Γ has been translated into the linear action of $B_{\bar{\Gamma}}$ on $\bar{\Gamma}$ - for functionals Γ satisfying the ghost equations of motion (5.90) (5.91).

In the α-gauge we shall proceed tentatively in the same way: namely assume that the ghost equations of motion hold

$$\frac{\delta\Gamma}{\delta c_-} = \frac{1}{128} \left(\frac{1}{\alpha}\right) \overline{DD}\, DD \frac{\delta\Gamma}{\delta\rho} \tag{5.97}$$

insert them into (5.84) and using

$$\bar{\Gamma} = \Gamma + \frac{1}{128} \cdot \frac{1}{\alpha} \mathrm{Tr} \int dV\, DD\Phi\, \overline{DD}\bar{\Phi} \tag{5.98}$$

we arrive at just the same result:

$$s(\Gamma) = \frac{1}{2} B_{\bar{\Gamma}} \bar{\Gamma} \tag{5.99}$$

How can we justify the use of the ghost equation of motion? A good hint is given by remembering that in this gauge the BRS-transformation was not completely nilpotent:

$$sc_- = \overline{DD}\Phi \qquad s^2c_- = s\,\overline{DD}\Phi = \overline{DD}Q_s = \overline{DD}\,\frac{\delta\Gamma}{\delta\rho} \tag{5.100}$$

This suggests that besides (5.84) there should also hold

$$\mathcal{s}^2(\Gamma) \equiv \mathrm{Tr} \int dV\, \frac{\delta\Gamma}{\delta\rho}\left(\frac{\delta\Gamma}{\delta c_-} + \frac{\delta\Gamma}{\delta \bar{c}_-}\right) = 0 \tag{5.101}$$

(here the first equation is not yet proved).

The solution of this condition is just

$$\frac{\delta\Gamma}{\delta c_-} \propto \overline{DD}\, DD\, \frac{\delta\Gamma}{\delta\rho} \tag{5.102}$$

due to the anti-commutativity of c_- and ρ. Thus, by fixing the proportionality constant - a normalization condition - we arrive at (5.97).

There remains again a question: How can (5.101) be proved to be the repeated non-linear application of $\mathcal{s}$ to $\mathcal{s}(\Gamma)$? This can be done by going over to connected Green's functions (since on those the BRS-transformations are linear), repeating $\mathcal{s}$ there and going back to Γ. The result is (5.101) and the necessary formulae are provided in App. B.

Let us recapitulate: the postulate $\mathcal{s}(\Gamma) = 0$ enforces the ghost equations of motion to hold

$$\begin{aligned} &- \text{ in the } \alpha\text{-gauge via } \mathcal{s}^2(\Gamma) = 0 \\ &- \text{ in the B-gauges via } \frac{\delta}{\delta B}\, \mathcal{s}(\Gamma) = 0 \end{aligned} \tag{5.103}$$

Having satisfied these we can get rid of all gauge dependence by utilizing $\bar{\Gamma}$ and arrive then at the linearized form of the Slavnov-identity

$$\mathcal{s}(\Gamma) = \frac{1}{2}\, B_{\bar{\Gamma}}\bar{\Gamma} = 0 \tag{5.99}$$

for all three gauges.

Let us note in passing relations which will be crucial in higher orders, namely, the functional differential operator B_γ satisfies

$$B_\gamma B_\gamma \gamma \equiv 0 \qquad \text{for any } \gamma \tag{5.104}$$

$$B_\gamma B_\gamma \quad = 0 \qquad \text{if } B_\gamma \gamma = 0 \tag{5.105}$$

The linear action of $B_{\bar{\Gamma}}$ on $\bar{\Gamma}$ permits us to interpret $B_{\bar{\Gamma}}$ as a transformation operator (like any one of the Ward-identity operators W_x before (3.24)).

For the special case $\bar{\Gamma} = \bar{\Gamma}^{(0)}$ (tree approximation, above) we give it a name:

$$\mathscr{b} \equiv B_{\bar{\Gamma}^{(o)}} \tag{5.106}$$

and note: $\mathscr{b}^2 = 0$ (5.107)

(due to 5.105 for $\gamma = \bar{\Gamma}^{(0)}$).

In fact,

$$\mathscr{b} \begin{cases} \Phi \\ c_+ \\ A \end{cases} = s \begin{cases} \Phi \\ c_+ \\ A \end{cases} \qquad \begin{aligned} \mathscr{b}\bar{\rho} &= \frac{\delta\bar{\Gamma}^{(o)}}{\delta\Phi} \\ \mathscr{b}\sigma &= \frac{\delta\bar{\Gamma}^{(o)}}{\delta c_+} \\ \mathscr{b}Y &= \frac{\delta\bar{\Gamma}^{(o)}}{\delta A} \end{aligned} \tag{5.108}$$

i.e. (5.99) defines first of all a transformation law for the fields, including the external ones and states then that $\bar{\Gamma}^{(0)}$ is an invariant functional under these transformations.

5.3 General solution of the Slavnov-identity

Since we want the Slavnov-identity (besides supersymmetry and rigid gauge invariance) to define our theory we have to find already at the tree level its (general) solution. Only if we can specify that do we have any hope of finding the continuation to higher orders.

We have seen in the last subsection that the ghost equations of motion

$$(\alpha) \qquad \frac{\delta\Gamma}{\delta c_-} = \frac{1}{128}\left(\frac{1}{\alpha}\right)\overline{DD}\,DD\,\frac{\delta\Gamma}{\delta\rho} \tag{5.98}$$

$$(\alpha,B) \qquad \frac{\delta\Gamma}{\delta c_-} = -\frac{1}{8}\overline{DD}\,DD\,\frac{\delta\Gamma}{\delta\rho} \tag{5.90}$$

$$(B) \qquad \frac{\delta\Gamma}{\delta c_-} = \frac{1}{128}\overline{DD}\,\frac{\delta\Gamma}{\delta\rho} \tag{5.91}$$

had to be satisfied (as a consequence of $\mathscr{s}^2(\Gamma) = 0$ for α = gauge, as a consequence of the gauge condition for the B-gauges). This means that the solution Γ we look for depends on c_-, $\bar{c}_-$ only via

$$(\alpha) \qquad \mathscr{r} = \rho + \frac{1}{128\alpha}(DDc_- + \overline{DD}\,\bar{c}_-) \tag{5.109}$$

$$(\alpha,B) \qquad \mathscr{r} = \rho - \frac{1}{8}(DDc_- + \overline{DD}\,\bar{c}_-) \tag{5.110}$$

$$(B) \qquad \mathscr{r} = \rho + \frac{1}{128}(c_- + \bar{c}_-) \tag{5.111}$$

i.e. we may eliminate all explicit dependence on c_-, $\bar{c}_-$ in Γ by performing the variable transformation $\rho \to \mathscr{r}$. In these variables the ghost equations simply read

$$\frac{\delta\Gamma(\mathscr{r})}{\delta c_-} = 0 \tag{5.112}$$

and the Slavnov identity becomes

$$\mathscr{s}(\Gamma) = \frac{1}{2}B_{\bar{\Gamma}}\bar{\Gamma} = 0 \tag{5.100}$$

with

$$B_{\bar{\Gamma}} = \frac{\delta\bar{\Gamma}}{\delta\mathscr{r}}\cdot\frac{\delta}{\delta\Phi} + \frac{\delta\bar{\Gamma}}{\delta\Phi}\cdot\frac{\delta}{\delta\mathscr{r}} + \frac{\delta\bar{\Gamma}}{\delta\sigma}\cdot\frac{\delta}{\delta c_+} + \frac{\delta\bar{\Gamma}}{\delta c_+}\frac{\delta}{\delta\sigma} + \frac{\delta\bar{\Gamma}}{\delta Y}\cdot\frac{\delta}{\delta A} + \frac{\delta\bar{\Gamma}}{\delta A}\cdot\frac{\delta}{\delta Y} \tag{5.113}$$

+ antichiral part

One way* to find the general solution Γ of (5.100) is to perform in a special solution, e.g. (5.83), all field substitutions compatible with all other requirements and to demand that the new Γ be again a solution of the Slavnov-identity [II.12,13]. The additional requirements are: naive dimensions must be respected, supersymmetry (in particular chirality) should be maintained; $\Phi\pi$-charge must be conserved; global gauge invariance should not be violated and finally R-invariance should not be broken. For the application of these postulates we have to assign quantum numbers to the fields (dimensions, R-weights, $\Phi\pi$-charges). This assignment is <u>not</u> unique and will be adapted for the different aims pursued. The one given in table 1 makes dilatation and conformal R-transformations commute with the BRS-transformation, hence is most suitable for considerations involving the superconformal group and, as it turns out, for discussion of the ultraviolet properties of the theory.

	θ	$\bar\theta$	D	$\bar D$	Φ	A	c_+	c_-	ρ,τ	Z	Y	$B_{(\alpha)}$	B
d	$-\frac{1}{2}$	$-\frac{1}{2}$	$\frac{1}{2}$	$\frac{1}{2}$	0	1	0	1	2	3	2	1	2
n	-1	1	1	-1	0	$-\frac{2}{3}$	0	-2	0	-2	$-\frac{4}{3}$	-2	0
$Q_{\Phi\pi}$	0	0	0	0	0	0	1	-1	-1	-2	-1	0	0

<u>Table 5.3.1</u>: Dimensions d, R-weights n and Faddeev-Popov charges $Q_{\Phi\pi}$. If a field Φ is assigned numbers $(d,n,Q_{\Phi\pi})$, its conjugate Φ has numbers $(d,-n,Q_{\Phi\pi})$. Fields of even (odd) Faddeev-Popov charge are commuting (anticommuting).

Dimensions, $\Phi\pi$-charge, conformal R-invariance and chirality permit:

$$\Phi \to \mathcal{F}(\Phi) \qquad \tau \to \mathcal{H}(\Phi,\tau)$$

$$c_+ \to z' c_+ \qquad \sigma \to z\,\sigma$$

$$A \to y' A \qquad Y \to y\,Y \tag{5.114}$$

---

* For another one cf. sect. 5.4.

here $\mathcal{F}$, $\mathcal{H}$ are arbitrary functions of Φ ($\mathcal{H}$ is linear in $\boldsymbol{\tau}$) and z', z, y', y are numerical matrices. Rigid gauge invariance yields further restrictions: requiring that the new fields c_+', c', A', Y' transform as the old ones means that the matrices z', z, y', y are equivalence transformations in their representation space. Since the τ's are irreducible z', z are just numbers (multiples of the unit matrix), but the y', y may be direct sums of multiples of unit matrices according to the reducibility property of T. We shall suppress this dependence and simply write numbers.

Writing components for $\mathcal{F}$ we see that

$$\mathcal{F}_i = c_1\Phi_i + \sum_{k\geq 2} \sum_{\omega=1}^{\Omega(k)} c_{\omega k}\, s^{\omega}_{i(i_1---i_k)}\, \Phi_{i_1}\cdots\Phi_{i_k} \tag{5.115}$$

where c_1 and $c_{\omega k}$ are arbitrary coefficients and s^{ω} invariant tensors which are completely symmetric in the indices $(i_1 \ldots i_k)$. $\Omega(k)$ is the number of such independent tensors for a given rank k + 1. $\mathcal{F}$ will be abbreviated by

$$\mathcal{F} = c_1\Phi + \sum_{k\geq 2} c_k(\Phi)^k \tag{5.116}$$

Analogous relations are valid for $\mathcal{H}$. We shall not write them since $\mathcal{H}$ will be a function of $\mathcal{F}$.

Let us introduce a still more condensed notation, where one index represents the Yang-Mills index and the superspace point:

$$\Phi_1 \equiv \Phi_{i_1}(z_1) \tag{5.117}$$

and the summation - integration convention:

$$\Phi_1\psi_1 \equiv \sum_{i_1} \int dz_1\, \Phi_{i_1}(z_1)\, \psi_{i_1}(z_1) \tag{5.118}$$

where dz = dV, dS, d$\bar{S}$ appropriately. In this notation the Slavnov-identity (5.100) reads

$$\frac{1}{2} B_{\bar{\Gamma}}\bar{\Gamma} = \frac{\delta\bar{\Gamma}}{\delta\boldsymbol{\tau}_1}\frac{\delta\bar{\Gamma}}{\delta\Phi_1} + \frac{\delta\bar{\Gamma}}{\delta\sigma_1}\frac{\delta\bar{\Gamma}}{\delta c_{+1}} + \frac{\delta\bar{\Gamma}}{\delta Y_1}\frac{\delta\bar{\Gamma}}{\delta A_1} + \frac{\delta\bar{\Gamma}}{\delta\bar{\sigma}_1}\frac{\delta\bar{\Gamma}}{\delta\bar{c}_{+1}} + \frac{\delta\bar{\Gamma}}{\delta\bar{Y}_1}\cdot\frac{\delta\bar{\Gamma}}{\delta\bar{A}_1} = 0 \tag{5.119}$$

We now perform in a solution $\bar{\Gamma}$ of (5.119) e.g. the one given by (5.83):

$$\bar{\Gamma} = \Gamma_{inv}\,(\Phi,A) + \mathfrak{t}_1\, Q_{s1}(\Phi,c_+) - \sigma_1(c_+c_+)_1 - Y_1(\tilde{c}_+A)_1 + \text{conj.} \tag{5.120}$$

the substitutions (5.114) and require that

$$\hat{\Gamma}(\Phi,A,c_+,\mathfrak{t},Y,Z) = \Gamma\,(\mathcal{F}\,(\Phi),\ y'A,\ z'c_+,\ \mathcal{H}(\Phi,\mathfrak{t}), yY,\ z\sigma) \tag{5.121}$$

be again a solution of (5.119). This yields

$$z\,z' = 1 \qquad y\,y' = 1 \tag{5.122}$$

$$\frac{\delta\,\mathcal{F}_2}{\delta\Phi_1}\,\frac{\delta\,\mathcal{H}_3}{\delta\mathfrak{t}_1} = \delta_{23}\ ;\ \frac{\delta\,\mathcal{H}_2}{\delta\Phi_1}\,\frac{\delta\,\mathcal{H}_3}{\delta\ \ 1} = \frac{\delta\,\mathcal{H}_3}{\delta\Phi_1}\,\frac{\delta\,\mathcal{H}_2}{\delta\ \ 1} \tag{5.123}$$

$(\delta_{23} = \delta_{i_2\,i_3}\ \delta_V\ (2,3))$.

The solutions of these equations are

$$z' = z^{-1} \qquad y' = y^{-1} \tag{5.124}$$

$$\mathcal{H}(2) = \frac{\delta}{\delta\hat{\Phi}(2)}\left[\mathfrak{t}_1\,\bar{\mathcal{F}}_1^{\,1}\,(\hat{\Phi})\right]\Big|_{\hat{\Phi}\ =\ \mathcal{F}(\Phi)}\ , \tag{5.125}$$

where z,y are constants and $\bar{\mathcal{F}}^{1}\,(\hat{\Phi})$ is the inverse of $\hat{\Phi} = \mathcal{F}(\Phi)$.

Performing in $\bar{\Gamma}$ (5.120) the substitutions

$$\begin{aligned} &c_+ \to z^{-1}\,c_+ && A \to y^{-1}\,A \\ &\sigma \to z\,\sigma && Y \to y\,Y \end{aligned} \tag{5.126}$$

is nothing but the same field amplitude renormalization in Γ. The substitution for Φ and $\mathfrak{t}$ yields:

$$\Gamma(\Phi,A,c_+,\mathfrak{t},Y,Z) = \Gamma_{inv}\ (\mathcal{F}\,(\Phi),A) + \hat{\mathfrak{t}}_1\ Q_{s1}(\mathcal{F}\,(\Phi),c_+) \tag{5.127}$$

$$- \sigma_1\ (c_+c_+)_1 - Y_1(\tilde{c}_+A)_1 + \text{conj.}$$

with

$$\hat{\mathfrak{r}} = \mathfrak{r}_1 \left.\frac{\delta \mathfrak{F}_1^{-1}(\hat{\Phi})}{\delta\hat{\Phi}}\right|_{\hat{\Phi} = \mathfrak{F}(\Phi)} \tag{5.128}$$

(Q is given by (5.6) with $i\Lambda \rightarrow c_+$; $-i\bar{\Lambda} \rightarrow -\bar{c}_+$.)

The special case

$$\mathfrak{F}(\Phi) = c_1\Phi \qquad \hat{\mathfrak{r}} \equiv \mathfrak{X}(\Phi_1\mathfrak{r}) = c_1^{-1}\mathfrak{r}, \qquad c_1 = \text{const.} \tag{5.129}$$

corresponds to a field amplitude renormalization $\Phi \rightarrow c_1\Phi$, $\mathfrak{r} \rightarrow \bar{c}_1\mathfrak{r}$ in $\bar{\Gamma}$. In Γ it is accompanied by a redefinition of the gauge parameter $\alpha \rightarrow c_1^{\,2}\alpha$ in the α gauge; $B \rightarrow c_1^{-1}B$, $\alpha \rightarrow c_1^{-2}\alpha$ in the (α, B) gauge; $B \rightarrow c_1^{-1}B$ in the B-gauge.

At vanishing external fields (5.127) shows that the substitution $\Phi \rightarrow \mathfrak{F}(\Phi)$ (5.115) leads from one invariant to another; for nonvanishing external fields one solution of the Slavnov identity is transported into another. This property solves therefore the problem we are left with from section (5.1), namely to find invariants, solutions of the corresponding Slavnov identity, for the general group transformations (5.32),(5.63) once we are able to relate the Q defined by

(5.26),(5.27),(5.28),(5.29) to $Q_2 \equiv \left.\frac{\delta\, \mathfrak{F}_2^{-1}(\hat{\Phi})}{\delta\hat{\Phi}_1}\right|_{\hat{\Phi} = \mathfrak{F}(1)\Phi} Q_{s1}(\mathfrak{F}(\Phi), c_+)$

The problem is that the BRS-transformations as determined self-consistently by (5.26) - (5.29) are "blind" to the requirements of rigid invariance which we have explicitly imposed in (5.64), i.e. we may still choose in which basis of products ... $c_+\Phi$ ... (with commutators and anti-commutators) we wish to expand the terms of $\hat{H}_{n+1}$ (5.63)*. We start therefore from (5.115)

$$\hat{\Phi}_i = \mathfrak{F}_i(\Phi) = c_1\Phi_i + \sum_{k\geq 2} \sum_{\omega=1}^{\Omega(k)} c_{\omega k} S^{\omega}_{i(i_1 \ldots i_k)}\, \Phi_{i_1} \cdots \Phi_{i_k} \quad (c_1 \neq 0)$$

---

* For example

$\Phi c_+\Phi = -\frac{1}{3}[\Phi[\Phi, c_+]] + \frac{1}{3} b_0\Phi^3$ constitutes a linear relation amongst 3 possible candidates for the basis at order 3.

and the special transformation law (5.6) ($i\Lambda \to c_+$) which we represent as

$$Q_s(\Phi,c_+) = c_+ - \bar{c}_+ + \frac{1}{2}[\Phi,c_+ + \bar{c}_+] + \frac{1}{12}[\Phi[\Phi\,,c_+ - \bar{c}_+]] + \ldots$$
$$= c_+ - \bar{c}_+ + \sum_{n\geq 1} h_n\, [\Phi[\Phi[\ldots[\Phi,c_+ - (-1)^n\bar{c}_+\underbrace{]\ldots]}_{n} \qquad (5.130)$$

Since we have to calculate $\hat{\mathfrak{r}}$ we need the inverse function

$$\Phi_i = \mathfrak{F}^{-1}(\Phi) = x_1\,\hat{\Phi}_1 + \sum_{k\geq 2}\sum_{\omega=1}^{\Omega(k)} x_{\omega k}\, S^{\omega}_{i(i_1\ldots i_k)}\,\hat{\Phi}_{i_1}\ldots\hat{\Phi}_{i_k} \qquad (5.131)$$

The coefficients $x_{\omega k}$ are computable functions of the $c_{\omega k}$:

$$x_1 = \frac{1}{c_1}\ ,\quad x_2 = -\frac{c_2}{c_1^3}\ ,\ \ldots \qquad (5.132)$$

$$x_{\omega k} = -\frac{c_{\omega k}}{c_1^{k+1}} + \bar{x}_{\omega k}\ (c_{\omega' k'}\ ;\ k' < k)$$

where the $\bar{x}_{\omega k}$ depend only on c-coefficients with $k' < k$. Hence $\hat{\mathfrak{r}}$ (5.128) is given by

$$\hat{\mathfrak{r}}_i = x_1\,\mathfrak{r}_i + \sum_{k\geq 2}\sum_{\omega=1}^{\omega(k)} k\, x_{\omega k}\,\mathfrak{r}_j\, s^{\omega}_{j(ii_1\ldots i_{k-1})}\,\hat{\Phi}_{i_1}\ldots\hat{\Phi}_{i_k} \qquad (5.133)$$

Considering now in the product $\hat{\mathfrak{r}}_i\, Q_{si}(\hat{\Phi},c_+)$ the first term

$$\hat{\mathfrak{r}}_i(c_+ - \bar{c}_+)_i = \mathfrak{r}_i(x_1(c_+ - \bar{c}_+)_i$$
$$+ \sum_{k\geq 2}\sum_{\omega=1}^{\Omega(k)} k\, x_{\omega k} s^{\omega}_{i(i_1\ldots i_k)}(c_+ - \bar{c}_+)_{i_1}\cdot\hat{\Phi}_{i_2}\ldots\hat{\Phi}_{i_k}) \qquad (5.134)$$

we find for it on substitution of $\hat{\Phi}$, the expression

$$= \tau_i \Bigg(\frac{1}{c_1} (c_+ - \bar{c}_+)_i + \sum_{k \geq 2} \sum_{\omega=1}^{\Omega(k)} \Big(-k \frac{c_{\omega k}}{c_1^2} + f_{\omega k}(c_{\omega' k'}; k' < k)\Big) s^{\omega}_{i(i_1 \ldots i_k)} \cdot (c_+ - \bar{c}_+)_{i_1} \Phi_{i_2} \ldots \Phi_{i_k} + \sum_{k \geq 2} \sum_{\alpha=1}^{A(k)} g_{\alpha k}(c_{\omega' k'}\,; k' < k)\, r^{\alpha}_{i[i_1 \ldots i_k)} (c_+ - \bar{c}_+)_{i_1} \Phi_{i_2} \ldots \Phi_{i_k} \Bigg) \tag{5.135}$$

I.e. there is a part which contains the completely symmetric tensor $s^{\omega}_{i(i_1 \ldots i_k)}$ multiplying $(c_+ - c_+)\, \Phi \ldots \Phi$ and a part which contains mixed tensors $r^{\alpha}_{i[i, \ldots i_k)}$ multiplying the same monomial. Indeed, each term of the expansion of Φ has the form

$$t_{i(i_2 \ldots i_k)}\, \tau_i (c_+ - \bar{c}_+)_{i_1} \Phi_{i_2} \ldots \Phi_{i_k} \tag{5.136}$$

and each such tensor t can be decomposed into parts s, r.

Now the commutator terms in $\hat{\tau}_i\, Q_{si}\, (\hat{\Phi}, c_+)$ are of the form

$$r_{i[i_1 \ldots i_k)}\, \tau_i (c_+ \pm \bar{c}_+)_{i_1} \Phi_{i_2} \ldots \Phi_{i_k} \tag{5.137}$$

with coefficients depending only on $c_{\omega' k'}$ with $k' < k$.

Hence the complete expression, which will define our new transformation law, is

$$\hat{\tau}_i\, Q_{si}(\hat{\Phi}, c_+) = \tau_i\, Q_i(\Phi, c_+\,; a_k) \tag{5.138}$$

$$= \tau_i \Bigg(\frac{1}{c_1} (c_+ - \bar{c}_+)_i + \sum_{k \geq 2} \sum_{\omega=1}^{\omega(k)} \Big(-k \frac{c_{\omega k}}{c_1^2} + f_{\omega k}(c_{\omega' k'}\,; k' < k)\Big) s^{\omega}_{i(i_1 \ldots i_k)} (c_+ - \bar{c}_+)_{i_1} \Phi_{i_2} \ldots \Phi_{i_k} + \sum_{k \geq 2} \sum_{\alpha=1}^{A(k)} h_{\alpha k}(c_{\omega' k'}\,; k' < k)\, r^{\alpha}_{i[i_1 \ldots i_k)} (c_+ \pm \bar{c}_+)_{i_1} \Phi_{i_2} \ldots \Phi_{i_k} \Bigg) \tag{5.139}$$

Putting now $c_1 = 1$ we note first of all that

$$- k\, c_{\omega k} + f_{\omega k}\, (c_{\omega' k'}\ ;\ k' < k) = a_{\omega k} \tag{5.140}$$

can be solved for $c_{\omega k}$ in terms of $a_{\omega k}$; second that after performing this replacement the content of the bracket multiplying η is exactly a Q as prescribed by (5.63) and (5.64): it solves $b\,Q = 0$ and has just the correct undetermined term $a_n\, b_0(\Phi^n)'$, which is thus shown to define the remainder of Q by the above construction.

5.4 Interpretation of the parameters a_k

In sections 5.1 and 5.3 we have found gauge transformations (BRS-transformations) forming a closed algebra and containing infinitely many parameters (to any finite order in the fields there is of course only a finite number of parameters). We now have to check whether these parameters have physical significance. Since they may be understood as determining the function $\mathcal{F}(\Phi)\big|_{\theta = 0} = \mathcal{F}(C)$ ($\mathcal{F}(\Phi)$ then being fixed by supersymmetry) upon which the theory depends and C is a pure gauge degree of freedom we do not expect any impact on the physical properties of the theory caused by the a_k. Indeed, what we want to show now is just that by a proper choice of gauge a theory with $a_k \neq 0$ is equivalent to a theory with $a_k = 0$. In other words, the a_k are a type of gauge parameters.

In order to derive this result we start in a generalized gauge of (α,B)-type*, given by (5.86) with Φ replaced by an arbitrary function $G(\Phi)$ lying in the adjoint representation. Its expression is similar to that of the function (5.115):

$$G_i(\Phi) = \alpha_1 \Phi + \sum_{k \geq 2} \sum_{\omega=1}^{\Omega(k)} \alpha_{\omega k}\, s^{\omega}_{i(i_1 \ldots i_k)}\, \Phi_{i_1} \ldots \Phi_{i_k} \tag{5.141}$$

Since G is a composite field it must be coupled to an external field $\mathcal{G} = \mathcal{G}_i \tau^i$ ($d = 2$, $Q_{\Phi\Pi} = 0$) and likewise its BRS variation H to an ex-

---

* See [II.13] for an equivalent discussion within an α-type gauge.

ternal field $\mathcal{K} = \mathcal{K}_i \tau^i$ (d = 0, $Q_{\Phi\pi}$ = -1). The generalization of the gauge condition (5.86) reads now

$$\frac{\delta\Gamma}{\delta B} = \overline{DD}\,\bar{B} + \overline{DD}\,DD\,\frac{\delta\Gamma}{\delta\mathcal{G}} \quad . \tag{5.142}$$

and, omitting matter fields which are irrelevant for the present discussion, we have as Slavnov identity

$$\begin{aligned} \mathcal{S}(\Gamma) \equiv \frac{\delta\Gamma}{\delta\rho_1}\frac{\delta\Gamma}{\delta\Phi_1} + B_1\frac{\delta\Gamma}{\delta c_{-1}} + \bar{B}_1\frac{\delta\Gamma}{\delta\bar{c}_{-1}} - \mathcal{G}_1\frac{\delta\Gamma}{\delta\mathcal{K}_1} \\ + \frac{\delta\Gamma}{\delta\sigma_1}\frac{\delta\Gamma}{\delta c_{+1}} + \frac{\delta\Gamma}{\delta\bar{\sigma}_1}\frac{\delta\Gamma}{\bar{c}_{+1}} = 0 \end{aligned} \tag{5.143}$$

The new ghost equation

$$\frac{\delta\Gamma}{\delta c_-} + \overline{DD}\,DD\,\frac{\delta\Gamma}{\delta\mathcal{K}} = 0 \tag{5.144}$$

follows as the old one (5.89) from the Slavnov identity (5.143) and the gauge condition (5.142).

We look now for the general solution of the Slavnov identity and of the gauge condition. As a consequence of the latter and of the ghost equation Γ has the form

$$\Gamma = B_1\bar{B}_1 + \bar{\Gamma}(\Phi, c_{+1}, \rho, \sigma, \mathcal{G}', \mathcal{K}') \tag{5.145}$$

where

$$\begin{aligned} \mathcal{G}' &= \mathcal{G} + DDB + \overline{DD}\,\bar{B} \\ \mathcal{K}' &= \mathcal{K} - DDc_- - \overline{DD}\,\bar{c}_- \end{aligned} \tag{5.146}$$

and $\bar{\Gamma}$ does not depend explicitly on B and c_- . Supersymmetry, rigid and R-invariance being taken into account, the most general form for $\bar{\Gamma}$ can be written as

$$\begin{aligned} \bar{\Gamma} = \Lambda(\Phi) + \rho_1 R_1(\Phi, c_+) + \mathcal{G}_1' G_1(\Phi) + \mathcal{K}_1' H_1(\Phi, c_+) \\ - \sigma_1(c_+c_+)_1 - \bar{\sigma}_1(\bar{c}_+\bar{c}_+)_1 \end{aligned} \tag{5.147}$$

The supersymmetric invariant Λ and the superfields R, G, H are arbitrary functions of their arguments. They have zero dimension and $\Phi\pi$-

charge 0, 1, 0, 1 respectively. The coefficient -1 in front of the last terms fixes the amplitudes of σ and $\bar{\sigma}$.

We calculate next the implication of BRS-invariance on (5.147). For this we have to insert (5.144) into (5.143):

$$\mathcal{s}(\Gamma) = \mathcal{B}(\bar{\Gamma}) \equiv \frac{\delta\Gamma}{\delta\rho_1}\frac{\delta\Gamma}{\delta\Phi_1} - \mathcal{G}_1^{\mathrm{T}}\frac{\delta\Gamma}{\delta\mathcal{X}_1} + \frac{\delta\Gamma}{\delta\sigma_1}\frac{\delta\Gamma}{\delta c_{+1}} + \frac{\delta\Gamma}{\delta\sigma_1}\frac{\delta\Gamma}{\delta\bar{c}_{+1}} = 0 \qquad (5.148)$$

and then to perform the action of B on Γ of (5.147). We thus find the following conditions on the functions appearing in $\bar{\Gamma}$:

$$\mathcal{b}R = 0 \qquad (5.149)$$

$$\mathcal{b}G = H \qquad (5.150)$$

$$\mathcal{b}H = 0 \qquad (5.151)$$

$$\mathcal{b}\Lambda = 0 \qquad (5.152)$$

where $\mathcal{b}$ is the linear operator acting on functionals of Φ, c_+ according to

$$\mathcal{b}\Phi = R, \quad \mathcal{b}c_+ = -c_+c_+ , \quad \mathcal{b}\bar{c}_+ = - \bar{c}_+\bar{c}_+ \qquad (5.153)$$

The solution of (5.149) is, as we know from sections 5.1 and 5.3, the function

$$R = Q^{\mathcal{F}}(\Phi,c_+) = \left.\frac{\delta\mathcal{F}^{-1}(\hat{\Phi})}{\delta\Phi_1} Q_{s1}(\hat{\Phi},c_+)\right|_{\hat{\Phi} = \mathcal{F}(\Phi)} \qquad (5.154)$$

$\mathcal{F}$ being given by (5.115) with $c_1 = 1$ and Q_s by (5.6).

Since $\mathcal{b}$ is nilpotent on functionals $\tau(\Phi,c_+)$

$$\mathcal{b}^2\,\tau = 0 \qquad (5.155)$$

(5.151) follows from (5.150) and H is then given by

$$H = \mathcal{S}G = R_1 \frac{\delta G}{\delta \Phi_1} = \frac{\delta \mathfrak{F}_1^{-1}(\hat{\Phi})}{\delta \hat{\Phi}_2} \; Q_{s2} \; (\hat{\Phi}, c_+) \; \frac{\delta G(\Phi)}{\delta \Phi_1}$$

$$= Q_{s2} \; (\hat{\Phi}, c_+) \frac{\delta}{\delta \hat{\Phi}_2} G \; (\mathfrak{F}^{-1}(\hat{\Phi}))\Big|_{\hat{\Phi} = \mathfrak{F}(\hat{\Phi})} \tag{5.156}$$

(5.152) is solved by

$$\Lambda(\Phi) = \Gamma_{YM} \; (\mathfrak{F}(\Phi)) \tag{5.157}$$

Γ_{YM} being given by (5.20).

Therefore the general solution of the Slavnov identity (5.143) reads

$$\Gamma(\Phi, c_+, \rho, \sigma, \mathcal{G}', \mathcal{H}') = \Big[\Gamma_{YM} \; (\hat{\Phi}) + \rho_1 \frac{\delta \mathfrak{F}_1^{-1}(\hat{\Phi})}{\delta \hat{\Phi}_2} Q_{s2} \; (\hat{\Phi}, c_+)$$

$$+ \mathcal{H}_1' \; Q_{s2} \; (\hat{\Phi}, c_+) \frac{\delta}{\delta \hat{\Phi}_2} G_1(\mathfrak{F}^{-1}(\hat{\Phi}) \Big]_{\hat{\Phi} = \mathfrak{F}(\Phi)} \tag{5.158}$$

$$- \sigma_1(c_+ c_+)_1 - \bar{\sigma}_1(\bar{c}_+ \bar{c}_+)_1 + \mathcal{G}_1' \; G_1(\Phi) + B_1 \bar{B}_1$$

Now, let us define a new gauge function $\hat{G}$ by

$$\hat{G}(\Phi) = G(\mathfrak{F}^{-1}(\hat{\Phi})) \tag{5.159}$$

$$\Gamma(\Phi, c_+, \rho, \mathcal{G}, \mathcal{H}') = \Big[\Gamma_{inv}(\hat{\Phi}) + \hat{\rho}_1 \; Q_{s2}(\hat{\Phi}, c_+) + \mathcal{H}_1' Q_{s2}(\hat{\Phi}, c_+) \; \frac{\delta}{\delta \Phi_2} \hat{G}_1(\hat{\Phi})$$

$$+ \mathcal{G}_1' \hat{G}_1(\hat{\Phi}) + DD\hat{G}_1(\hat{\Phi}) \; \overline{DD\hat{G}_1(\hat{\Phi})} \Big]_{\hat{\Phi} = \mathfrak{F}(\Phi)} \tag{5.160}$$

$$- \sigma_1(c_+ c_+)_1 - \bar{\sigma}_1(\bar{c}_+ \bar{c}_+)_1$$

with

$$\hat{\rho} = \rho_1 \frac{\delta \mathfrak{F}_1^{-1}(\hat{\Phi})}{\delta \hat{\Phi}}\Bigg|_{\hat{\Phi} = \mathfrak{F}(\Phi)} \tag{5.161}$$

This shows that the solution Γ (5.158) corresponding to a function $\mathfrak{F}(\Phi)$ and a gauge function $G(\Phi)$ is equivalent, modulo the canonical transformation

$$\Phi \to \hat{\Phi} = \mathfrak{F}(\Phi) \qquad \rho \to \hat{\rho}(\rho,\Phi) \tag{5.162}$$

to the solution corresponding to $\mathfrak{F}(\Phi) = \Phi$ and gauge function $\hat{G} = G(\mathfrak{F}^{-1}(\Phi))$.

5.5 Gauge independence

In the last subsection, we have shown that the parameters a_k implied in the BRS-transformation law for the gauge field can be interpreted as gauge parameters. Our purpose is now to check explicitly their non-physical character and to provide a formulation which can be extended to higher orders. We shall in fact show [II.14] that for any gauge parameter p there exists an insertion Δ_p such that

$$\partial_p Z(j) = \mathcal{S}\Delta_p \cdot Z(j) \tag{5.163}$$

Here Z is the generating functional for the Green's functions, obtained in the tree approximation from the classical action Γ by Legendre transformation (see Appendix B). j stands for the sources of the elementary fields Φ, c_+, c_-, B and A. $\Delta_p \cdot Z$ denotes the generating functional for the Green's functions with the insertion Δ_p - an integrated field polynominal -, and $\mathcal{S}$ the Slavnov operator (B.23). $\mathcal{S}$ being linear in the sources j the right-hand-side of (5.163) vanishes on shell, i.e. at j = o, which shows that the scattering matrix is indeed gauge parameter independent. This equation can be generalized for the generating functional Z (j, q) of Green's functions involving gauge invariant operators Q_i, the latter being coupled in the classical action to the BRS-invariant external fields q_i. Eq. (5.163) at j = o then shows that the Green's functions of gauge invariant operators

$$< T\, Q_1 \ldots Q_n > = i^{-n} \frac{\delta^n}{\delta_{q_1} \ldots \delta_{1_n}} Z(j,q) \Big|_{j=0,q=0} \tag{5.164}$$

are independent of the gauge parameters, too.

We shall prove (5.163) in the linear (α,B)-gauge defined by (5.86), the gauge parameters being α and the a_k. To do this we impose a generalization of the Slavnov identity (5.85) (we suppress trace as well as integration symbols, and we omit the matter fields):

$$\mathcal{S}(\Gamma) \equiv \frac{\delta\Gamma}{\delta\rho}\frac{\delta\Gamma}{\delta\Phi} + B\frac{\delta\Gamma}{\delta c_-} + \bar{B}\frac{\delta\Gamma}{\delta\bar{c}_-} + \frac{\delta\Gamma}{\delta\sigma}\frac{\delta\Gamma}{\delta c_+} + \frac{\delta\Gamma}{\delta\bar{\sigma}}\frac{\delta\Gamma}{\delta\bar{c}_+} + \chi\partial_\alpha\Gamma + \sum_{\omega,k} x_{\omega k}\,\partial_{a_{\omega k}}\Gamma = 0 \tag{5.165}$$

The terms which are derivatives with respect to α, a_k mean that we allow the gauge parameters p to transform under BRS into Grassmann (anticommuting)parameters z which are BRS-invariant. The notation is

$$p = (p_1) = (\alpha,\ a_{\omega k}) \qquad\qquad z = (z_1) = (\chi,\ x_{\omega k}) \tag{5.166}$$

The gauge condition (5.86) is modified, too:

$$\begin{aligned} \frac{\delta\Gamma}{\delta B} &= \alpha\,\overline{DD}\,\bar{B} + \frac{1}{8}\,\overline{DD}DD\,\Phi + \frac{1}{2}\,\chi\overline{DD}\,\bar{c}_- \\ \frac{\delta\Gamma}{\delta\bar{B}} &= \alpha\,DD\,B + \frac{1}{8}\,DD\overline{DD}\,\Phi + \frac{1}{2}\,\chi DD\,c_- \end{aligned} \tag{5.167}$$

It is clear that the theory defined by (5.165) and (5.167) coincides at z = o with that constructed in sections 5.2 and 5.3. To see that the Slavnov identity (5.165) implies the condition of gauge independence (5.163), let us translate it in terms of the functional Z:

$$\mathcal{S}Z \equiv \left\{ -J\frac{\delta}{\delta\rho} + \xi_+\frac{\delta}{\delta J_B} - \bar{\xi}_+\frac{\delta}{\delta\bar{J}_B} + \xi_-\frac{\delta}{\delta\sigma} - \bar{\xi}_-\frac{\delta}{\delta\bar{\sigma}} + \sum_1 z_1\,\partial_{p_1}\right\} Z = 0 \tag{5.168}$$

and differentiate it with respect to z_p. This yields the equation

$$\partial_{p_1} Z = \mathcal{S}\partial_{z_1} Z \tag{5.169}$$

which taken at z = o is indeed (5.163) with $\Delta_p \cdot Z \equiv \partial_z Z$.

It remains to find the general solution of the Slavnov identity (5.165) and of the gauge condition (5.167). We first notice that, like in section 5.2, they imply ghost equations

$$\begin{aligned} \mathcal{G}\Gamma &\equiv \left[\frac{\delta}{\delta c_-} + \frac{1}{8}\,\overline{DD}DD\,\frac{\delta}{\delta\rho}\right]\Gamma = -\frac{1}{2}\,\chi\overline{DD}\,\bar{B} \\ \bar{\mathcal{G}}\Gamma &\equiv -\left[\frac{\delta}{\delta\bar{c}_-} + \frac{1}{8}\,DD\overline{DD}\,\frac{\delta}{\delta\rho}\right]\Gamma = \frac{1}{2}\,\chi DD\,B \end{aligned} \tag{5.170}$$

which are now inhomogeneous. Inserting them into the Slovnov identity yields

$$\mathcal{S}(\Gamma) = \mathcal{B}(\bar{\Gamma}) \equiv \frac{\delta\bar{\Gamma}}{\delta\eta}\frac{\delta\bar{\Gamma}}{\delta\Phi} + \frac{\delta\bar{\Gamma}}{\delta\sigma}\frac{\delta\bar{\Gamma}}{\delta c_+} + \frac{\delta\bar{\Gamma}}{\delta\bar{\sigma}}\frac{\delta\bar{\Gamma}}{\delta\bar{c}_+} + \sum_l z_l\, \partial_{p_l}\, \bar{\Gamma} = 0 \tag{5.171}$$

Γ is defined by

$$\Gamma\,(\Phi,c_+,c_-,B,\rho,\sigma,p,z) = \bar{\Gamma}(\Phi,c_+,\eta,\sigma,p,z) + B\, DD\, \Phi + \bar{B}\, \overline{DD}\, \Phi + \alpha\, B\bar{B} + \frac{1}{2}\, \chi(c_- B + \bar{c}_- B) \tag{5.172}$$

and depends on c_- and ρ only through the combination (5.110):

$$\eta = \rho - \frac{1}{8}\,(DD\, c_- + \overline{DD}\, \bar{c}_-) \tag{5.173}$$

since it fulfills the homogeneous ghost equations

$$\mathcal{G}\bar{\Gamma} = 0 \qquad \bar{\mathcal{G}}\bar{\Gamma} = 0 \tag{5.174}$$

A special p- and z-independent solution of (5.171) is obviously given by

$$\bar{\Gamma}_s(\Phi,c_+,\eta,\sigma) = \Gamma_{YM}(\Phi) + \eta\, Q_s(\Phi,c_+) - \sigma c_+ c_+ - \bar{\sigma}\bar{c}_+\bar{c}_+ \tag{5.175}$$

with Γ_{YM} and Q_s given in (5.20) and (5.6). Γ_s is nothing else than the special solution ($a_k = o$) given in section 5.2. The search for the general solution begins like in section 5.3: the substitutions

$$\begin{aligned} \hat{\Phi} &= \mathcal{F}(\Phi,p) \\ \hat{\eta} &= \frac{\delta}{\delta\hat{\Phi}}\, \mathrm{Tr}\!\int\! dV\, \eta\, \mathcal{F}^{-1}(\hat{\Phi},p)\Big|_{\hat{\Phi} = \mathcal{F}(p)} \\ \hat{c}_+ &= t(p)c_+ \qquad \hat{\sigma} = \frac{1}{t(p)}\,\sigma \end{aligned} \tag{5.176}$$

- with $\mathcal{F}$ and t arbitrary functions of their arguments - into $\bar{\Gamma}_s$ preserve the Slavnov identity (5.171) at z = o and thus provide the z-independent terms of the general solution. We can, therefore, write the latter as

$$\bar{\Gamma}(\Phi,c_+,\eta,\sigma,p,z) = \bar{\Gamma}\,(\hat{\Phi},\hat{c}_+,\hat{\eta},\hat{\sigma})$$

$$+ \sum_1 z_1\,[\eta\, G_1(\Phi,p) + h_1(\sigma c_+ + \bar{\sigma}\bar{c}_+)] \tag{5.177}$$

where the unknown dimensionless functions G_p and h_p are determined by the full Slavnov identity. The result is

$$G_1(\Phi,p) = - \frac{\partial}{\partial p_1}\, \mathfrak{F}^{-1}(\hat{\Phi},p)\Big|_{\hat{\Phi} = \mathfrak{F}(\Phi,p)} \tag{5.178}$$

$$h_1\,(p) = - \frac{1}{t(p)}\, \partial_{p_1}\, t(p)$$

The function $\mathfrak{F}$ has the general form (5.115) with p-dependent coefficients c_1, $c_{\omega k}$. Hence the general solution (5.177) depends on the gauge coupling constant g, on p_1 and z_1, as well as on the arbitrary functions t(p), $c_1(p)$ and $c_{\omega k}(p)$. It reduces at z = o to the general solution given in section 5.3, parametrized with g and p_1, if we choose

$$t(p) = 1$$

$$c_1(p) = 1, \quad c_{\omega k}(p) = c_{\omega k}(a_{\omega' k'};k' \leq k) \tag{5.179}$$

$c_{\omega k}$ (a) being the solution of the system of equations (5.140). Inspection of (5.139) shows that this may be achieved by imposing the normalization conditions

$$\Gamma_{\sigma_i^F\, c_{+j}^A\, c_{+k}^A} = - 4\, if_{ijk} \tag{5.180}$$

$$\Gamma_{\eta_i^D\, c_{+j}^A} = - 4\, \delta_{ij} \tag{5.181}$$

$$S_k^\omega\, \Gamma_{\eta^D\, c_+^A(\Phi^C)^{k-1}} = - 4\,(k-1)a_{\omega k}\,, \quad k \geq 2 \tag{5.182}$$

where S_k^ω denotes the projection onto $s^\omega_{i(i_1 \dots i_k)}$, the symmetric invariant tensor appearing in the expansion of the coupling $\eta_i Q_i$ according to (5.139). Condition (5.180) fixes the amplitude of c_+ and Z (5.181), that of Φ and η and (5.182) defines the gauge parameters $a_{\omega k}$. No normalization condition is needed for α which is already defined by the gauge condition (5.167). Finally the condition

$$\Gamma_{D_i D_j} = \frac{1}{4g^2} \delta_{ij} \tag{5.183}$$

defines the gauge coupling constant in accordance with the expression (5.14) for the SYM action (5.20). If matter fields are present their amplitudes, masses and self-coupling constants are defined by suitable normalization conditions which we shall not write down here.

5.6 Summary of section 5

As a supersymmetric extension of non-Abelian gauge transformations we have exhibited the transformation law

$$\begin{aligned} \delta\Phi &= i(\Lambda-\bar{\Lambda}) + \frac{i}{2}[\Phi,\Lambda+\bar{\Lambda}] + \frac{i}{12}[\Phi,[\Phi,\Lambda-\bar{\Lambda}]] + \ldots \\ &= i\,Q_s(\Phi,\Lambda) \end{aligned} \tag{5.6}$$

arising from

$$e^{\Phi} \rightarrow e^{-i\bar{\Lambda}}\, e^{\Phi}\, e^{i\Lambda} \tag{5.5}$$

for infinitesimal $\Lambda,\bar{\Lambda}$. It has then been established that this is only a special case of the more general law (5.63), (5.64) which can be explicitly obtained from (5.138), (5.128):

$$\delta\Phi_i = i\,Q_i(\Phi,\Lambda) = i\,\frac{\delta\mathcal{F}_i^{-1}(\hat{\Phi})}{\delta\hat{\Phi}_j}\, Q_{sj}(\hat{\Phi},\Lambda)\Bigg|_{\hat{\Phi}=\mathcal{F}(\Phi)} \tag{5.184}$$

with

$$\Phi_i \rightarrow \hat{\Phi}_i = \mathcal{F}_i(\Phi) = \Phi_i + \sum_{k\geq 2}\sum_{\omega=1}^{\Omega(k)} c_{\omega k}\, s^{\omega}_{i(i_1\ldots i_k)}\, \Phi_{i_1}\ldots\Phi_{i_k} \tag{5.115}$$

As for the special gauge transformation law there exist also for the general one BRS-transformations and, in fact, it is the general solution of the Slavnov identity which forces one to consider the general law. The desire of rigid invariance fixes the peculiar appearance of symmetric invariant tensors (with respect to the gauge group G) in

(5.115) resp. in (5.63) (5.64). The parameters $c_{\omega k}$ (5.115) (in 1-1 correspondence with $a_{\omega k}$) are seen to be gauge-type parameters.

Supersymmetric Yang-Mills theory can thus in the tree approximation be characterized by

- supersymmetry: $W_\alpha \Gamma = 0$ $\qquad \bar{W}_{\dot\alpha}\Gamma = 0$

- rigid gauge invariance: $W_\omega \Gamma = 0$

- BRS-invariance: Slavnov-identity (5.84) (5.85) (including gauge conditions (4.34) (4.43))

- (softly broken) conformal R-invariance $W_R \Gamma = W_R$ (mass terms)

The parameters of the model g, $a_{\omega k}$ as well as the field amplitudes can be fixed by the normalization conditions (5.180) to (5.183).

With these symmetry requirements and normalization conditions the action is uniquely determined.

6. Supercurrents

6.1 Generalities

The models presented in sections 3 to 5 are characterized by Ward-identities. First of all by the one of supersymmetry, then by R-invariance, internal symmetries and gauge invariance. All of the rigid symmetries are described by Ward-identity operators of the type

$$W^X \equiv -i \int dz\ \delta^X\psi \frac{\delta}{\delta\psi} \tag{6.1}$$

where X stands for the symmetry in question, the integration measure dz is appropriate for the type of field (dS for chiral, dV for vector superfield) and a sum over all fields ψ_i $i = 1,\ldots$ is understood if there are several fields present. The variations $\delta^X\psi$ contain information on the algebra of symmetry transformations and - together with the normalization factors chosen in (6.1) - imply that the functional differential operators W^X satisfy just the algebra of the charges associated with the symmetries. So, for instance, in the case of supersymmetry we find

$$\{W_\alpha, \bar{W}_{\dot{\alpha}}\} = 2\,\sigma^\mu_{\alpha\dot{\alpha}}\, W^P_\mu \qquad , \tag{6.2}$$

$$\{W_\alpha, W_\beta\} = 0 = \{\bar{W}_{\dot{\alpha}}, \bar{W}_{\dot{\beta}}\} , \tag{6.3}$$

where W^P_μ denotes the Ward-identity operator of translations:

$$W^P_\mu \equiv -i \int dz\, \partial_\mu \psi \frac{\delta}{\delta\psi} \qquad . \tag{6.4}$$

I.e. we have reproduced (1.5), (1.6). Acting with the operators W on Γ, the action in the tree approximation or the generating functional of vertex functions in higher orders, we have represented the symmetry algebra on an object which is tractable by perturbation theoretic means.

The proper subject of the present section is the study of the currents associated with the symmetries of the superconformal group and specifically their covariance with respect to supersymmetry [II.15]. As suggested by the analogy to Lorentz covariance one might expect that some currents combine to superfields. Obvious candidates are the currents for R-invariance, supersymmetry and translation invariance since their charges form already superfields as shown in section 2, where we used them to collect individual Ward-identities. Let us consider the "longest" superfield $\hat{W}^R$ - see (2.9), (A.27), (A.28) - and a theory Γ which satisfies:

$$\hat{W}^R \Gamma = 0 \quad . \tag{6.5}$$

Writing

$$\hat{W}^R = \int dx\; w \tag{6.6}$$

we have as a consequence of (6.5)

$$w\Gamma = -i\,\partial^\mu V_\mu \quad , \tag{6.7}$$

i.e. the r.h.s. is a total derivative, the factor -i is chosen for convenience. If w possesses supercovariance, $\partial^\mu V_\mu$ has it and then - hopefully - V_μ. But if V_μ is a superfield it makes sense to rewrite ∂^μ covariantly (use (A.10))

$$w\Gamma = -\frac{1}{4}\sigma^{\mu}_{\alpha\dot\alpha}\{D^{\alpha},\bar D^{\dot\alpha}\}V_{\mu}$$

$$= D^{\alpha}(-\frac{1}{4}\sigma^{\mu}_{\alpha\dot\alpha}\bar D^{\dot\alpha}V_{\mu}) + \bar D_{\dot\alpha}(-\frac{1}{4}\bar\sigma^{\dot\alpha\alpha}D_{\alpha}V_{\mu}) , \tag{6.8}$$

which suggests that the l.h.s. too permits such a decomposition

$$w = D^{\alpha}w_{\alpha} - \bar D_{\dot\alpha}\bar w^{\dot\alpha} . \tag{6.9}$$

In fact, it will turn out in all of our examples that even more is possible: one can "integrate over D_{α}" and establish separate equations, namely

$$w_{\alpha}\Gamma = -\frac{1}{4}\sigma^{\mu}_{\alpha\dot\alpha}\bar D^{\dot\alpha}V_{\mu} + \frac{1}{2}B_{\alpha} \tag{6.10}$$

and its conjugate

$$\bar w_{\dot\alpha}\Gamma = -\frac{1}{4}\sigma^{\mu}_{\alpha\dot\alpha}D^{\alpha}V_{\mu} + \frac{1}{2}\bar B_{\dot\alpha} . \tag{6.11}$$

Here B_{α}, $\bar B_{\dot\alpha}$ are constrained by

$$D^{\alpha}B_{\alpha} - \bar D_{\dot\alpha}\bar B^{\dot\alpha} = 0 . \tag{6.12}$$

A close analysis of all possible contributions B_{α} will also reveal that in our examples V_{μ} has amongst its components an axial current, associated with R-transformations, a supersymmetry current and an energy-momentum tensor. We shall now go through our standard examples and then come back to general considerations in subsection 6.5. Before actually doing so let us make two comments:

(i) The above procedure is a slightly unconventional way to establish local Ward-identities from given global ones. Usually one makes the parameter of the respective symmetry transformation space-time dependent, differentiates with respect to them and thus isolates the total divergence which vanishes on shell. We bypass this orthodox construction since we want to avoid the complicated algebra of local R-, susy- and translational transformations which is nothing but the algebra of supergravity transformations. (For a different approach compare [III.24] and section 6.6.)

(ii) The transition from (6.5) to (6.7) requires the existence of an algebraic Poincaré-Lemma in the space of all field polynomials. A general proof of it has only recently been given [III.25].

6.2 Chiral models*

In order to derive local Ward-identities from global ones while preserving supercovariance as sketched in the preceding subsection let us look at the space-time integrand of the R-Ward-identity operator (s. (A.29))

$$W^R \Gamma \equiv -i \int dS\, i(n+\Theta\partial_\Theta)\psi \frac{\delta\Gamma}{\delta\psi} -i \int d\bar{S}\, i(-n+\bar{\Theta}\partial_{\bar{\Theta}})\bar{\psi} \frac{\delta\Gamma}{\delta\bar{\psi}} \tag{6.13}$$

namely at $-DD((n+\Theta\partial_\Theta)\psi j) - \bar{D}\bar{D}((-n+\bar{\Theta}\,\partial_{\bar{\Theta}})\bar{\psi}\bar{j})$. (Here we have written $-j$ instead of $\delta\Gamma/\delta\psi$.) Due to the explicit appearance of Θ's and ∂_Θ's these terms are obviously not supersymmetric. But they hint to a possible supersymmetric form. This becomes even clearer if we recall (6.9) and try to identify the contributions to w_α and $\bar{w}_{\dot{\alpha}}$. Because of dimensions and the form j there are in fact only two supersymmetric candidates as terms of w_α:

$$w_\alpha\, \Gamma = a\, D_\alpha(\psi j) + b\, D_\alpha\, \psi j \quad , \tag{6.14}$$

(here a and b are numbers to be determined). With the help of the formula

$$\int dx\, D_\alpha\, U = -\frac{1}{2}\,\Theta_\alpha \int dSU + \frac{1}{2}\int dS\, \Theta_\alpha\, U \tag{6.15}$$

one can fix the coefficients a,b by forming $D^\alpha w_\alpha$, integrating over space-time and comparing with the desired part of W_R:

$$\begin{aligned}\int dx\, D^\alpha w_\alpha &= a\int dS\, \psi j + b\int dx\, D^\alpha(D_\alpha \psi j)\\ &= a\int dS\, \psi j + \frac{1}{2}\, b\int dS\, \Theta^\alpha \partial_\alpha\, \psi j + \frac{1}{2}\,\Theta^\alpha \int dS\, D_\alpha \psi j + \dots\end{aligned}$$

$$\text{i.e.}\quad a = -n \; , \quad b = -2 \tag{6.16}$$

Encouraged by this result we go ahead and compute

$$\int dx\, w\Gamma = \int dx(D^\alpha w_\alpha \Gamma - \bar{D}_{\dot{\alpha}} \bar{w}_{\dot{\alpha}} \Gamma)$$

with formulas corresponding to (6.15)

---

*Maintaining parity

$$\int dx\ U = -\frac{1}{4}\int dS\Theta^2 U + \frac{1}{2}\Theta^\alpha \int dS\Theta_\alpha U - \frac{1}{4}\Theta^2 \int dS\ U,$$

$$\int dx\ U = -\frac{1}{4}\int d\bar{S}\bar{\Theta}^2 U + \frac{1}{2}\bar{\Theta}_{\dot{\alpha}}\int d\bar{S}\bar{\Theta}^{\dot{\alpha}} U - \frac{1}{4}\bar{\Theta}^2 \int d\bar{S}\ U\ , \tag{6.17}$$

$$\int dx\bar{D}_{\dot{\alpha}} U = \frac{1}{2}\int d\bar{S}\bar{\Theta}_{\dot{\alpha}} U - \frac{1}{2}\bar{\Theta}_{\dot{\alpha}}\int d\bar{S}\ U\ .$$

The result is

$$\int dx\ w\Gamma = \hat{W}^R\ \Gamma\ . \tag{6.18}$$

I.e. the contact terms

$$w_\alpha = n\ D_\alpha (A\frac{\delta}{\delta A}) + 2D_\alpha\ A\frac{\delta}{\delta A}\ ,$$
$$\bar{w}_{\dot{\alpha}} = n\ \bar{D}_{\dot{\alpha}}(\bar{A}\frac{\delta}{\delta \bar{A}}) + 2\bar{D}_{\dot{\alpha}}\ \bar{A}\frac{\delta}{\delta \bar{A}}\ , \tag{6.19}$$

which are themselves supercovariant on the local level give rise to the covariant rigid contact terms which we wanted to reproduce. These contact terms will now serve us to derive supercovariant local Ward-identities in concrete models.

6.2.1 Massless Wess-Zumino model

We choose $n = -2/3$, the conformal R-weight, and the corresponding R-invariant massless action ((3.26) with $\lambda = 0$, $m = 0$)

$$\Gamma = \frac{1}{16}\int dV\ \bar{A}A + \frac{g}{48}\left(\int dS\ A^3 + \int d\bar{S}\ \bar{A}^3\right)\ . \tag{6.20}$$

The derivatives needed for (6.19) read

$$\frac{\delta\Gamma}{\delta A} = -\ j = \frac{1}{16}\bar{D}\bar{D}\ \bar{A} + \frac{g}{16}A^2\ ,$$
$$\frac{\delta\Gamma}{\delta \bar{A}} = -\ \bar{j} = \frac{1}{16}DD\ A + \frac{g}{16}\bar{A}^2 \tag{6.21}$$

They yield what we shall call the trace equations

$$-2\ w_\alpha\ \Gamma = \bar{D}^{\dot{\alpha}}\ V_{\alpha\dot{\alpha}},$$
$$-2\ \bar{w}_{\dot{\alpha}}\ \Gamma = D^\alpha\ V_{\alpha\dot{\alpha}}, \tag{6.22}$$

with

$$V_{\alpha\dot{\alpha}} = -\frac{1}{6}(D_\alpha A\, \bar{D}_{\dot{\alpha}}\bar{A} - A\, D_\alpha\bar{D}_\alpha\, \bar{A} + \bar{A}\, \bar{D}_\alpha D_\alpha\, A). \tag{6.23}$$

As a consequence we derive

$$w\,\Gamma \equiv D^\alpha w_\alpha\, \Gamma - \bar{D}_{\dot{\alpha}}\bar{w}^{\dot{\alpha}}\, \Gamma = -i\, \partial^\mu V_\mu \quad , \tag{6.24}$$

$$V_{\alpha\dot{\alpha}} = \frac{1}{2}\sigma^\mu_{\alpha\alpha}\, V_\mu \quad , \quad V_\mu = \sigma^{\alpha\dot{\alpha}}_\mu\, V_{\alpha\dot{\alpha}} \quad , \tag{6.25}$$

i.e. the superfield V_μ is strictly conserved on shell.

Due to (6.18) we may write

$$w = w^R - i\,\Theta^\alpha w_\alpha{}^Q + i\,\bar{\Theta}_\alpha \bar{w}^{\bar{Q}\alpha} - 2\Theta\sigma^\mu\bar{\Theta} w_\mu{}^P + \text{total derivatives} \tag{6.26}$$

and expect that the zeroth Θ-component of V_μ contains the R-current, the first components $\Theta(\bar{\Theta})$ the supersymmetry currents $Q_\mu{}^\alpha(\bar{Q}_\mu{}^{\dot{\alpha}})$, the $\Theta\bar{\Theta}$-component the energy-momentum tensor $T_{\mu\nu}$. This is in fact true (and will be shown in detail in section 6.5). We therefore call V_μ <u>supercurrent</u>. It is <u>the</u> supercurrent because all other currents of the superconformal group may be obtained from it by taking covariant derivatives and space-time moments. Let us give the respective definitions. (The name of the current refers to the symmetry of the $\Theta = 0$ component of it, cf. sect. 6.5.)

$$\hat{R}_\mu = V_\mu \quad , \tag{6.27a}$$

$$\hat{Q}_{\mu\alpha} = i\, D_\alpha\, V_\mu, \tag{6.27b}$$

$$\hat{\bar{Q}}_{\mu\dot{\alpha}} = -i\bar{D}_{\dot{\alpha}}\, V_\mu\ . \tag{6.27c}$$

$$\hat{T}_{\mu\nu} = -\frac{1}{16}(V_{\mu\nu} + V_{\nu\mu}), \quad V_{\mu\nu} = [D^\beta, \bar{D}^{\dot{\beta}}]V_{\alpha\dot{\alpha}}\sigma_{\mu\beta\dot{\beta}}\sigma_\nu^{\alpha\dot{\alpha}} \quad , \tag{6.27d}$$

$$\hat{M}_{\mu\nu\rho} = x_\rho\, \hat{T}_{\mu\nu} - x_\nu\, \hat{T}_{\mu\rho}\ . \tag{6.27e}$$

$$\hat{D}_\mu = x^\nu\, \hat{T}_{\mu\nu} \quad , \tag{6.27f}$$

$$\hat{K}_{\mu\nu} = (2x_\nu x^\lambda - g^\lambda{}_\nu x^2)\, \hat{T}_{\mu\lambda} \quad , \tag{6.27g}$$

$$\hat{S}_{\mu\alpha} = i\, x_\nu\, \sigma^\nu_{\alpha\dot{\alpha}}\, \hat{\bar{Q}}_\mu{}^{\dot{\alpha}} \quad , \tag{6.27h}$$

$$\hat{\bar{S}}_{\mu\dot{\alpha}} = -i\, Q^\alpha_\mu\, \sigma^\nu_{\alpha\dot{\alpha}}\, x_\nu \quad . \tag{6.27i}$$

The x-moment currents are uniquely determined by requiring that they be strictly conserved (s. App. B of [I.6]). The most convenient expression of their conservation properties is in terms of the trace identities of the supercurrent.

$$\partial^\mu \hat{R}_\mu = \partial^\mu V_\mu = -\frac{i}{2}(\bar{D}^{\dot\alpha}D^\alpha V_{\alpha\dot\alpha} + D^\alpha\bar{D}^{\dot\alpha}V_{\alpha\dot\alpha}), \tag{6.28a}$$

$$\partial^\mu \hat{Q}_{\mu\alpha} = i\, D_\alpha \partial^\mu V_\mu , \tag{6.28b}$$

trace identiy: $$\hat{Q}^\alpha_\mu\, \sigma^\mu_{\alpha\dot\alpha} = 2i\, D^\alpha V_{\alpha\dot\alpha} , \tag{6.28c}$$

$$\partial^\mu \hat{\bar{Q}}_{\mu\dot\alpha} = -i\, \bar{D}_{\dot\alpha}\partial^\mu V_\mu \tag{6.28d}$$

trace identiy: $$\sigma^\mu_{\alpha\dot\alpha}\, \hat{\bar{Q}}^{\dot\alpha}_\mu = -2i\, \bar{D}^{\dot\alpha} V_{\alpha\dot\alpha} , \tag{6.28e}$$

$$\partial^\mu \hat{T}_{\mu\nu} = \frac{i}{32}(DD\bar{D}_{\dot\alpha}\bar{D}^{\dot\beta}V_{\alpha\dot\beta}\,\sigma^{\alpha\dot\alpha}_\nu + \bar{D}\bar{D}\, D_\alpha D^\beta V_{\beta\dot\alpha}\,\sigma^{\alpha\dot\alpha}_\nu) - \frac{1}{16}\sigma_{\nu\beta\dot\beta}\,[D^\beta,\bar{D}^{\dot\beta}]\,\partial^\mu V_\mu \tag{6.28f}$$

trace identiy:

$$\hat{T}_\lambda{}^\lambda = -\frac{i}{8}(\bar{D}^{\dot\alpha}\hat{Q}^\alpha_\mu\,\sigma^\mu_{\alpha\dot\alpha} + D^\alpha\sigma^\mu_{\alpha\dot\alpha}\,\hat{\bar{Q}}^{\dot\alpha}_\mu) = -\frac{1}{4}[D^\alpha,\bar{D}^{\dot\alpha}]V_{\alpha\dot\alpha} \tag{6.28g}$$

$$\partial^\mu \hat{M}_{\mu\nu\rho} = (\hat{T}_{\rho\nu} - \hat{T}_{\nu\rho}) + x_\rho\partial^\mu\hat{T}_{\mu\nu} - x_\nu\partial^\mu\hat{T}_{\mu\rho} , \tag{6.28h}$$

$$\partial^\mu \hat{D}_\mu = \hat{T}_\lambda{}^\lambda + x^\nu\partial^\mu\hat{T}_{\mu\nu} , \tag{6.28i}$$

$$\partial^\mu \hat{K}_{\mu\nu} = 2x^\lambda(\hat{T}_{\nu\lambda} - \hat{T}_{\lambda\nu}) + 2x_\nu\hat{T}^\lambda{}_\lambda + (2x_\nu x^\lambda - g_\nu{}^\lambda x^2)\partial^\mu\hat{T}_{\mu\lambda} , \tag{6.28j}$$

$$\partial^\mu \hat{S}_{\mu\alpha} = i\sigma^\mu_{\alpha\dot\alpha}\,\hat{\bar{Q}}^{\dot\alpha}_\mu + i\, x_\nu\sigma^\nu_{\alpha\dot\alpha}\partial^\mu\,\hat{\bar{Q}}^{\dot\alpha}_\mu , \tag{6.28k}$$

$$\partial^\mu \hat{\bar{S}}_{\mu\dot\alpha} = -i\,\hat{Q}^\alpha_\mu\,\sigma^\mu_{\alpha\dot\alpha} - i\, x_\nu\sigma^\nu_{\alpha\dot\alpha}\partial^\mu\,\hat{Q}^\alpha_\mu . \tag{6.28l}$$

Hence, as long as (6.22) holds all currents (6.27) are conserved. They express the superconformal invariance of the massless Wess-Zumino model in the classical approximation. The currents (6.27) and the identities (6.28) will also be useful in other models, s.b., and will be of great importance even in the case when superconformal invariance is broken. In order to prepare for higher orders where in general breaking by anomalies is to be expected let us study already in the classical approximation the breaking induced by a mass term.

6.2.2 Massive Wess-Zumino model

We add a mass term

$$\Gamma_m = \frac{m}{8}\left(\int dS\, A^2 + d\bar{S}\, \bar{A}^2\right) \tag{6.29}$$

to the action (6.20) and break thereby explicitly conformal R-invariance

$$\hat{W}^R\, \Gamma = \frac{m}{12}\left(\int dS\, A^2 - \int d\bar{S}\, \bar{A}^2\right). \tag{6.30}$$

Since supersymmetry and translation invariance are both unbroken, (6.30) is correct as it stands. The mass term contributes to the equation of motion

$$\frac{\delta\Gamma}{\delta A} = \frac{1}{16}\bar{D}\bar{D}\,\bar{A} + \frac{m}{4}A + \frac{g}{16}A^2 \quad , \tag{6.31}$$

and thus to the trace equation

$$-2\, w_\alpha\, \Gamma = \bar{D}^{\dot{\alpha}} V_{\alpha\dot{\alpha}} - 2\, D_\alpha\left(\frac{1}{12} mA^2\right) , \tag{6.32}$$

and to the current conservation equation

$$\partial^\mu V_\mu = iw\Gamma - i\left(DD\left(\frac{m}{12}A^2\right) - \bar{D}\bar{D}\left(\frac{m}{12}\bar{A}^2\right)\right) . \tag{6.33}$$

Obviously there is no redefinition of V_μ by which one could absorb the breaking term. This is, of course, alright since conformal R is broken, (6.30). But looking at the conservation of $Q_{\mu\alpha}$ as defined by (6.27) we find

$$\partial^\mu\, \hat{Q}_{\mu\alpha} = -\, D_\alpha\, w\Gamma - D_\alpha\bar{D}\bar{D}\left(\frac{m}{12}\bar{A}^2\right) , \tag{6.34}$$

a non-conserved supersymmetry current. Since the mass term Γ_m, (6.29), was supersymmetric there should exist a strictly conserved supersymmetry current and, indeed, the disturbing term in (6.34) is a total derivative (use (A.10))

$$\partial^\mu(\hat{Q}_{\mu\alpha} + 4i\ \sigma_{\mu\alpha\dot{\alpha}}\ \bar{D}^{\dot{\alpha}}\ (\frac{m}{12}\ \bar{A}^2)) = -D_\alpha\ w\Gamma. \tag{6.35}$$

To the cost of a total derivative contribution to the contact terms this new supersymmetry current can also be obtained directly from the supercurrent by the definition

$$\hat{Q}_{\mu\alpha} = i(D_\alpha V_\mu - (\sigma_\mu \bar{\sigma}^\nu D)_\alpha V_\nu)\ . \tag{6.27b'}$$

The corresponding redefinition for $\hat{T}_{\mu\nu}$ reads

$$\hat{T}_{\mu\nu} = -\frac{1}{16}\ (V_{\mu\nu} + V_{\nu\mu} - 2\ g_{\mu\nu} V_\lambda{}^\lambda) \tag{6.27d'}$$

and shows that we have been manipulating trace contributions i.e. improvement terms for the supersymmetry current and the energy-momentum tensor. The conservation and trace relations associated with these definitions are the following

$$\partial^\mu\ \hat{Q}_{\mu\alpha} = i(D_\alpha \partial^\mu V_\mu + 2\ \partial_\mu \sigma^\mu{}_\alpha{}^{\dot{\alpha}} D^\beta V_{\beta\dot{\alpha}}), \tag{6.28b'}$$

$$\text{trace identity:}\quad \hat{Q}_\mu{}^\alpha\ \sigma^\mu_{\alpha\dot{\alpha}} = -\ 6i\ D^\alpha V_{\alpha\dot{\alpha}}\ , \tag{6.28c'}$$

$$\begin{aligned}\partial^\mu\ \hat{T}_{\mu\nu} = \frac{i}{32}\ (DD\ \bar{D}_{\dot{\alpha}} \bar{D}^{\dot{\beta}} V_{\alpha\dot{\beta}}\ \sigma_\nu{}^{\alpha\dot{\alpha}} + \bar{D}\bar{D}\ D_\alpha D^\beta V_{\beta\dot{\alpha}} \sigma_\nu{}^{\alpha\dot{\alpha}}\ )\\ -\frac{1}{16}\ \sigma_{\nu\beta\dot{\beta}}\ [D^\beta,\bar{D}^{\dot{\beta}}]\partial^\mu V_\mu + \frac{1}{4}\ [D^\alpha,\bar{D}^{\dot{\alpha}}]V_{\alpha\dot{\alpha}}\ ,\end{aligned} \tag{6.28f'}$$

$$\text{trace identity:}\quad \hat{T}_\lambda{}^\lambda = \frac{3}{4}\ [D^\alpha,\bar{D}^{\dot{\alpha}}]V_{\alpha\dot{\alpha}}\ . \tag{6.28g'}$$

All other definitions of moment currents, (6.27), remain unchanged. The breaking of superconformal symmetry by the mass term of a chiral field shows therefore up via trace terms in the conservation equation for the R-current, dilatations, conformal transformations and special supersymmetry transformations.

6.2.3 General chiral model

The preceding considerations can easily be generalized to any chiral model, (3.20),

$$\Gamma = \frac{1}{16}\int dV\,\bar{A}_k A_k + \int dS(\lambda_k A_k + m_{kk'}A_k A_{k'} + g_{kk'k''}A_k A_{k'} A_{k''})$$

$$+ \int d\bar{S}(\lambda_k \bar{A}_k + m_{kk'}\bar{A}_k \bar{A}_{k'} + g_{kk'k''}\bar{A}_k \bar{A}_{k'} \bar{A}_{k''}) \tag{6.36}$$

since the respective transformations are diagonal in the fields. We choose contact terms $w_\alpha^{(k)}$ which are diagonal in the fields

$$w_\alpha \Gamma \equiv \sum_k w_\alpha^{(k)}\Gamma = \sum_k (n_k D_\alpha (A_k \frac{\delta D}{\delta A_k}) + 2\, D_\alpha A_k \frac{\delta D}{\delta A_k}) \tag{6.37}$$

and find as trace equation

$$-2\, w_\alpha \Gamma = \bar{D}^{\dot\alpha} V_{\alpha\dot\alpha} + 2\, D_\alpha\, S - B_\alpha \tag{6.38}$$

where the supercurrent V is also a sum of the individual supercurrents

$$V \;=\; \sum_k V^{(k)}\,, \tag{6.39}$$

$$V^{(k)}_{\alpha\dot\alpha} = -\frac{1}{2}\,(n_k + 1)\, D_\alpha A_k \bar{D}_{\dot\alpha}\bar{A}_k + \frac{n_k}{4}(-\, A_k D\bar{D}_{\dot\alpha}\bar{A}_k + \bar{A}_k \bar{D}_{\dot\alpha} D_\alpha A_k)$$

and the breaking terms are of the form

$$S = -\, g_{kk'k''}\,(n_k + n_{k'} + n_{k''} + 2)\, A_k A_{k'} A_{k''}$$
$$-\, m_{kk'}\,(n_k + n_{k'} + 2)\, A_k A_{k'} - \lambda_k (n_k + 2)\, A_k \quad , \tag{6.40}$$

$$B_\alpha = -\frac{1}{4}\,(\frac{3}{2}\, n_k + 1)\, \bar{D}\bar{D}\, D_\alpha (A_k \bar{A}_k)\,. \tag{6.41}$$

Hence the current conservation equation reads

$$\partial^\mu V_\mu = iw\Gamma + i(DD\, S - \bar{D}\bar{D}\,\bar{S})\,. \tag{6.42}$$

For conformal weights, $n_k = -2/3$, we just recover the analogs of the Wess-Zumino model treated before. For R-weights leading to a non-conformal R-symmetry we consider as specific example the O'Raifeartaigh model, cf. sect. 3.3, where $n_0 = n_2 = -2$, $n_1 = 0$. The corresponding breaking S vanishes and B reads

$$B_\alpha = \frac{1}{4} \bar{D}\bar{D}\, D_\alpha(2\, A_0\bar{A}_0 - A_1\bar{A}_1 + 2\, A_2\bar{A}_2) \tag{6.43}$$

and the R-current is strictly conserved

$$\partial^\mu V_\mu = iw\Gamma. \tag{6.44}$$

The breaking term B_α is hard, hence the energy-momentum tensor not improved. This goes hand in hand with the fact that for O'Raifeartaigh weights the superconformal algebra does already not close on the fields A_k (cf. sect. 2). There is thus no point in going into the moment construction of conformal currents.

6.3 Abelian gauge theory

In a gauge theory observables are gauge invariant quantities. The supercurrent - containing an axial current and the energy-momentum tensor - should therefore be constructed as a gauge invariant operator if possible. This is straightforward in the classical approximation [II.15] where one can discard gauge fixing terms and just write down gauge invariant extensions of the matter terms (6.39) and corresponding expressions for the gauge vector field. But for higher orders a more systematic construction is needed which includes contact terms i.e. permits to go off-shell [II.16]. Hence our first task is to find the contact terms for the gauge field. In order to do so we first observe that the matter field contact terms (6.37)

$$w_\alpha\ \Gamma \equiv \sum_k n_k D_\alpha(A_k \frac{\delta}{\delta A_k}) + 2\, D_\alpha A_k \frac{\delta}{\delta A_k} \tag{6.45}$$

have quite specific covariance properties (besides being superfields). In fact, under conformal R-transformations they transform as densities

$$[W_R, w_\alpha] = -i(-1 + \Theta\partial_\Theta)w_\alpha = -i\ \delta_R^{(-1)}\ w_\alpha \tag{6.46}$$

of weight -1. Similarly, under abelian gauge transformations (4.68) - we think of SQED -

$$\begin{aligned} w_\Lambda &= \bar{D}\bar{D}\, \frac{\delta}{\delta\Phi} - g\ A_+ \frac{\delta}{\delta A_+} + g\ A_- \frac{\delta}{\delta A_-} \\ w_{\bar{\Lambda}} &= DD\, \frac{\delta}{\delta\Phi} + g\ \bar{A}_+ \frac{\delta}{\delta\bar{A}_+} - g\ \bar{A}_- \frac{\delta}{\delta\bar{A}_-} \end{aligned} \tag{6.47}$$

they transform as

$$[w_\Lambda(1), w_\alpha(2)] = 2\, D_\alpha^{(2)} \delta_S\, (2,1)\, w_\Lambda(2),$$
$$[w_{\bar\Lambda}(1), w_\alpha(2)] = 0 \; . \tag{6.48}$$

This suggests to postulate that also the contact terms of the gauge vector fields have the same covariance. (6.46) permits for a real field Φ the following contributions

$$\begin{aligned} w_\alpha = {} & a_1\, \bar D\bar D D_\alpha \Phi \delta_\Phi + a_2\, \bar D\bar D \Phi D_\alpha \delta_\Phi + a_3\, \bar D \Phi \bar D D_\alpha \delta_\Phi \\ & + a_4\, \bar D \Phi D_\alpha \bar D \delta_\Phi + a_5\, \Phi \bar D \bar D D_\alpha \delta_\Phi + a_6\, D_\alpha \bar D \bar D \Phi \delta_\Phi \\ & + a_7\, D_\alpha \bar D \Phi \bar D \delta_\Phi + a_8\, \bar D D_\alpha \Phi \bar D \delta_\Phi + a_9\, D_\alpha \Phi \bar D \bar D \delta_\Phi \\ & + a_{10} \Phi D_\alpha \bar D \bar D \delta_\Phi , \end{aligned} \tag{6.49}$$

$(\delta_\Phi \equiv \delta/\delta\Phi)$.

The requirement (6.48) of gauge covariance fixes

$$a_9 = - a_1 ,$$
$$a_k = \;\; 0 \qquad k \neq 1,9 \tag{6.50}$$

and leaves a_1 arbitrary. It is fixed and found to be

$$a_1 = - 2 \tag{6.51}$$

by demanding - in accordance with (6.18) -

$$\int dx (D^\alpha w_\alpha - \bar D_{\dot\alpha} \bar w^{\dot\alpha}) \Gamma \Big|_{\Theta=0} = W^R\, \Gamma \tag{6.52}$$

where $\delta^R \Phi = (\Theta\partial_\Theta + \Theta\partial_\Theta)\Phi$.

(A real vector field Φ has R-weight $n(\Phi) = 0$, cf. (A.28).) Hence in an abelian gauge theory we may use as contact terms for the gauge vector field Φ [II.16]

$$w_\alpha \equiv - 2\, \bar D\bar D D_\alpha \Phi \frac{\delta}{\delta\Phi} + 2\, D_\alpha \Phi \bar D \bar D \frac{\delta}{\delta\Phi} \; . \tag{6.53}$$

A somewhat lengthy calculation (repeated use of (6.15) and (6.17) has to be made) shows that these contact terms yield

$$W^D\Gamma = \hat{W}^D\Gamma\Big|_{\tilde{\Theta}=0}$$

$$\hat{W}^D\Gamma = \int dx\, w^D$$

$$w^D = w^P_{trace} + x^\nu w^P_\nu$$

$$w^P_{trace} = \frac{3}{2} i\,(D^\alpha w_\alpha + \bar{D}_{\dot\alpha} w^{\dot\alpha})$$

$$w^P_\nu = \frac{1}{16}\,\Big(-(DD\bar{D}_{\dot\alpha} w_\alpha + \bar{D}\bar{D}D_\alpha \bar{w}_{\dot\alpha})\sigma_\nu{}^{\alpha\dot\alpha}$$

$$- \sigma_\nu{}^{\beta\dot\beta}\{D_\beta,\bar{D}_{\dot\beta}\}(D^\alpha w_\alpha - \bar{D}_{\dot\alpha}\bar{w}^{\dot\alpha}) + 8i\,\partial_\nu(D^\alpha w_\alpha + \bar{D}_{\dot\alpha}\bar{w}^{\dot\alpha}) \qquad (6.54)$$

with $\delta_D\Phi = (d + \frac{1}{2}\,\Theta\partial_\Theta - \frac{1}{2}\,\bar{\Theta}\partial_{\bar{\Theta}})\Phi$

and $d(\Phi) = 0.$

This result is encouraging because it indicates that the moment construction of sect. 6.2.1, 6.2.2 might again yield all currents of the superconformal group.

After these preparations it has become an algebraic exercise to calculate the explicit form of the supercurrent. For the SQED action in the α-gauge (4.66), (4.67)

$$\Gamma = \frac{1}{128}\int dV\, \Phi(D\bar{D}\bar{D}D - \frac{1}{2\alpha}\{DD,\bar{D}\bar{D}\} + 8M^2)\Phi$$

$$+ \frac{1}{16}\int dV\,(\bar{A}_+ e^{g\Phi} A_+ + A_- e^{-g\Phi}\bar{A}_-)$$

$$- \frac{m}{4}\,(\int dS\, A_+A_- + \int d\bar{S}\,\bar{A}_-\bar{A}_+)\ . \qquad (4.67)$$

We need the equations of motion; they read:

$$\frac{\delta\Gamma}{\delta\Phi} = \frac{1}{64}\,(D\bar{D}\bar{D}D - \frac{1}{2\alpha}\{DD,\bar{D}\bar{D}\} + 8M^2)\Phi + \frac{g}{16}\,(\bar{A}_+ e^{g\Phi} A_+ - A_- e^{-g\Phi}\bar{A}_-),$$

$$\frac{\delta\Gamma}{\delta A_\pm} = \frac{1}{16}\,\bar{D}\bar{D}(\bar{A}_\pm e^{\pm g\Phi}) - \frac{m}{4}\,A_\mp\ ,$$

$$\frac{\delta\Gamma}{\delta\bar{A}_\pm} = \frac{1}{16}\,DD(e^{\pm g\Phi}A_\pm) - \frac{m}{4}\,\bar{A}_\mp\ . \qquad (6.55)$$

We have to combine them as the contact terms w_α prescribe.

$$w_\alpha \equiv -2\,\bar{D}\bar{D}D_\alpha\Phi\,\frac{\delta}{\delta\Phi} + 2\,D_\alpha\Phi\bar{D}\bar{D}\,\frac{\delta}{\delta\Phi}$$

$$+\,n\,D_\alpha(A_\pm\,\frac{\delta}{\delta A_\pm}) + 2\,D_\alpha\,A_\pm\,\frac{\delta}{\delta A_\pm} \tag{6.56}$$

The result is the following:

$$\bar{D}^{\dot{\alpha}}V_{\alpha\dot{\alpha}} = -2\,w_\alpha\Gamma - 2\,D_\alpha S - 6(n+\tfrac{2}{3})\,\bar{D}\bar{D}D_\alpha I_5 \tag{6.57}$$

with

$$V = V^{inv} + M^2V^M + \frac{1}{\alpha}V^g \quad ,$$

$$V^{inv} = (-\frac{1}{4}nV^1 - \frac{1}{2}V^2 + \frac{1}{16}V^3) \; ,$$

$$V^1_{\alpha\dot{\alpha}} = [D_\alpha,\bar{D}_{\dot{\alpha}}](\bar{A}_+e^{g\Phi}A_+ + A_-e^{-g\Phi}\bar{A}_-) \quad ,$$

$$V^2_{\alpha\dot{\alpha}} = D_\alpha(A_-e^{-g\Phi})e^{g\Phi}\bar{D}_{\dot{\alpha}}(e^{-g\Phi}\bar{A}_-)$$

$$-\,\bar{D}_{\dot{\alpha}}(\bar{A}_+e^{g\Phi})e^{-g\Phi}D_\alpha(e^{g\Phi}A_+) \; ,$$

$$V^3_{\alpha\dot{\alpha}} = \bar{D}\bar{D}D_\alpha\Phi\; DD\bar{D}_{\dot{\alpha}}\Phi \; ,$$

$$V^M_{\alpha\dot{\alpha}} = -\,D_\alpha\Phi\bar{D}_{\dot{\alpha}}\Phi + \frac{1}{6}[D_\alpha,\bar{D}_{\dot{\alpha}}]\Phi^2 \; ,$$

$$V^g_{\alpha\dot{\alpha}} = -\frac{1}{48}(\Phi D\bar{D}\bar{\psi} - \Phi\bar{D}D\psi - D\Phi\bar{D}\bar{\psi} + \bar{D}\Phi D\psi$$

$$+\,D\bar{D}\Phi\psi - \bar{D}D\Phi\bar{\psi} + D\bar{D}\Phi\bar{\psi} - \bar{D}D\Phi\psi$$

$$-\,D\bar{D}\bar{D}\Phi\bar{D}DD\Phi + \bar{D}\bar{D}\Phi D\bar{D}DD\Phi - DD\Phi\bar{D}D\bar{D}\bar{D}\Phi)_{\alpha\dot{\alpha}} \; ,$$

$$\psi \equiv \bar{D}\bar{D}DD\Phi \quad ,$$

$$S = \frac{(n+1)}{2}\,m\,(A_+A_-) - \frac{1}{24}M^2\,DD(\Phi^2) + S_o \; ,$$

$$S_o = \frac{1}{\alpha}\frac{1}{192}\,\bar{D}\bar{D}(\Phi(\bar{\psi}-\psi)) \quad ,$$

$$I_5 = \frac{1}{16}(\bar{A}_+e^{g\Phi}A_+ + A_-e^{-g\Phi}\bar{A}_-) \; .$$

This is the supercurrent written in the α-gauge.
It is further to be noted that we have permitted general R-weights $n = n(A_+) = n(A_-)$ for the matter fields. Two cases will turn out to be particularly interesting: $n = -2/3$, the conformal value of the weights,

and n = ∞. We discuss them separately.

$n = -\frac{2}{3}$

The trace equation (6.57) reads now

$$\bar{D}^{\dot{\alpha}} V^c_{\alpha\dot{\alpha}} = -2\, w^c_\alpha \Gamma - 2\, D_\alpha S^c \tag{6.58}$$

and the current conservation equation

$$\partial^\mu V^c_\mu = i w^c \Gamma + i(DDS^c - \bar{D}\bar{D}\bar{S}^c) \tag{6.59}$$

(upper index c for conformal) and has a form which is comparable with (6.32). An important difference is represented by the term S_0 in S^c. It has dimension three i.e. is a hard breaking term, so is a priori a candidate for a hard breaking term in the R- and D-Ward-identity. From the construction of the supercurrent it is clear that

$$W^R \Gamma = -\int dSS + \int d\bar{S}\bar{S} \tag{6.60}$$

i.e. concretely

$$W^R \Gamma = \frac{-1}{6} m \left(\int dSA_+A_- - \int d\bar{S}\bar{A}_-\bar{A}_+\right) + \frac{M^2}{24}\left(\int dS\bar{D}\bar{D}(\Phi^2) - \int d\bar{S}DD(\Phi^2)\right)$$
$$+ \frac{1}{192\alpha}\left(\int dS\bar{D}\bar{D}(\Phi(\psi-\bar{\psi})) - \int d\bar{S}DD(\Phi(\bar{\psi}-\psi))\right),$$

hence

$$W^R \Gamma = -\frac{m}{6}\left(\int dSA_+A_- - \int d\bar{S}\bar{A}_-\bar{A}_+\right). \tag{6.61}$$

The breaking in the R-Ward-identity reduces to the soft term as it must. Analogously to (6.59) one can show, cf. sect. 6.5, that

$$W^D \Gamma = -\frac{3}{2} i \left(\int dSS + \int d\bar{S}\bar{S}\right) . \tag{6.62}$$

Again, due to

$$\int dS\, S_0 = 0 \tag{6.63}$$

one finds that

$$W^D\Gamma = -\frac{im}{4}\left(\int dS A_+ A_- + \int d\bar{S}\bar{A}_-\bar{A}_+\right) + \frac{iM^2}{8}\int dV\Phi^2 \tag{6.64}$$

i.e. the breaking of the dilational Ward-identity is caused by the soft mass terms only.
But nevertheless it is to be noted that here, due to S_0, the energy-momentum tensor is not improved off-shell. It is, however, improved on-shell since between physical states S_0 vanishes.

$n = \infty$

The trace equation (6.57) reduces in this limit to

$$-4\bar{D}^{\dot\alpha}[D_\alpha,\bar{D}_{\dot\alpha}]I_5 = -2D_\alpha(A_\pm \frac{\delta}{\delta A_\pm}) - m\,D_\alpha(A_+A_-) - 6\,\bar{D}\bar{D}D_\alpha I_5 \tag{6.65}$$

which suggests rewriting it into the "D_α-integrated" form (use (A.11))

$$\bar{D}\bar{D}\,I_5 = A_+\frac{\delta\Gamma}{\delta A_+} + A_-\frac{\delta\Gamma}{\delta A_-} + \frac{1}{2}m\,A_+A_- . \tag{6.66}$$

Similarly, the current conservation equation becomes

$$[DD,\bar{D}\bar{D}]I_5 = DD(A_\pm\frac{\delta\Gamma}{\delta A_\pm}) - \bar{D}\bar{D}(\bar{A}_\pm\frac{\delta\Gamma}{\delta\bar{A}_\pm}) + \frac{m}{2}(DD(A_+A_-) - \bar{D}\bar{D}(\bar{A}_-\bar{A}_+)). \tag{6.67}$$

Integrating over space-time we obtain

$$W_5\Gamma = i\,\frac{m}{2}\left(\int dS A_+A_- - \int d\bar{S}\bar{A}_-\bar{A}_+\right) , \tag{6.68}$$

$$W_5\Gamma \equiv -i\int dS\,(A_+\frac{\delta\Gamma}{\delta A_+} + A_-\frac{\delta\Gamma}{\delta A_-}) + i\int d\bar{S}\,(\bar{A}_+\frac{\delta\Gamma}{\delta\bar{A}_+} + \bar{A}_-\frac{\delta\Gamma}{\delta\bar{A}_-}) . \tag{6.69}$$

This result clears up what we were doing: the contact terms (6.69) represent chiral transformations which commute with supersymmetry since they act equally on all components of the superfields $A_\pm$; the Ward-identity (6.68) shows that the matter mass term breaks this symmetry; (6.67) is indeed the current conservation equation associated with these transformations and (6.66), the "trace" equation indicates that there exists a simpler form for it from which it can be derived. Using

$$[DD,\bar{D}\bar{D}] = 4i(D\sigma^\mu\bar{D} - \bar{D}\bar{\sigma}^\mu D)\partial_\mu \tag{6.70}$$

(cf. (A.12)) we see that the $\Theta\bar{\Theta}$-component of I_5 is the physical one:

it contains the axial current belonging to the transformations (6.69) in components. All other components of I_5 do not generate symmetry transformations.

6.4 Non-abelian gauge theory

In non-abelian gauge models the role of gauge invariance is taken over by BRS-invariance (cf. sect. 5). The supercurrent will therefore be constructed as a BRS-invariant operator [II.17]. The realization of this task in higher orders requires again going off-shell, i.e. calculating the contributions of all ghost and external fields. Hence an analysis of the contact terms is once more necessary. Since BRS-transformations act non-linearly on Γ, but linearly on Z - the functional for general Greens functions - we shall for the time being work with Z. We demand - cf. (6.46) - that the contact terms w_α transform as density with weight -1 under R-transformations

$$[W_R, w_\alpha] = -i\delta_R^{(-1)} w_\alpha \quad , \tag{6.71}$$

that they give rise to the correct R-weights for the fields

$$\int dx (D^\alpha w_\alpha - \bar{D}_{\dot\alpha} w^{\dot\alpha})\Big|_{\Theta=0} = W_R \quad , \tag{6.72}$$

and - the new requirement - that

$$[w_\alpha, \mathfrak{s}]\, Z = 0 \quad . \tag{6.73}$$

Here w_α are the sought contact terms in Legendre transformed form and $\mathfrak{s}$, the BRS-operator, is given explicitly in (B.21). It turns out that all chiral fields come as their R-weight n predicts (cf. (6.19) and table 5.3.1, page 62).

$$w_\alpha(\text{chiral})\Gamma = n(\varphi) D_\alpha(\varphi \frac{\delta\Gamma}{\delta\varphi}) - 2\, D_\alpha\varphi \frac{\delta\Gamma}{\delta\varphi} \quad , \varphi \text{ chiral} \tag{6.74}$$

(everything already transformed back on Γ). For vector fields we have the list of terms corresponding to (6.49) and now commutation with $\mathfrak{s}$, (6.73), leads to

$$w_\alpha(\text{vector})\Gamma = 2\Phi\,\bar{D}\bar{D}_\alpha \frac{\delta\Gamma}{\delta\Phi} - 2\,\bar{D}\bar{D}\Phi D_\alpha \frac{\delta\Gamma}{\delta\Phi}$$

$$-2\,\bar{D}\bar{D}_\alpha\rho \frac{\delta\Gamma}{\delta\rho} + 2\,D_\alpha\rho\bar{D}\bar{D}\frac{\delta\Gamma}{\delta\rho} \qquad (6.75)$$

(everything transformed back on Γ). It is important to note the difference to the abelian case, (6.53), where the contact terms for Φ had the form they have here for ρ. Both forms lead to vanishing R-weight:

$$n(\Phi) = n(\rho) = 0\ , \qquad (6.76)$$

but they differ in their result for associated dimensions. The moment construction (cf. section 6.5) i.e.

$$W'^D\Gamma = \int dx\; w^D\Gamma \qquad (6.77)$$

given by (6.54) yields indeed a dilatational Ward-identity operator, but one which associates $d(\Phi) = 2$, $d(\rho) = 0$ as dimensions to the fields Φ, ρ. Similarly the R-weight of the Faddeev-Popov ghost c_-, $n(c_-) = -2$, leads to $d(c_-) = 3$. These three dimensions are not the ones which we had given beforehand to those fields, namely $d(\Phi) = 0$, $d(\rho) = 2$, $d(c_-) = 1$, hence W'^D is <u>not</u> the Ward-identity operator of the physical dilatations. It is clear that the true dilatational Ward-identity operator is given by

$$W^D = W'^D - (-i)\int dV\left(2\Phi\frac{\delta}{\delta\Phi} - 2\rho\frac{\delta}{\delta\rho}\right)$$

$$- (-i)\left(\int dS\; 2c_-\frac{\delta}{\delta c_-} + \int d\bar{S}\; 2\bar{c}_-\frac{\delta}{\delta\bar{c}_-}\right)\ . \qquad (6.78)$$

Comparing with the abelian gauge case we conclude that the moment construction for supercurrents will not work automatically, but has to be supplemented by "corrections" like (6.78).

Let us now go ahead and use the contact terms we found, (6.74), (6.75):

$$\frac{1}{2} w_\alpha = \mathrm{Tr}(\Phi\bar{D}\bar{D}D_\alpha\delta_\Phi - \bar{D}\bar{D}\Phi D_\alpha\delta_\Phi - \bar{D}\bar{D}D_\alpha\rho\delta_\rho + D_\alpha\rho\bar{D}\bar{D}\delta_\rho$$

$$+ D_\alpha c_+\delta_{c_+} - \sigma D_\alpha\delta_\sigma - c_- D_\alpha\delta_{c_-}) + \frac{1}{2} w_\alpha(\mathrm{matter})$$

$$\frac{1}{2} w_\alpha(\mathrm{matter}) = -\frac{1}{3} D_\alpha(A_+\delta_{A_+}) + D_\alpha A_+\delta_{A_+} + \frac{1}{3} D_\alpha(Y_+\delta_{Y_+}) \tag{6.79}$$

$$- Y_+ D_\alpha\delta_{Y_+} + (+ \to -)$$

$$+ \mathrm{Tr}(-\frac{1}{3} D_\alpha(A\delta_A) + D_\alpha A\delta_A + \frac{1}{3} D_\alpha(Y\delta_Y) - YD_\alpha\delta_Y)$$

(here $\delta_\varphi \equiv \delta/\delta\varphi$).

In order to construct a supercurrent we have to deal with the equations of motion. For a general massless supersymmetric Yang-Mills theory which maintains parity and contains matter fields in the adjoint and some other unitary representation the action is given in the α-gauge by (cf. (5.20), (5.11), (4.33), (5.78), (5.83)):

$$\Gamma = \Gamma_{inv} + \Gamma_{g.f.} + \Gamma_{\Phi\Pi} + \Gamma_{ext.f.}$$

$$\Gamma_{inv} = \Gamma_{YM} + \Gamma_{matter}$$

$$\Gamma_{YM} = \frac{-1}{256}\frac{1}{g^2}\mathrm{Tr}(\int dS F^\alpha F_\alpha + \int d\bar{S}\bar{F}_{\dot\alpha}\bar{F}^{\dot\alpha}),\quad F_\alpha = \bar{D}\bar{D}(e^{-\Phi}D_\alpha e^\Phi)$$

$$\Gamma_{matter} = \int dV(z_1(\bar{A}_+ e^{\tilde\Phi}A_+ + A_-\bar{e}^{\tilde\Phi}\bar{A}_-) + z_2\,\mathrm{Tr}\,\bar{e}^{\Phi}\bar{A}e^{\Phi}A) \tag{6.80}$$

$$+ h\,\mathrm{Tr}(\int dS A^3 + \int d\bar{S}\bar{A}^3)$$

$$\Gamma_{g.f.} = \frac{-1}{128\alpha}\mathrm{Tr}\int dV\, DD\Phi\bar{D}\bar{D}\Phi$$

$$\Gamma_{\Phi\Pi} = \frac{1}{128\alpha}\mathrm{Tr}\int dV(DDc_- Q_s + \bar{D}\bar{D}\bar{c}_- Q_s)$$

$$\Gamma_{ext.f.} = \mathrm{Tr}\int dV\rho Q_s - \int dS(\mathrm{Tr}\,\sigma c_+ c_+ + YP) - \int d\bar{S}(\mathrm{Tr}\,\bar{\sigma}\bar{c}_+\bar{c}_+ + \bar{Y}\bar{P})$$

$$YP \equiv Y_-\tilde{c}_+ A_+ + A_-\tilde{c}_+ Y_+ + \mathrm{Tr}\,Y[c_+, A]$$

Let us recall some of the notation.

$\Phi \equiv \Phi^i \tau^i$, τ^i generates the fundamental representation of the gauge group G;

$\tilde{\Phi} \equiv \Phi^i T^i$, T^i generates the unitary representation of G under which $A_\pm$ transform;

$A_\pm \equiv (A_{\pm a})$ cf. (5.19);

$A \equiv A^i \tau^i$ transforms under the adjoint representation (cf. (5.19a));

$c_\pm \equiv c_\pm^i \tau^i$; $\tilde{c}_\pm \equiv c_\pm^i T^i$;

$Q_s = Q(a_k = 0) = c_+ - \bar{c}_+ + \frac{1}{2}[\Phi, c_+ + \bar{c}_+] + \ldots$ cf. (5.6), (5.65).

We shall in fact use below the following decomposition of Q_s:

$$Q_s(\Phi, c_+, \bar{c}_+) = R\{c_+\} - \bar{R}\{\bar{c}_+\} \tag{6.81}$$

where

$$R^{-1}\{\Lambda\} \equiv \sum_{n=0}^{\infty} \frac{(-1)^n}{(2n+1)!}[\Phi,[\ldots[\Phi,\Lambda]\ldots] ,$$

$$\bar{R}^{-1}\{\bar{\Lambda}\} \equiv \sum_{n=0}^{\infty} \frac{1}{(2n+1)!}[\Phi,[\ldots[\Phi,\bar{\Lambda}]\ldots] ,$$

$$R \{\Lambda\} \equiv \sum_{n=0} a_n [\Phi,[\ldots[\Phi,\Lambda]\ldots] ,$$

$$\bar{R} \{\bar{\Lambda}\} \equiv \sum_{n=0} (-1)^n a_n [\Phi,[\ldots[\Phi,\bar{\Lambda}]\ldots] ,$$

and the coefficients a_n are determined by

$$R\{R^{-1}\{\Lambda\}\} = \Lambda .$$

The equations of motion for the action (6.80) are as follows:

$$\begin{aligned}\frac{\delta\Gamma}{\delta\Phi} &= \frac{1}{64g^2} R^{-1}\{D(\Psi F \Psi^{-1}) - \frac{1}{128\alpha}\{DD,\bar{D}\bar{D}\}\Phi + Q' \\ &+ z_1 \tau^i (\bar{A}_+ e^{\tilde{\Phi}} R^{-1}\{T^i\} A_+ + A_- e^{-\tilde{\Phi}} \bar{R}^{-1}\{T^i\}\bar{A}_-) \\ &+ z_2 R^{-1}\{ - \bar{A}\Psi A \Psi^{-1} + \Psi A \Psi^{-1}\bar{A}\} \\ &\equiv \frac{\delta\bar{\Gamma}}{\delta\Phi} - \frac{1}{128\alpha}\{DD,\bar{D}\bar{D}\}\Phi ,\end{aligned} \tag{6.82}$$

$$Q' \equiv \sum_{n=0}^{\infty} a_n \sum_{k=1}^{n} ((-1)^k [\Phi^{n-k}\{c_+\}, \Phi^{k-1}\{\eta\}]_+ - (-1)^{n-k}[\Phi^{n-k}\{\bar{c}_+\}, \Phi^{k-1}\{\eta\}]_+) , \qquad (6.82 \text{ cont'd})$$

$$\Psi \equiv e^{\Phi} \ , \quad \Psi^{-1} \equiv e^{-\Phi} \ , \quad \eta \equiv \rho + \frac{1}{128\alpha} (DDc_- + \bar{D}\bar{D}\bar{c}_-) \quad (\text{cf. } (5.109)),$$

$$\frac{\delta\Gamma}{\delta c_-} = \frac{1}{128\alpha} \bar{D}\bar{D}DD \, Q_s \ ,$$

$$\frac{\delta\Gamma}{\delta c_+} = - \bar{D}\bar{D} \, \bar{R}\{\eta\} + [\sigma, c_+] + Y_- TA_+ - A_- TY_+ - [Y, A],$$

$$\frac{\delta\Gamma}{\delta\rho} = Q_s \ , \quad \frac{\delta\Gamma}{\delta\sigma} = -c_+ c_+ \ ,$$

$$\frac{\delta\Gamma}{\delta A_+} = z_1 \, \bar{D}\bar{D}(\bar{A}_+ e^{\tilde{\Phi}}) - Y_- \tilde{c}_+ \ ,$$

$$\frac{\delta\Gamma}{\delta A_-} = z_1 \, \bar{D}\bar{D}(e^{-\tilde{\Phi}} A_-) - \tilde{c}_+ Y_+ \ ,$$

$$\frac{\delta\Gamma}{\delta A} = z_2 \, \bar{D}\bar{D}(\Psi^{-1} A \Psi) + 3h(A^2 - \frac{1}{N} \text{Tr } A^2) - \{Y, c_+\} \ ,$$

$$\frac{\delta\Gamma}{\delta Y_+} = A_- \tilde{c}_+ \ , \quad \frac{\delta\Gamma}{\delta Y_-} = -\tilde{c}_+ A_+ \ , \quad \frac{\delta\Gamma}{\delta Y} = [A, c_+] \ .$$

The supercurrent is now to be found by decomposing

$$-2 \, w_\alpha \Gamma = \Delta_\alpha = \sum_{k=1}^{6} \Delta_\alpha^k \qquad (6.83)$$

into

$$\Delta_\alpha = \bar{D}^{\dot{\alpha}} V_{\alpha\dot{\alpha}} + 2 \, D_\alpha \, S \qquad (6.84)$$

with V and S being BRS-invariant, V_μ an axial current vector superfield and S a chiral superfield related to the breaking of the conformal symmetries by the gauge fixing procedure. The list of the terms Δ_α^k is the following:

$$- \frac{1}{4} \Delta_\alpha^1 = \Phi \bar{D}\bar{D} D_\alpha \frac{\delta\bar{\Gamma}}{\delta\Phi} - \bar{D}\bar{D} D_\alpha \eta Q_s$$

$$- \frac{1}{4} \Delta_\alpha^2 = - \bar{D}\bar{D}\Phi D_\alpha \frac{\delta\bar{\Gamma}}{\delta\Phi} + D_\alpha \eta \bar{D}\bar{D} Q_s$$

$$- \frac{1}{4} \Delta_\alpha^3 = - D_\alpha c_+ \, \bar{D}\bar{D}\bar{R}\{\eta\} - Y_- D_\alpha \tilde{c}_+ A_+ - A_- D_\alpha \tilde{c}_+ Y_+ - D_\alpha c_+ [Y, A] \ ,$$

$$-\frac{1}{4}\Delta_\alpha^4 = -\frac{1}{128\alpha}(c_-D_\alpha LQ_s + D_\alpha\bar{D}\bar{D}\bar{c}_-\bar{D}\bar{D}Q_s - \bar{D}\bar{D}\Phi D_\alpha L\Phi) ,$$

$$-\frac{1}{4}\Delta_\alpha^5 = -\frac{1}{3}z_1 D_\alpha\bar{D}\bar{D}(\bar{\mathfrak{a}}_+A_+ + A_-\bar{\mathfrak{a}}_-) - \frac{1}{3}z_2 D_\alpha\bar{D}\bar{D}(A\bar{\mathfrak{a}}),$$

$$-\frac{1}{4}\Delta_\alpha^6 = z_1(\bar{D}\bar{D}\bar{\mathfrak{a}}_+D_\alpha A_+ + D_\alpha A_-\bar{D}\bar{D}\bar{\mathfrak{a}}_-)$$
$$+ (\bar{D}\bar{D}\bar{\mathfrak{a}}_+\tilde{f}_\alpha A_+ - A_-\tilde{f}_\alpha\bar{D}\bar{D}\bar{\mathfrak{a}}_-)$$
$$-\mathfrak{s}(f_\alpha(Y_-TA_+ - A_-TY_+)) + z_2 D_\alpha A\bar{D}\bar{D}\bar{\mathfrak{a}} + D_\alpha c_+[Y,A] .$$

Again we have introduced a number of abbreviations:

$$L \equiv \{DD,\bar{D}\bar{D}\}$$

$$\mathfrak{a} \equiv \Psi A \Psi^{-1} , \quad \mathfrak{a}_+ \equiv \tilde{\Psi} A_+ , \quad \mathfrak{a}_- \equiv A_-\tilde{\Psi}^{-1} ,$$

$$\bar{\mathfrak{a}} \equiv \Psi^{-1}\bar{A}\Psi , \quad \bar{\mathfrak{a}}_+ \equiv \bar{A}_+\tilde{\Psi}^{-1} , \quad \bar{\mathfrak{a}}_- \equiv \tilde{\Psi}\bar{A}_- ,$$

$$\tilde{\Psi} \equiv e^{\tilde{\Phi}} , \quad f_\alpha \equiv \Psi^{-1}D_\alpha\Psi, \quad \Psi \equiv e^{\Phi} .$$

One should also note that Tr symbols are omitted.

For the decomposition (6.83) we use two guiding principles: first we try to rewrite Δ_α^k as $\mathfrak{s}$-variations, because then their BRS-invariance is obvious; second we use the limit to the abelian case where we know already all terms. At this stage one obtains

$$\begin{aligned}
-\frac{1}{4}\Delta_\alpha^1 &= \mathfrak{s}(\bar{D}\bar{D}D_\alpha\eta\Phi) , \\
-\frac{1}{4}\Delta_\alpha^2 &= -\mathfrak{s}(D_\alpha\eta\bar{D}\bar{D}\Phi) , \\
-\frac{1}{4}\Delta_\alpha^3 &= \mathfrak{s}(-f_\alpha\bar{D}\bar{D}\bar{R}\{\eta\} + f_\alpha(Y_-TA_+ - A_-TY_+) - f_\alpha[Y,A]) , \\
-\frac{1}{4}\Delta_\alpha^4 &= -\frac{1}{128\alpha}(c_-DLQ_s + D_\alpha\bar{D}\bar{D}\bar{c}_-\bar{D}\bar{D}Q_s - \bar{D}\bar{D}\Phi D_\alpha L\Phi), \\
\mathfrak{s}\,\Delta_\alpha^4 &= 0 , \\
-\frac{1}{4}\Delta_\alpha^5 &= -\frac{1}{3}D_\alpha\mathfrak{s}(Y_-A_+ + A_-Y_+ + YA) + hD_\alpha A^3, \\
\mathfrak{s}\,\Delta_\alpha^5 &= 0 ,
\end{aligned} \tag{6.86}$$

$$-\frac{1}{4}\Delta^6_\alpha = z_1(\bar D\bar D\bar{a}_+ D_\alpha A_+ + D_\alpha A_-\bar D\bar D\bar{a}_- + \bar D\bar D\bar{a}_+\tilde f_\alpha A_+ - A_-\tilde f_\alpha \bar D\bar D\bar{a}_-)$$

$$-s(f_\alpha(Y_-TA_+ - A_-TY_+) - YD_\alpha A) \qquad (6.86\ \text{cont'd})$$

$$-h\, D_\alpha A^3 \ ,$$

$$s\Delta^6_\alpha = 0 \ .$$

Similarly, it is useful to have a - possibly only partial - list for the terms contributing to S and V as they are characterized by their quantum numbers. Discarding matter terms for the moment we have as possible terms for S where S is of the form*)

$$S = \bar D\bar D B \ ,$$

$$\bar B = B \ ,$$

(quantum numbers of B : dim:2, R-weight:0, $Q_{\Phi\Pi}$:0, parity:+) the following ones:

$$B^1 = s(\eta\Phi),$$

$$B^2 = DD\Phi\bar D\bar D\Phi + \bar D\bar D Q_s\bar c_- + DDQ_s c_- \ , \qquad (6.87)$$

$$sB^2 = 0 \ .$$

The current terms are characterized by dim:3, R-weight:0, $Q_{\Phi\Pi}$:0, parity:- . Hence we expect certainly the following terms*):

$$V^Y_{\alpha\dot\alpha} = F_\alpha\Psi^{-1}\bar F_{\dot\alpha}\Psi,$$

$$V^1_{\alpha\dot\alpha} = D_\alpha Q_s\bar D_{\dot\alpha}\eta - \bar D_{\dot\alpha}Q_s D_\alpha\eta + D_\alpha\Phi\bar D_{\dot\alpha}\frac{\delta\bar\Gamma}{\delta\Phi} - \bar D_{\dot\alpha}\Phi D_\alpha\frac{\delta\bar\Gamma}{\delta\Phi}$$

$$= s(D_\alpha\Phi\bar D_{\dot\alpha}\eta - \bar D_{\dot\alpha}\Phi D_\alpha\eta) \equiv s\,\hat V^1_{\alpha\dot\alpha}, \qquad (6.88)$$

$$V^2_{\alpha\dot\alpha} = D_\alpha\bar D\bar D\Phi\bar D_{\dot\alpha}DD\Phi + D_\alpha\bar D\bar D Q_s\bar D_{\dot\alpha}\bar c_- - \bar D_{\dot\alpha}DDQ_s D_\alpha c_- \ ,$$

$$sV^2_{\alpha\dot\alpha} = 0 \ ,$$

---

*)Tr symbols omitted

$$V^3_{\alpha\dot\alpha} = D_\alpha\bar D_{\dot\alpha}DD\Phi\bar D\bar D\Phi - \bar D_{\dot\alpha}D_\alpha\bar D\bar D\Phi DD\Phi + D_\alpha\bar D_{\dot\alpha}DDQ_s c_-$$

$$- \bar D_{\dot\alpha}D_\alpha\bar D\bar D Q_s\bar c_- + \bar D\bar D Q_s D_\alpha\bar D_{\dot\alpha}\bar c_- - DDQ_s\bar D_{\dot\alpha}D_\alpha c_- \,, \qquad (6.88\text{ cont.'d})$$

$$\delta V^3_{\alpha\dot\alpha} = 0 \,.$$

All these terms are generalizations of corresponding abelian ones, so they must occur in the current.
In the first step we try to establish a relation amongst the non-variations, i.e. Δ^4_α, B^2, V^2, V^3 and find indeed

$$16\alpha\Delta^4_\alpha = \bar D\bar D D_\alpha B^2 - \bar D^{\dot\alpha}(V^2_{\alpha\dot\alpha} + V^3_{\alpha\dot\alpha}) \,. \qquad (6.89)$$

In the second step we link the terms which are variations, i.e. Δ^1, Δ^2, Δ^3, B^1 and V^1 and introduce, again in analogy to the abelian case,

$$-\frac{1}{4}C_\alpha \equiv \sum_{k=1}^{3}\Delta^k_\alpha - \bar D\bar D D_\alpha B^1 + 2\bar D^{\dot\alpha}V^1_{\alpha\dot\alpha} \,. \qquad (6.90)$$

We know that C_α is a variation

$$-\frac{1}{4}C_\alpha = -\frac{1}{4}\delta\hat C_\alpha$$

$$-\frac{1}{4}\hat C_\alpha = -\bar D\bar D D_\alpha\Phi\eta + D_\alpha\Phi\bar D\bar D\eta - \Psi^{-1}D_\alpha\Psi\bar D\bar D\bar R\{\eta\} \qquad (6.91)$$

and that this reduces in the abelian limit to

$$-\frac{1}{4}\hat C^{\text{abelian}}_\alpha = -\bar D\bar D D_\alpha\Phi\eta$$

$$= -\bar D\bar D D_\alpha\Phi\,\frac{\delta\bar\Gamma_{\text{abelian}}}{\delta\Phi} = -\bar D\bar D D_\alpha\Phi\bar D_{\dot\alpha}DD\bar D^{\dot\alpha}\Phi \qquad (6.92)$$

$$= -\bar D^{\dot\alpha}\,(F_\alpha\bar F_{\dot\alpha})_{\text{abelian}}$$

This suggests the introduction of

$$-\frac{1}{4}E_\alpha \equiv \bar D^{\dot\alpha}\Big(-\frac{1}{64g^2}V^{YM}_{\alpha\dot\alpha}\Big) \qquad (6.93)$$

in the non-abelian case and to search for a $\hat E_\alpha$ whose variation yields E_α:

$$E_\alpha = \mathfrak{G}\hat{E}_\alpha$$
$$-\frac{1}{4}\hat{E}_\alpha = -F_\alpha\bar{R}\{n\} \quad . \tag{6.94}$$

All of this construction was to find out that

$$\hat{C}_\alpha - \hat{E}_\alpha = -2\bar{D}^{\dot{\alpha}}\hat{V}^4_{\alpha\dot{\alpha}} \ , \tag{6.95}$$

where $\hat{V}^4_{\alpha\alpha}$ has the complicated structure

$$\hat{V}^4_{\alpha\dot{\alpha}} \equiv R^{-1\prime}\{\bar{D}_{\dot{\alpha}}\Phi, D_\alpha\Phi\}\bar{R}\{n\},$$
$$R^{-1\prime}\{A,B\} \equiv \sum_{n=1}^{\infty} \frac{(-1)^n}{(2n+1)!} \sum_{k=1}^{n} \Phi^{k-1}\{[A,\Phi^{n-k}\{B\}]_\mp\} \tag{6.96}$$

(Tr symbols omitted).

$V^4_{\alpha\dot{\alpha}} = \mathfrak{G}\hat{V}^4_{\alpha\dot{\alpha}}$ is the last element of the basis of V needed in the tree approximation without matter.

The decomposition problem has been pushed by now to the following stage:

$$-\frac{1}{4}\Delta_\alpha = \frac{-1}{64\alpha}\ \bar{D}\bar{D}D_\alpha B^2 + \frac{1}{64\alpha}\ \bar{D}^{\dot{\alpha}}(V^2+V^3)_{\alpha\dot{\alpha}}$$
$$+\ \bar{D}\bar{D}D_\alpha B^1 - 2\bar{D}^{\dot{\alpha}}(V^1+V^4)_{\alpha\dot{\alpha}} - \frac{1}{64g^2}\ \bar{D}^{\dot{\alpha}}V^{YM}_{\alpha\dot{\alpha}} \tag{6.97}$$

Let us next incorporate the matter contributions. To be dealt with are Δ^5, Δ^6 and an additional term in Δ^3. The abelian limit (6.57), the Wess-Zumino model (6.23), respectively suggest

$$V^5_{\alpha\dot{\alpha}} = z_1\ [D_\alpha,\bar{D}_{\dot{\alpha}}](A_-\ \bar{a}_- + \bar{A}_+\ a_+) + z_2[D_\alpha,\bar{D}_{\dot{\alpha}}](\Psi^{-1}\bar{A}\Psi A) \tag{6.98}$$

$$V^6_{\alpha\alpha} = z_1\ D_\alpha(A_-\tilde{\Psi}^{-1})\tilde{\Psi}\bar{D}_{\dot{\alpha}}(\tilde{\Psi}^{-1}\bar{A}_-) + z_1\ \bar{D}_{\dot{\alpha}}(\bar{A}_+\tilde{\Psi})\tilde{\Psi}^{-1}D_\alpha(\tilde{\Psi}A_+)$$
$$+\ z_2\ \Psi^{-1}D_\alpha(\Psi A\Psi^{-1})\Psi\bar{D}_{\dot{\alpha}}(\Psi^{-1}\bar{A}\ ) \tag{6.99}$$

as further basis elements of the current and actual calculation of $\bar{D}^{\dot{\alpha}}V_{\alpha\dot{\alpha}}$ shows that $\frac{2}{3}\ \bar{D}^{\dot{\alpha}}V^5_{\alpha\dot{\alpha}} - 2\bar{D}^{\dot{\alpha}}V^6_{\alpha\dot{\alpha}}$ accounts for the matter contributions Δ^5, Δ^6 and in Δ^3. Collecting these partial results we arrive finally at

$$\bar{D}^{\dot{\alpha}} V'_{\alpha\dot{\alpha}} = -2w_\alpha \Gamma + \bar{D}\bar{D}D_\alpha B, \tag{6.100}$$

where

$$B = 4(B^1 - \frac{1}{64\alpha} B^2) ,$$

$$V' = \frac{1}{16g^2} V^{YM} - \frac{1}{16\alpha}(V^2 + V^3) + 8(V^1 + V^4) + \frac{8}{3} V^5 - 8V^6 .$$

It is noteworthy that the supercurrent V' and the breaking B are both BRS-invariant. From the form (6.100) we find that the supercurrent is strictly conserved

$$\partial^\mu V'_\mu = iw\Gamma. \tag{6.101}$$

It will turn out in higher orders that not all terms B contributing to the trace equation are BRS-invariant, but DDB ($\bar{D}\bar{D}B$) are. We therefore transform already here in the tree approximation current and breaking into a more appropriate form. We define a new supercurrent

$$V_{\alpha\dot{\alpha}} = V'_{\alpha\dot{\alpha}} - \frac{2}{3}[D_\alpha, \bar{D}_{\dot{\alpha}}]B \tag{6.102}$$

and then find that it satisfies as trace equation

$$\bar{D}^{\dot{\alpha}} V_{\alpha\dot{\alpha}} = -2w_\alpha \Gamma - 2D_\alpha S \tag{6.103}$$

with

$$S = \frac{1}{6} \bar{D}\bar{D}B \tag{6.104}$$

and as current conservation equation

$$\partial^\mu V_\mu = iw\Gamma + i(DDS - \bar{D}\bar{D}\bar{S}) \quad . \tag{6.105}$$

Let us now use the existence of a supercurrent for the discussion of R- and dilatational symmetry. From (6.101) it is obvious that R-invariance holds, as it must, since the action (6.80) is invariant under conformal R-transformations (due to the absence of masses and our choice of weights, table 5.3.1). Dilatational invariance, however, is not guaranteed since the trace of the energy-momentum tensor does not vanish (cf. definition (6.27d) and (6.28i)):

$$\hat{T}_\mu{}^\mu = -\frac{3}{2}\, i\, (DDS + \bar{D}\bar{D}\bar{S}) \neq 0 \tag{6.106}$$

and the dilatational Ward-identity as obtained from the moment construction is given by

$$W'^D\Gamma = \int dx\, \hat{T}_\mu{}^\mu = -\frac{3}{2}\, i\, (\int dSS + \int d\bar{S}\bar{S}) \,. \tag{6.107}$$

But above, (6.78), we have already identified the physical dilatational Ward-identity operator:

$$W^D = W'^D + 2i\,(N_\Phi - N_\rho + N_-) \tag{6.108}$$

(N_φ is the counting operator for the field φ). And, indeed, one can verify that the breaking terms $-\frac{3}{2}\, i \int dV \frac{1}{6} \cdot 2B$ yield precisely this difference:

$$\begin{aligned} \mathrm{Tr} \int dV \mathcal{B}(\eta\Phi) &= (N_\Phi - N_\eta + 2\alpha\partial_\alpha)\Gamma(\eta,\ldots) \,, \\ \int dVB^2 &= 128\alpha^2\partial_\alpha\Gamma + \mathrm{Tr} \int dV(\bar{D}\bar{D}Q_s\bar{c}_- + DDQ_s c_-) \,. \end{aligned} \tag{6.109}$$

When we rewrite Γ in the variables ρ and c_- this becomes

$$\begin{aligned} \mathrm{Tr} \int dV \mathcal{B}(\eta\Phi) &= (N_\Phi - N_\rho + N_- + 2\alpha\partial_\alpha)\Gamma(\rho,c_-,\ldots) \\ \int dVB^2 &= 128\alpha^2\partial_\alpha\Gamma(\rho,c_-,\ldots) \end{aligned} \tag{6.110}$$

hence

$$-\frac{i}{2} \int dVB = -2i(N_\Phi - N_\rho + N_-) \,. \tag{6.111}$$

Let us summarize these results: for supersymmetric Yang-Mills theories there exists a supercurrent off-shell; it is BRS-invariant and contains an axial current, the supersymmetry currents and a supersymmetric energy-momentum tensor. The latter is however not improved as a consequence of the gauge fixing procedure. Still, the supercurrent contains all informations on the superconformal group.

6.5 The identification of component currents

In the preceding sections we have introduced moment currents, (6.27), (6.36), (6.37), and already made partial use of the knowledge where and in which supercurrent component currents belonging to the superconformal symmetries are situated. We present now a systematic investigation of this problem (cf. [I.6] App. C).

The starting point is the trace equation for which we assume the form

$$- 2w_\alpha \Gamma = \bar{D}^{\dot{\alpha}} V_{\alpha\dot{\alpha}} + 2D_\alpha S - B_\alpha \tag{6.112}$$

(and its conjugate).
Here $V_{\alpha\dot{\alpha}}$ is an axial vector superfield, S a chiral field and B_α constrained by

$$D^\alpha B_\alpha - \bar{D}_{\dot{\alpha}} \bar{B}^{\dot{\alpha}} = 0 \; . \tag{6.113}$$

Together with (6.112) holds therefore the current conservation equation

$$iw\Gamma = \partial^\mu V_\mu - i\,(DDS - \bar{D}\bar{D}\bar{S}) \; . \tag{6.114}$$

(We shall actually prove in section 16.3 that (6.112) is the most general possibility for all models, i.e. contact terms treated in this book, hence we are dealing with the most general case.)
The expansion of the contact terms is given by

$$w = w^R - i\theta^\alpha w_\alpha{}^Q + i\bar{\theta}_{\dot{\alpha}} \bar{w}^{Q\dot{\alpha}} - 2\theta\sigma^\mu\bar{\theta} w_\nu^P \; + \text{total derivatives} \tag{6.115}$$

(integration yields $\hat{W}^R$, (6.18)). Similarly we expand the superfields V_μ, B_α and S.

$$V_\mu = C_\mu + \theta\chi_\mu + \bar{\theta}\bar{\chi}_\mu + \tfrac{1}{2}\theta^2 M_\mu + \tfrac{1}{2}\bar{\theta}^2 \bar{M}_\mu \tag{6.116a}$$
$$+ \theta\sigma^\nu\bar{\theta} v_{\mu\nu} + \tfrac{1}{2}\theta^2\bar{\theta}\bar{\lambda}_\mu + \tfrac{1}{2}\bar{\theta}^2\theta\lambda_\mu + \tfrac{1}{4}\theta^2\bar{\theta}^2 D_\mu \; ,$$

$$B_\alpha = C_\alpha + \theta^\beta\chi_{\alpha\beta} + \bar{\theta}_{\dot{\beta}}\chi_\alpha{}^{\dot{\beta}} + \tfrac{1}{2}\theta^2 M_\alpha + \tfrac{1}{2}\bar{\theta}^2 N_\alpha \tag{6.116b}$$
$$+ \theta\sigma^\nu\bar{\theta} v_{\alpha\nu} + \tfrac{1}{2}\theta^2\bar{\theta}_{\dot{\beta}}\lambda_\alpha{}^{\dot{\beta}} + \tfrac{1}{2}\bar{\theta}^2\theta^\beta\lambda_{\alpha\beta} + \tfrac{1}{2}\theta^2\bar{\theta}^2 D_\alpha ,$$

$$S = A + \Theta\Psi + \Theta^2 F \ . \tag{6.116c}$$

We shall first show that if both S and B_α are non-zero then it is impossible to construct any conserved energy-momentum tensor $T_{\mu\nu}$ or conserved supersymmetry current $Q_{\mu\alpha}$ from the components of V_μ. Consider $Q_{\mu\alpha}$ and assume that it is given by some combination of components of V_μ. Dimensional considerations (dim V_μ = 3, dim $Q_{\mu\alpha}$ = 7/2) as well as Lorentz covariance restrict this combination to be

$$Q_\mu = a\chi_\mu + b\sigma_\mu\bar{\sigma}_\nu\chi^\nu . \tag{6.117}$$

(6.114) yields

$$\partial_\mu\chi^\mu = 4\partial_\mu(\sigma^\mu\bar{\psi}) - iw^Q\Gamma + \text{t.d.c.t.} \tag{6.118}$$

(the abbreviations c.t. = contact terms and t.d.c.t. = total derivative contact terms) whereas (6.112) yields the "trace" relation

$$\bar{\sigma}_\nu\chi^\nu = 4\bar{\psi} - 2\bar{C} + \text{c.t.} \tag{6.119}$$

so that

$$\partial_\mu Q^\mu = 4(a + b)\sigma^\mu\partial_\mu\bar{\psi} - 2b\sigma^\mu\partial_\mu\bar{C} + w^Q\Gamma + \text{t.d.c.t.} \tag{6.120}$$

There are only two possibilities for having a conservation euqation

$$\partial_\mu Q^\mu = w^Q\Gamma + \text{t.d.c.t.} \tag{6.121}$$

giving rise upon integration to the supersymmetry WI

$$W^Q\Gamma \equiv \int dx w^Q\Gamma = 0 \ . \tag{6.122}$$

The first one ist

$$\bar{C}_{\dot{\alpha}} = 0 \quad (\text{which implies } B_\alpha = \bar{B}_{\dot{\alpha}} = 0) \tag{6.123}$$

and

$b = - a = - i$.

In this case

$$\partial_\mu V^\mu = iw\Gamma + i(DDS - \bar{D}\bar{D}\bar{S})$$

$$D^\alpha V_{\alpha\dot{\alpha}} = - 2\bar{w}_{\dot{\alpha}}\Gamma - 2\bar{D}_{\dot{\alpha}}\bar{S}$$

$$Q_\mu = i(\chi_\mu - \sigma_\mu\bar{\sigma}_\nu\chi^\nu)$$

$$\partial_\mu Q^\mu = w^Q\Gamma + \text{t.d.c.t.} \tag{6.124}$$

and the trace relation becomes

$$Q^\mu\sigma_\mu = -3i\chi^\mu\sigma_\mu = 12i\bar{\psi} + \text{c.t.} \quad . \tag{6.125}$$

A precise examination of all components of equation (6.112) with $\bar{B} = 0$ yields the following identification of components and trace identities (care must be taken concerning parity quantum numbers; recall that $PV_\mu P^{-1} = - V_\mu$ and $PSP^{-1} = \bar{S}$):

1) $C_\mu = R_\mu$,

$\partial_\mu R^\mu = - 4i(F - \bar{F}) + w^R\Gamma + \text{t.d.c.t.}$

2) $\chi_\mu = - i(Q_\mu - \frac{1}{3} Q^\nu\sigma_\nu\bar{\sigma}_\mu)$,

$\partial_\mu Q^\mu = w^Q\Gamma + \text{t.d.c.t.}$

trace identity: $Q_\mu\sigma^\mu = 12i\bar{\psi} + \text{c.t.}$ (6.126)

3) $v_{(\mu\nu)} = - 2(T_{\mu\nu} - \frac{1}{3} g_{\mu\nu}T^\lambda{}_\lambda)$,

$\partial^\mu T_{\mu\nu} = w^P_\nu\Gamma + \text{t.d.c.t.}$

trace identity: $T^\lambda{}_\lambda = 6(F + \bar{F}) + \text{c.t.}$

4) $v_{[\mu\nu]} = - \frac{1}{2} \varepsilon_{\mu\nu\rho\sigma}\partial^\rho c^\sigma + \text{c.t.}$

5) $M_\mu = - 8i\partial_\mu\bar{A} + \text{c.t.}$

6) $\lambda_\mu = - i\sigma^\nu\partial_\mu\bar{\chi}_\nu + \varepsilon_{\mu\lambda\nu\rho}\sigma^\rho\partial^\lambda\bar{\chi}^\nu + c.t.$ (6.126 cont'd)

7) $D_\mu = \partial^2 c_\mu - 2\partial_\mu\partial_\nu c^\nu + c.t.$.

It is remarkable that if S is a mass term (vanishing with the mass), then $Q_{\mu\alpha}$ and $T_{\mu\nu}$ are improved conserved currents, as the two trace identities above show. On the other hand R_μ is not conserved, but its non-conservation is also a mass term. Renormalization will bring anomalous higher order contributions (non-vanishing with the mass) to S, giving rise to anomalous trace identities and an anomalous Ward-identity for R_μ.

The second possibility is given by

$$\bar{\Psi} = 0 \quad \text{(which implies that } S = \bar{S} = 0) \tag{6.127}$$

and

$$b = 0, \quad a = + i \quad .$$

Thus

$$\partial_\mu V^\mu = iw\Gamma$$

$$D^\alpha V_{\alpha\dot{\alpha}} = - 2\bar{w}_{\dot{\alpha}}\Gamma + \bar{B}_{\dot{\alpha}}$$

$$Q_\mu = i\chi_\mu$$

$$\partial_\mu Q^\mu = w^Q\Gamma + \text{t.d.c.t.}$$

$$\text{trace identity:} \quad Q^\alpha_\mu\sigma^\mu_{\alpha\dot{\alpha}} = 2i\bar{c}_{\dot{\alpha}} + c.t. \; . \tag{6.128}$$

Similarly the other currents are found to be

1) $R_\mu = C_\mu$,

$$\partial_\mu R^\mu = w^R\Gamma + \text{t.d.c.t.}$$

2) $T_{\mu\nu} = - \frac{1}{2} v_{\mu\nu}$,

$$\partial^\mu T_{\mu\nu} = w^P_\nu + \text{t.d.c.t.}$$

$$\text{trace identity:} \quad T^\lambda{}_\lambda = \frac{1}{2} \chi^\alpha{}_\alpha + c.t. \tag{6.129}$$

So all three currents are conserved. If $\bar{B}$ has the special form $DD\bar{D}\Phi$, with $\phi = \bar{\Phi}$, then it can be shown that $\partial^{\mu} v_{(\mu\nu)} = \text{c.t.}$; so that a symmetric energy-momentum tensor can be defined $T_{\mu\nu} = -\frac{1}{2} v_{(\mu\nu)}$; but this is not true in general. The very peculiar circumstance is realized in the massless limit of the Wess-Zumino model (to all orders) and in the massless limits of SQED and SYM in the tree approximation.* Then there exists a strictly conserved R-current as a component of a supercurrent. This term $DD\bar{D}\Phi$ shows up in the traces as an anomaly (if it comes with higher orders). If either one of the above cases is realized for an operator V_{μ} satisfying (6.112) we shall speak of it as "supercurrent" or "supercurrent with interpretation". Case (6.123) we shall call "S-type breaking", case (6.127) will be called "B-type breaking".

Proceed now to the identification of the components of the other supercurrents where for the first case the definitions (6.27') are adopted and in the second case the old definitions (6.27) are used. By explicit computation of the lowest Θ-components, it is found that the supercurrents begin with the component currents whose name they bear. The higher Θ-components are made from other component currents (compare with the corresponding $\hat{W}^{A}$ (2.7)), trace terms and improvement terms (i.e. contributions identically vanishing upon taking the divergence). The results are

1) $\hat{Q}_{\mu\alpha} = Q_{\mu\alpha} + \ldots$

2) $\hat{T}_{\mu\nu} = T_{\mu\nu} + \ldots$

3) $\hat{D}_{\mu} = D_{\mu} + \ldots$

with $D_{\mu} = x^{\nu} T_{\mu\nu}$

4) $\hat{M}_{\mu\nu\rho} = M_{\mu\nu\rho} + \ldots$

with $M_{\mu\nu\rho} = x_{\nu} T_{\mu\rho} - x_{\rho} T_{\mu\nu}$

5) $\hat{K}_{\mu\nu} = K_{\mu\nu} + \ldots$

with $K_{\mu\nu} = (2x_{\nu} x^{\lambda} - g_{\nu}{}^{\lambda} x^2) T_{\mu\lambda}$ (6.130)

---

*s.b. sects. 16.2.3, 16.3.4

6) $\hat{S}_{\mu\alpha} = S_{\mu\alpha} + \ldots$

$$\text{with } S_{\mu\alpha} = ix_\lambda \sigma^\lambda_{\alpha\dot{\alpha}} \bar{Q}^{\dot{\alpha}}_\mu \quad . \qquad (6.130 \text{ cont'd.})$$

6.6 Superfield form of internal symmetry currents

Suppose a supersymmetric theory is also invariant under a rigid internal symmetry group (cf. (5.19)) G

$$\begin{aligned} \delta_\omega A_j &= -i\, \omega^a T^a_{jk} A_k \\ \delta_\omega \bar{A}_j &= +i\, \bar{A}_k T^a_{kj} \omega^a \end{aligned} \qquad (6.131)$$

where the Hermitean matrices T^a generate a unitary representation of G on the chiral fields A. Then the associated current j^a_μ lies also in a supermultiplet and the question is simply: in which one? Let us follow the conventional Noether procedure for finding it. The global Ward-identity

$$W_\omega \Gamma \equiv -i \int dS(-i)\omega^a T^a_{jk} A_k \frac{\delta\Gamma}{\delta A_j} - i \int d\bar{S}(+i)\omega^a \bar{A}_k T^a_{kj} \frac{\delta\Gamma}{\delta\bar{A}_j} = 0 \qquad (6.132)$$

yields the current conservation equation if we let ω^a become space-time dependent and then differentiate with respect to it. For an invariant action

$$\Gamma = \frac{1}{16} \int dV\, \bar{A}_i A_i + [\int dS(m_{ij} A_i A_j + g_{ijk} A_i A_j A_k) + c.c.] \qquad (6.133)$$

we find

$$W_\omega \Gamma = -\frac{1}{16} \int dS\, \omega^a(x) \bar{D}\bar{D}(\bar{A} T^a A) + \frac{1}{16} \int d\bar{S}\, \omega^a(x) DD(\bar{A} T^a A) \qquad (6.134)$$

If $\omega^a(x)$ has no susy-covariance we shall obtain after differentiation with respect to it the non-supersymmetric version of the internal symmetry current. For the supersymmetric form we thus better make ω^a to a chiral field in the dS contribution (6.134), respectively to an anti-chiral field in the $d\bar{S}$ part of (6.134). Differentiation leads then to two conjugate equations

$$w^a_\omega\Gamma = -\frac{1}{16}\bar{D}\bar{D}(\bar{A}T^aA)$$

$$\bar{w}^{\bar{a}}_{\bar{\omega}}\Gamma = +\frac{1}{16}DD(\bar{A}T^aA)$$

where we have defined

$$w^a_\omega \equiv \frac{\delta}{\delta\omega^a(x)}W_\omega\Gamma = -T^a_{jk}A_k\frac{\delta}{\delta A_j}$$
$$\bar{w}^{\bar{a}}_{\bar{\omega}} \equiv \frac{\delta}{\delta\bar{\omega}^a(x)}W_\omega\Gamma = \bar{A}_kT^a_{kj}\frac{\delta}{\delta\bar{A}_j} \tag{6.136}$$

They correspond to the trace equations for the supercurrent (6.22), (6.58), (6.103) and coincide in form with (6.66) which is a special case of this general statement. The current conservation equation is deduced from them by differentiating and adding:

$$DDw^a_\omega\Gamma + \bar{D}\bar{D}w^{\bar{a}}_{\bar{\omega}}\Gamma = -\frac{1}{16}[DD,\bar{D}\bar{D}](\bar{A}T^aA)$$
$$= +\frac{i}{4}(\bar{D}\bar{\sigma}^\mu D - D\sigma^\mu\bar{D})\partial_\mu(\bar{A}T^aA) \tag{6.137}$$

(to be compared with (6.24), (6.59), (6.105) and, in particular, with (6.67)).

The zeroth Θ-component of this equation is the current conservation equation of the rigid symmetry which we are looking for. The other components do not correspond to symmetry <u>currents</u> of the original action (6.133) because this action is simply not invariant under the other transformations generated by ω^a and $\bar{\omega}^a$, they constitute "accidental" identities following from the equations of motion.

Of course, these "accidental" identities are readily recognized: we enlarge the kinetic term in the action

$$\frac{1}{16}\int dV\,\bar{A}A \rightarrow \frac{1}{16}\int dV\,\bar{A}e^{\tilde{\Phi}}A \quad , \quad \tilde{\Phi} \equiv \Phi^aT^a \quad , \tag{6.138}$$

and look for transformations of $\tilde{\Phi}$ rendering the enlarged term invariant: (recall that ω^a is chiral)

$$W_\omega\Gamma = -\frac{1}{16}\int dV\,\bar{A}e^{\tilde{\Phi}}T^a\omega^aA \quad = -\frac{i}{16}\int dV\,\bar{A}\,(-\delta_\omega e^{\tilde{\Phi}})A,$$
$$\bar{W}_{\bar{\omega}}\Gamma = \frac{1}{16}\int dV\,\bar{A}\bar{\omega}^aT^ae^{\tilde{\Phi}}A \quad = -\frac{i}{16}\int dV\,\bar{A}\,(-\delta_{\bar{\omega}}e^{\tilde{\Phi}})A \quad . \tag{6.139}$$

Hence the transformations on $\tilde{\Phi}$ are those of (6.81):

$$\delta_\omega e^{\tilde{\Phi}} = e^{\tilde{\Phi}}(i\tilde{\omega}) \ , \ \delta_\omega \tilde{\Phi} = R_{\tilde{\Phi}}^{-1}\{i\tilde{\omega}\} \ , \quad \tilde{\omega} \equiv T^a\omega^a \ ,$$
$$\delta_{\bar{\omega}} e^{\tilde{\Phi}} = -i\tilde{\bar{\omega}}\, e^{\tilde{\Phi}} \ , \ \delta_{\bar{\omega}} \tilde{\Phi} = \bar{R}_{\tilde{\Phi}}^{-1}\{i\tilde{\bar{\omega}}\} \ , \qquad (6.140)$$

and (6.139) can be recast into the form of local Ward-identities

$$-\bar{D}\bar{D}\, R_{\tilde{\Phi}}\{T^a\}\, \frac{\delta\Gamma}{\delta\Phi} + w^a_\omega\Gamma = 0 \ ,$$
$$DD\, \bar{R}_{\tilde{\Phi}}\{T^a\}\, \frac{\delta\Gamma}{\delta\Phi} + \bar{w}^a_{\bar{\omega}}\Gamma = 0 \ . \qquad (6.141)$$

They express the invariance of Γ under the local, non-abelian gauge transformations (6.131), (6.140). We had introduced these additional transformations at the point where we enlarged $\omega^a(x)$ to the chiral superfields

$$\omega^a(x) = \omega^a_A(x) + \Theta\omega^a_\psi(x) + \Theta^2\omega^a_F(x) \qquad (6.142)$$

and its conjugate. The non-current terms in (6.137) correspond to the local Ward-identities (6.141) associated with ω and $\bar{\omega}$.

CHAPTER III

PERTURBATION THEORY IN SUPERSPACE

In the axiomatic approach to perturbative quantum field theory (cf. for instance [R.12]) which we follow here one defines in one way or the other perturbative Green's functions and shows then that they satisfy the required axioms - in particular those of relativistic covariance, causality and unitarity - in the sense of perturbation theory. Perturbation is usually performed in powers of a coupling constant or in the powers of $\hbar$ which counts the number of loops of associated Feyman diagrams. When, as is often the case in supersymmetry, fields of canonical dimension zero are present, one expands also in the number of fields. If mass generation occurs, the expansion will be in $\sqrt{\hbar}$ ln $\hbar$-arising in a well-defined way from an $\hbar$-expansion. All of these expansions are considered as formal ones, i.e. questions of convergence of the series considered are not answered, in fact, not even posed.

The problem which remains even in this rather restrictive framework is the mathematically consistent definition of Green's functions. A complete treatment for the general case can be found in [III.11]. The present chapter is devoted to its solution in the context of supersymmetric theories by a generalization of the ordinary momentum subtraction scheme to superspace such that covariance with respect to supersymmetry is manifestly preserved. In the cases to be treated later where supersymmetry is broken we shall work in component fields in ordinary space-time. As subtraction scheme one may choose conventional BPHZ [III.1,4,13] whose rules can be abstracted from the ones given here by just leaving off all Θ-variables. The broken supersymmetry Ward-identities have to be checked then with the help of a theorem given in chapter IV (section 13.3).

Section 7. A simple example

Let $\{\Phi_i\}$ be a set of free fields where the index i collects the dependence on a point in space-time, internal and Lorentz indices. Then the Green's functions $G(1,\ldots,n)$ are defined by the (modified) Gell-Mann-Low formula

$$G(1,\ldots,n) \equiv \langle T(\Phi_1\ldots\Phi_n) \rangle$$

$$= R \frac{\langle T(\Phi_1\ldots\Phi_n\, e^{i\int\mathcal{L}_{int}}) \rangle_{(0)}}{\langle T\, e^{i\int\mathcal{L}_{int}} \rangle_{(0)}} \tag{7.1}$$

(no attempt is made to "prove" this formula, in our approach it is a basic definition of the theory). Here T denotes time ordering, the subscript (o) means "free fields", $\mathcal{L}_{int}$ is a function of the free fields Φ_i and the symbol R ("renormalization") stands for an operation to render meaningful in a way explained below the formal expressions one obtains by expanding the exponential in (7.1). This expansion is governed by Wick's theorem and yields the Feynman rules of the theory in question: all terms of the type $\langle T\, \Phi_1\ldots\Phi_n i\int dx_1\, \mathcal{L}_{int}\ldots\, i\int dx_m \mathcal{L}_{int} \rangle$ can be reduced to sums of products of 2-point functions $\langle T\, \Phi_a\Phi_b\rangle_{(0)}$ which are the propagators derived from a bilinear Lagrangian for the fields Φ_a, Φ_b. The diagrammatic representation of this procedure consists in drawing lines for the propagators, vertices for the factors $\int\mathcal{L}_{int}$ and in associating with these lines and vertices calculational prescriptions such that the terms in the Wick expansion are precisely recovered.

For the supersymmetric theories presented in chapter II we have already derived the free propagators, hence we may interpret (7.1) immediately in superspace, with Φ_i now being a superfield and the integration measure in $\mathcal{L}_{int}$, according to the case, scalar or vectorial.

In order to motivate definitions and procedures to come in this chapter we shall work out a simple example in detail.

Choose in (7.1)

$$\Gamma_{int} \equiv \int dS\, \mathcal{L}_{int} + \int d\bar{S}\, \bar{\mathcal{L}}_{int} = \frac{g}{48} (\int dS\, A^3 + \int d\bar{S}\, \bar{A}^3 \tag{7.2}$$

(cf. (3.26) !) and write in the expansion of the exponential of (7.1) the g^2-term for a 2-point Green's function.*

It reads:

$$G_{\tilde{1}\tilde{2}}(g^2) = \frac{1}{2}\left(\frac{ig}{48}\right)^2 \langle T(\tilde{A}(1)\tilde{A}(2)(\int d\bar{S}_3\bar{A}^3 + \int dS_3A^3)(\int dS_4A^3 + \int d\bar{S}_4\bar{A}^3))\rangle \quad (7.3)$$

Wick's theorem tells us to perform all contractions:

$$\begin{aligned} G_{\tilde{1}\tilde{2}}(g^2) = \frac{(2\cdot 3)^2}{2}\left(\frac{ig}{48}\right)^2 (&\int dS_3dS_4 \ \overline{\tilde{A}(1)A(3)}\ (\overline{A(3)A(4)})^2\ \overline{A(4)\tilde{A}(2)} \\ &+ \int d\bar{S}_3d\bar{S}_4 \ \overline{\tilde{A}(1)\bar{A}(3)}\ (\overline{\bar{A}(3)\bar{A}(4)})^2\ \overline{\bar{A}(4)\tilde{A}(2)} \\ &+ \int dS_3d\bar{S}_4 \ \overline{\tilde{A}(1)A(3)}\ (\overline{A(3)\bar{A}(4)})^2\ \overline{\bar{A}(4)\tilde{A}(2)} \\ &+ \int d\bar{S}_3dS_4 \ \overline{\tilde{A}(1)\bar{A}(3)}\ (\overline{\bar{A}(3)A(4)})^2\ \overline{A(4)\tilde{A}(2)}) \end{aligned} \quad (7.4)$$

(the contraction symbol ⌐¬ is only a shorthand:

$$\overline{\Phi(1)\Phi(2)} \equiv \langle T(\Phi(1)\Phi(2))\rangle_{(0)}$$

The result we may represent graphically by <u>Fig. 7.1</u>

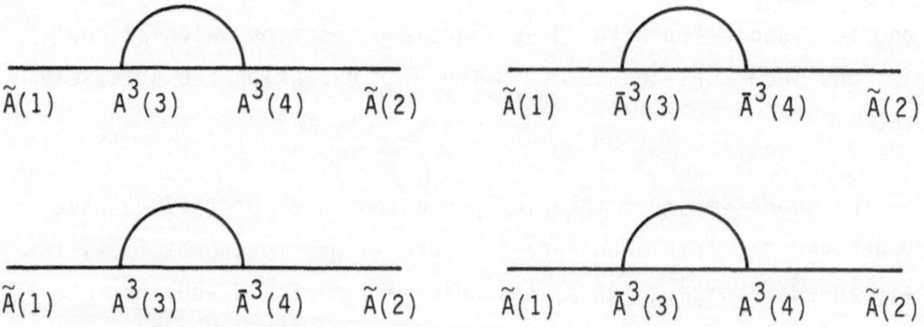

and analytically by

$$G_{\tilde{1}\tilde{2}}(g^2) = G^{(o)}_{\tilde{1}\tilde{a}}\ \Gamma^{(1)}_{\tilde{a}\tilde{b}}\ G^{(o)}_{\tilde{b}\tilde{2}} \quad (7.5)$$

with summation and integration over $\tilde{a}$, $\tilde{b}$ understood, the upper indices denoting: free theory with (o); order $\hbar$ ≡ one loop with (1), respec-

---

*$\tilde{A}$ stands for <u>either</u> A or $\bar{A}$.

tively, and

$$\Gamma^{(1)}_{ab} \equiv \begin{pmatrix} \Gamma^{(1)}_{A\bar{A}} & \Gamma^{(1)}_{AA} \\ \Gamma^{(1)}_{\bar{A}\bar{A}} & \Gamma^{(1)}_{\bar{A}A} \end{pmatrix} = \frac{(2.3)^2}{2}\left(\frac{ig}{48}\right)^2 \begin{pmatrix} (\overline{A(3)\bar{A}(4)})^2 & (\overline{A(3)A(4)})^2 \\ (\overline{\bar{A}(3)\bar{A}(4)})^2 & (\overline{\bar{A}(3)A(4)})^2 \end{pmatrix} \tag{7.6}$$

Using (3.18) (3.19) we calculate now the vertex functions:

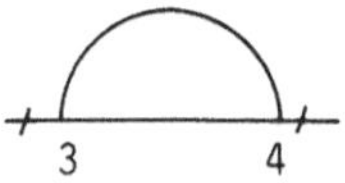

$$: \Gamma^{(1)}_{AA} = \chi \; (\overline{A(3)A(4)})^2$$

$$= \chi \, \frac{4im\,\delta_s(3,4)}{\Box + m^2} \; \frac{4im\,\delta_s(3,4)}{\Box + m^2} = 0 \tag{7.7}$$

since

$$\delta_s(3,4) = -\frac{1}{4}\,\Theta^2_{34} \tag{7.8}$$

in the chiral-chiral basis, since $\delta_s(3,4)\;\delta_s(3,4) \equiv 0$.

$$\left(\chi \equiv \frac{(2.3)^2}{2}\left(\frac{ig}{48}\right)^2\right).$$

Analogously: $\Gamma^{(1)}_{\bar{A}\bar{A}} = 0$

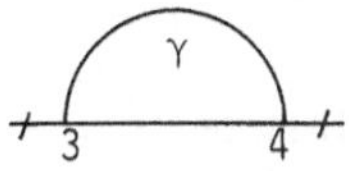

$$: \Gamma^{(1)}_{A\bar{A}} = \chi \; (\overline{A(3)\bar{A}(4)})^2 \tag{7.9}$$

$$= \chi \frac{1}{(2\pi)^4}\int dp\; e^{i(x_3-x_4)p} \frac{1}{(2\pi)^4}\int dk \, \frac{i\,DD_4\;\delta_s(3,4;k)}{k^2 - m^2} \; \frac{i\,DD_4\;\delta_s(3,4;p-k}{(p-k)^2 - m^2}$$

where the product of the translation invariant propagators in x-space yielded the convolution in momentum space.

With the help of suitable shifts (cf. section 1) one can show that from (7.8) follows

$$DD_4\;\delta_s\;(3,4;k) = \overline{DD}_3\;\delta_{\bar{s}}\;(3,4;k) = e^{E_{34}k} \tag{7.10}$$

with

$$E_{34} = \Theta_3\sigma\bar{\Theta}_3 + \Theta_4\sigma\bar{\Theta}_4 - 2\,\Theta_3\sigma\bar{\Theta}_4$$

Therefore:

$$I_\gamma(p,k,\Theta) \equiv \frac{i\,DD_4\,\delta_s(3,4;k)}{k^2 - m^2} \cdot \frac{i\,DD_4\,\delta_s(3,4;p-k)}{(p-k)^2 - m^2} = \frac{-e^{E_{34}p}}{(k^2-m^2)((p-k)^2-m^2)}$$

$$= -e^{E_{34}p}\, I_\gamma(p,k) \tag{7.11}$$

(γ refers to the diagram representing $\Gamma^{(1)}_{A\bar{A}}$).

But we note that $\int d^4k\; I_\gamma\;(p,k) = \infty$!

I.e. formula (7.1) in its naive form (we have plainly expanded the exponential) does not make sense. We therefore define:

$$R_\gamma(p,K) := (1-t^o_p)\; I_\gamma(p,k) = I_\gamma(p,k) - I_\gamma(o,k) \tag{7.12}$$

(t^o_p: Taylor operator with respect to p of order o) and find that $\int dk\; R_\gamma(p,k)$ exists, provided we take into account a suitable ε-prescription in the denominators for circumventing the poles (e.g. denominator $= k^2 - m^2 + i\,\varepsilon\,(\underline{k}^2 + m^2)$). *

This is the operation R of (7.1) in our simple example.

Let us collect the result following from this definition:

$$\Gamma^{(1)}_{A\bar{A}} = -\frac{\chi}{(2\pi)^4}\int dp\; e^{i(x_3-x_4)p}\; e^{(\Theta_3\sigma\bar{\Theta}_3+\Theta_4\sigma\bar{\Theta}_4-2\Theta_3\sigma\bar{\Theta}_4)p}\; \frac{1}{(2\pi)^4}\int dk\; R_\gamma(p,k) \tag{7.13}$$

i.e. $\Gamma^{(1)}_{A\bar{A}}\;(x_3,x_4)$ is a function of the variable

---

* This result is in fact known as power counting theorem [111.1]: the subtracted integrand has negative dimension on all hyperplanes in the 4-dimensional integration space, hence the integral exists and is absolutely convergent.

$$x \equiv x_3 - x_4 - i\,\Theta_3\sigma\bar{\Theta}_3 - i\,\Theta_4\sigma\bar{\Theta}_4 + 2\,i\,\Theta_3\sigma\bar{\Theta}_4 \tag{7.14}$$

This peculiar combination is annihilated by a similarly peculiar differential operator:

$$\delta f \equiv \left(\frac{\partial}{\partial\Theta_3} + i\,\sigma\bar{\Theta}_3 \frac{\partial}{\partial x_3} + \frac{\partial}{\partial\Theta_4} + i\,\sigma\bar{\Theta}_4 \frac{\partial}{\partial x_4} \right) f(x) = 0 \tag{7.15}$$

This is nothing but twice the differential operator occurring in the supersymmetry variation. Indeed: consider the functional

$$\Gamma^{(1)} := \int dS_3 A(x_3,\Theta,\bar{\Theta}) \int d\bar{S}_4 \bar{A}(x_4,\Theta,\bar{\Theta})\, \Gamma^{(1)}_{A\bar{A}}(3,4) \tag{7.16}$$

(A and $\bar{A}$ in the real basis)

and act on it with the functional differential operator (3.24)

$$W_\alpha \equiv -i \left(\int dS\ \delta^Q_\alpha A \frac{\delta}{\delta A} + \int d\bar{S}\ \delta^Q_\alpha \bar{A} \frac{\delta}{\delta \bar{A}} \right). \tag{7.17}$$

One obtains

$$W_\alpha \Gamma^{(1)} = -\, i \int dS_3 d\bar{S}_4 [\delta^Q A_3 \bar{A}_4 + A_3 \delta^Q_\alpha \bar{A}_4] \Gamma^{(1)}_{A\bar{A}}(3,4)$$

i.e. by partial integration

$$= \quad i \int dS_3 \int d\bar{S}_4\ A_3 \bar{A}_4\ \delta\Gamma^{(1)}_{A\bar{A}}(3,4) = 0 \tag{7.18}$$

And what we have checked with this calculation is nothing but the compatibility of the operation R with supersymmetry: the subtraction (7.12) left intact the Θ-exponential structure of $I_\gamma(p,k,\Theta)$ (7.11) which according to the above calculation suffices for supersymmetry. It modified only a Lorentz-scalar coefficient function. In the tree approximation this function of the external momentum p was just a constant.

Still another observation can be made in connection with the subtraction (7.12). The term we subtract is independent of p, hence contributes in (7.13) a $\delta(x_3-x_4)$ - term, non-vanishing at coincident arguments only: it is a local term in (7.16), hence our operation R has not violated the locality of the theory. That we do indeed have the freedom, to modify the theory by such a local term is seen from the propagators (at order $\hbar$): they contribute to and change the T-product of the field operators A and $\bar{A}$ precisely where it has from the very beginning a lack of definiteness: at coinciding time arguments. (This argument is somewhat hand-waving in that it uses an infinite quantity: the subtraction term; but it is basically correct and can be made rigorous [III.2] .)

The last requirement we had, was relativistic covariance. It is not easily demonstrated with the above ε-prescription [III.1,3], so we only quote the result: the limit $\varepsilon \to o$ exists and defines a covariant distribution in the case of Green's functions, a covariant complex function in the case of vertex functions. - In any case it is clear that this proof does not interfere with supersymmetry: the ε-limit properties of the fully supersymmetric quantity G_{ab} are those of the scalar "kernel" in Γ_{ab}.

Let us now summarize what we learned from our simple example:

- G_{ab} was divergent due to divergence of Γ_{ab}, (7.5), generalized: once the one-particle-irreducible Green's function (cf. App. B) are made finite all other Green's functions are finite.

- The subtraction prescription R (7.12) rendered Γ_{ab} finite; look for a generalization to all orders.

- R maintained supersymmetry because it respected the exponential Θ-structure coming from the propagators; (R maintained also locality and relativistic covariance in the limit $\varepsilon \to o$).

- $\Gamma^{(1)}_{ab} = \Gamma^{(1)}_{\bar{a}\bar{b}} = 0$ (7.7) was a specific consequence of the form of the chiral-chiral propagators ("non-renormalization theorem").

- $\Gamma^{(1)}_{a\bar{b}}$ needed only one subtraction: since in (7.11) the k dropped out in the numerator the resulting integral was only logarithmically divergent and not quadratically (as it could have been) ("cancellation of divergences").

In the following section we set up a subtraction scheme (give an operation R) which generalizes the above results - to the extent to which it is possible - to all orders of perturbation theory.

Section 8. Feynman rules and power counting

In order to compute Green's or vertex functions according to (7.1) we perform the expansion in Feynman diagrams with the following rules (in momentum space). We associate

with	the line	and the analytic expression
$\langle T\, A(1)A(2)\rangle_{(0)}$	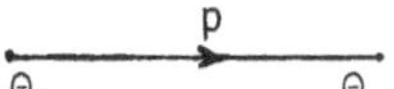 Θ_1 ... Θ_2	$\dfrac{4i\, m\, \delta_s(1,2)}{p^2 - m^2 + i\varepsilon}$
$\langle T\, \bar{A}(1)\bar{A}(2)\rangle_{(0)}$	 $\bar{\Theta}_1$ ... $\bar{\Theta}_2$	$\dfrac{4i\, m\, \delta_{\bar{s}}(1,2)}{p^2 - m^2 + i\varepsilon}$
$\langle T\, A(1)\bar{A}(2)\rangle_{(0)}$	 Θ_1 ... $\bar{\Theta}_2$	$\dfrac{iDD_2\delta_2(1,2)}{p^2-m^2 + i\varepsilon} = \dfrac{i\overline{DD}_1\delta_{\bar{s}}(1,2)}{p^2-m^2 + i\varepsilon}$
$\langle T\, \bar{A}(1)A(2)\rangle_{(0)}$	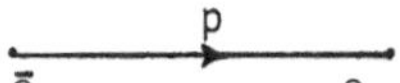 $\bar{\Theta}_1$ ... Θ_2	$\dfrac{i\overline{DD}_2\delta_{\bar{s}}(1,2)}{p^2-m^2 + i\varepsilon} = \dfrac{iDD_1\delta_s(1,2)}{p^2-m^2 + i\varepsilon}$
$\langle T\, \Phi(1)\Phi(2)\rangle_{(0)}$	 1 ... 2	$\dfrac{-8\, i\, \delta_V(1,2)}{p^2 - M^2(s-1)^2 + i\varepsilon}$

(SQED; α-gauge with $\alpha = 1$; cf.(4.32))

$\langle T\Phi_{i_1}(1)\Phi_{i_2}(2)\rangle_{(0)}$	 1 ... 2	$\dfrac{-8\, ig^2\, \delta_{i_1 i_2}\delta_V(1,2)}{p^2 - M^2(s-1)^2 + i\varepsilon}$

(SYM; α-gauge with $\alpha = g^2$; cf. (5.20)) (8.1)

(We write explicitly only Feynman gauge for gauge fields since all other gauges require a precise definition of terms like $1/(k^2)^2$ which we postpone until chapter V. The ε-terms should be read as $\varepsilon \equiv \varepsilon(\underline{P}^2 + m^2)$, it is the choice [III.1,4] guaranteeing majorization of Minkowski integrals by Euclidean ones, yielding absolute convergence.)

The δ-functions occurring in the propagators are understood to be in momentum space. Their explicit form depends on the basis chosen for the fields.*

If all fields are taken in the real basis, 1 and 2 stand for a pair of superspace points and if p denotes the momentum flowing from 1 to 2 we have

$$\delta_v(1,2) = \frac{1}{16}\,\Theta^2_{12}\,\bar{\Theta}^2_{12}$$

$$\delta_s(1,2) = \overline{DD}_1\delta_v(1,2) = -\frac{1}{4}\,\Theta^2_{12}\;e^{-\bar{\Theta}_1\gamma\Theta_2 p}$$

$$\delta_{\bar{s}}(1,2) = DD_1\delta_v(1,2) = -\frac{1}{4}\,\bar{\Theta}^2_{12}\;e^{-\bar{\Theta}_1\gamma\Theta_2 p}$$

$$DD_1\delta_s(1,2) = e^{-E_{21}p} \qquad \overline{DD}_1\delta_{\bar{s}}(1,2) = e^{E_{12}p}$$

$$\bar{\Theta}_a\gamma\Theta_b \equiv \Theta_a\sigma\bar{\Theta}_b - \Theta_b\sigma\bar{\Theta}_a = -\bar{\Theta}_b\gamma\Theta_a$$

$$E_{ab} \equiv \Theta_a\sigma\bar{\Theta}_a + \Theta_b\sigma\bar{\Theta}_b - 2\Theta_a\sigma\bar{\Theta}_b = -\bar{\Theta}_a\gamma\Theta_b + \Theta_{12}\sigma\bar{\Theta}_{12}$$

$$\Theta_{ab} \equiv \Theta_a - \Theta_b \tag{8.2}$$

Before proceeding further we have to generalize the above [III.11]. We shall incorporate massless particles into the subtraction scheme in general by using an auxiliary mass [III.12,13]: it is given by an ordinary supersymmetric mass term the parameter m simply being replaced

---

* It is mainly here that authors differ in their Feynman rules; another source of difference is the treatment of D-factors in the vertices. [II.2, III.5-10]

by m(s-1). The parameter s varies between zero and one and will play the role of an additional subtraction variable (like an external momentum p). At the end of all calculations it is to be put equal to one for all massless fields. At $s \neq 1$ such a mass term will describe a massive field (with off-shell normalization conditions, cf. below).* The general form

$$\begin{aligned} &< T\, \Phi_a(1)\, \Phi_b(2)>_{(0)} \equiv \Delta_{ab}(1,2) = \Delta_{ab}(p,s,\tilde{\Theta}_1,\tilde{\Theta}_2) \\ &= Q_{ab}(p,s,\tilde{\Theta}_{12})\, e^{-\bar{\Theta}_1 \gamma \Theta_2 p} \prod_i [p^2 - m^i_{ab}(s)^2 + i\varepsilon(\underset{\sim}{p}^2 + m^i_{ab}(s)^2]^{-1} \end{aligned} \tag{8.3}$$

emerges for the propagators (notation: $\tilde{\Theta} = \Theta$ or $\bar{\Theta}$). The Θ-structure can in fact be seen to follow from supersymmetry (c.f. (7.15 - 7.18)). $m^i_{ab}(s)^2$ is a quadratical polynomial in the mass parameters and in the parameter s. The factor Q_{ab} depends on the Θ's only via their differences Θ_{12}, $\bar{\Theta}_{12}$. This function is further restricted if Φ_a and/or Φ_b are chiral fields:

$$q_{ab}(p,s,\tilde{\Theta}_{12}) = \begin{cases} e^{\Theta_{12}\sigma\bar{\Theta}_{12}p}\, \bar{Q}_{ab}(p,s,\Theta_{12}) & (\Phi_a \text{ chiral}) \\ e^{\Theta_{12}\sigma\bar{\Theta}_{12}p}\, \bar{Q}_{ab}(p,s) & (\Phi_a \text{ chiral, } \Phi_b \text{ antichiral}) \\ \Theta^2_{12}\, \bar{Q}_{ab}(p,s) & (\Phi_a \text{ and } \Phi_b \text{ chiral}) \end{cases} \tag{8.4}$$

The vertices of the Feynman graphs correspond to the terms of Γ_{int} and, more generally, if one computes Green's functions of composite operators, to local superfield monomials - <u>local insertions</u> -

---

* In the context of the example of section 7 we prescribe:

$$I_\gamma(p,k,s) = \frac{1}{(k^2 - m^2(s-1)^2)\, ((k-p)^2 - m^2(s-1)^2)} \tag{7.11'}$$

be replaced by

$$R_\gamma(p,k,s) = I_\gamma(p,k,s) - I(o,k,o) \tag{7.12'}$$

in order to avoid a spurious infrared divergence caused by subtracting a massless propagator at zero external momentum. Masslessness of the field A will be guaranteed by further subtracting Γ_{AA} at $p = o$, $s = 1$.

$$Q_i(x,\tilde{\Theta}) = (s-1)^{a_i} (\frac{\partial}{\partial x})^{(\mu_i)} (\tilde{D})^{(\nu_i)} [\prod_a \Phi_a^{c_{ai}} (x,\tilde{\Theta})] \tag{8.5}$$

or <u>integrated insertions</u> (e.g. terms of Γ_{int})

$$\Delta_i = \int dx \;\; Q_i(x,\tilde{\Theta}) \tag{8.6}$$

where Q_i in (8.6) includes the covariant derivatives defining Θ-space integration. μ_i, ν_i are multi-indices, the differential operators $\partial/\partial x$ and $\tilde{D} = D, \bar{D}$ may be spread within the various factors of the bracket; $\Phi_a^{c_{ai}}$ denotes the c_{ai}-th power of the superfield Φ_a. Vertices associated with local insertions (8.5) will be called <u>external</u>: like for external lines, external momentum enters the diagram at these vertices. They will be called <u>internal</u> in the case of integrated insertions (8.6). The integration over x-space in the internal vertices leads in momentum space to integration over closed loops and an overall momentum conservation δ-function for each connected diagram. If derivatives are present at the vertices they have to be taken into account in accordance with Wick's theorem (namely by considering all possible contractions of the fields carrying derivatives). We shall not try to write down explicitly all numerical factors arising from Wick's theorem, but refer to the theorem.

Wick's theorem will yield a formal expression $\int dk \; I_\gamma(p^\gamma,k^\gamma,s^\gamma,\tilde{\Theta}^\gamma)$ associated with a diagram or subdiagram γ. Integration is extended over the closed loops of γ; p^γ, k^γ denote a basis of the external, resp. loop momenta; s^γ the s-variables and $\tilde{\Theta}^\gamma$ the Θ-variables of γ. The integrand I_γ is a product of propagators or covariant derivatives of propagators which all have the general form (8.2, 8.3). It can then be shown:

$$I_\gamma(p,k,s,\tilde{\Theta}) = e^{E(p,\tilde{\Theta})} \; \bar{I}_\gamma(p,k,s,\tilde{\Theta}..) \tag{8.7}$$

$\bar{I}_\gamma$ depends only on the differences $\tilde{\Theta}_{ij} = \tilde{\Theta}_i - \tilde{\Theta}_j$; moreover it does not depend on $\bar{\Theta}_{ij}$ if either i or j or both represent chiral external vertices or legs, and not on $\bar{\Theta}_{ij}$ in the antichiral case. $E(p,\tilde{\Theta})$ is a homogeneous function, linear in the external momenta, bilinear in the variables Θ_i and $\bar{\Theta}_i$:

$$E(p,\tilde{\Theta}) = \sum_{i=1}^{I} E_{iL}\, p_i - \sum_{k=1}^{K} E_{Lk}\, p_k - \sum_{l=1}^{L-1} \bar{\Theta}_l\, \gamma\Theta_L\, p_l \tag{8.8}$$

$$E^{\mu}_{ij} \equiv -\bar{\Theta}_i\gamma^{\mu}\Theta_j + \Theta_{ij}\sigma^{\mu}\bar{\Theta}_{ij}$$

the indices i = 1 ... I, k = 1 ... K and l = 1 ... L denote the external vertices and legs of chiral, antichiral and general type respectively. If they are only chiral and antichiral objects (L = 0) then

$$E(p,\tilde{\Theta}) = \sum_{i=1}^{I} E_{ik}\, p_i - \sum_{k=1}^{K-1} [\bar{\Theta}_k\gamma\Theta_K + (\Theta_{k1} + \Theta_{K1})\sigma\bar{\Theta}_{kK}]p_k \tag{8.9}$$

Finally in the purely chiral case (K = L = 0)

$$E(p,\tilde{\Theta}) = \sum_{i=1}^{I-1} E_{iI}\, p_i\,. \tag{8.10}$$

The expansion of (8.7) in powers of $\tilde{\Theta}_{..}$

$$I_{\gamma}(p.k,s,\tilde{\Theta}) = e^{E(p,\tilde{\Theta})} \sum_{|\omega|=0}^{\Omega} (\tilde{\Theta}..)^{(\omega)} I_{\gamma}^{(\omega)}(p,k,s) \tag{8.11}$$

stops at a finite degree Ω due to the anticommutativity of the Θ's. Ω can be computed by counting the number of independent $\tilde{\Theta}..$ components:

Ω = 2 (number of (anti-) chiral vertices and legs)
+4 (" " vector " " ")
-4 (8.12)

If, however, only chiral or only antichiral objects are present,

$$\Omega_{(\text{chiral})} = 2 \text{ (number of vertices and legs) } -2 \tag{8.13}$$

In (8.11) (ω) is a multiindex $(\omega_1\bar{\omega}_1 \ldots \omega_n\, \bar{\omega}_n)$ with

$$|\omega| = \sum_{m=1}^{n} (\omega_m + \bar{\omega}_m)$$
$$(\tilde{\Theta}..)^{(\omega)} = \prod_{m=1}^{n} \Theta_m^{\ \omega}\bar{\Theta}_m^{\ \omega} \tag{8.14}$$

(the index m = 1 ... n represents pairs (ij) as well as spinor indices $\alpha,\dot{\alpha}$). The summation in (8.11) extends over all monomials (8.14) of degree $|\omega|$ in the $\tilde{\Theta}_{ij}$ components, for each $|\omega|$ comprised between 0 and Ω.

We shall now exploit the Θ-structure of the integrand (8.11) to estimate its ultraviolet (UV) and infrared (IR) dimensions. The superficial UV and IR degrees of divergence for the component integrand $I_\gamma^{(\omega)}$ associated with a graph or subgraph γ are defined as

$$d^{(\omega)}(\gamma) = \overline{\deg}_{p,k,s}\, I_\gamma^{(\omega)}(p,k,s) + 4m(\gamma)$$
$$r^{(\omega)}(\gamma) = \underline{\deg}_{p,k,(s-1)}\, I_\gamma^{(\omega)}(p,k,s) + 4m(\gamma) \tag{8.15}$$

where $m(\gamma)$ is the number of loops of graph γ, and $\overline{\deg}_x$ means the asymptotic degree for $x \to \infty$, whereas $\underline{\deg}_x$ means the asymptotic degree for $x \to o$. It is clear from the expression (8.11) that $d(\gamma) \equiv d^{(o)}(\gamma)$ and $r(\gamma) \equiv r^{(o)}(\gamma)$ represent the UV and IR dimensions of the complete integrand I_γ and moreover that

$$d^{(\omega)}(\gamma) = d(\gamma) + \frac{1}{2}|\omega|$$
$$r^{(\omega)}(\gamma) = r(\gamma) + \frac{1}{2}|\omega| \tag{8.16}$$

This follows from homogeneity and the assignment -1/2 for the dimension of Θ. We now want to compute $d(\gamma)$ and $r(\gamma)$. We assign UV and IR dimensions d_a and r_a to the superfield Φ_a and define, analogously to (8.15), the UV and IR degrees of the free propagator Δ_{ab}, the dimension of Θ being taken into account:

$$d_{ab} = \overline{\deg}_{p,s(\Theta)}\, \Delta_{ab}(p,s,\Theta)$$
$$r_{ab} = \underline{\deg}_{p,s-1(\Theta)}\, \Delta_{ab}(p,s,\Theta) \tag{8.17}$$

The starting assumption will be that the superfield's dimensions d_a and r_a are chosen to fulfill the inequalities

$$4 + d_{ab} \leq d_a + d_b$$
$$4 + r_{ab} \geq r_a + r_b \tag{8.18}$$

The reason of this requirement is that it will lead to estimates independent of the detailed structure of the graph, in particular of its order.

We suppose that the graph γ consists of m loops, I internal lines, of which I_{ab} are of the Φ_a-Φ_b type, V vertices with V_i corresponding to the monomials Q_i, i = 1, ..., V, (8.4); N amputated external lines, of which N_S are chiral, N_V are vectorial, or, counting in another way, N_a are of type Φ_a, n_{ai} are of type Φ_a and are connected to the ith vertex. Clearly

$$I = (1/2) \sum_{a,b}^{a \neq b} I_{ab} + \sum_a I_{aa} \equiv \sum_{a,b}^{a \leq b} I_{ab} \tag{8.19}$$

$$m = I - V + 1 \tag{8.20}$$

$$N_s + N_v = \sum_a N_a \quad , \quad N_a = \sum_i n_{ai} \tag{8.21}$$

$$\sum_b^{b \neq a} I_{ab} + 2\, I_{aa} = \sum_i (c_{ai} - n_{ai}) = \sum_i c_{ai} - N_a \tag{8.22}$$

where the c_{ai} is the power of Φ_a in the ith vertex (see (8.4)).

A direct application of the Feynman rules yields the degrees

$$\left.\begin{matrix} d(\gamma) \\ r(\gamma) \end{matrix}\right\} = 4m + \sum_{a,b}^{a \leq b} I_{ab} \left\{\begin{matrix} d_{ab} \\ r_{ab} \end{matrix}\right. + \sum_{i=1}^{V} [a_i + |\mu_i| + (1/2)|\nu_i|]$$

$$- N_s - 2N_V$$

Abbreviating with

$$D_i \equiv a_i + |\mu_i| + (1/2)|\nu_i| \tag{8.23}$$

and using (8.20), we find

$$\left.\begin{matrix} d(\gamma) \\ r(\gamma) \end{matrix}\right\} = 4 - N_S - 2N_V + \sum_{a,b}^{a \leq b} I_{ab} \left[\left\{\begin{matrix} d_{ab} \\ r_{ab} \end{matrix}\right\} + 4 \right] + \sum_{V_i \in \gamma} (D_i - 4) \tag{8.24}$$

The inequalities (8.18), together with (8.22), yield the bounds

$$
\begin{aligned}
d(\gamma) &\leq 4 - N_s - 2N_V + \sum_{a,b}^{a\leq b} I_{ab}(d_a + d_b) + \sum_{V_i\epsilon\gamma} (D_i - 4) = \\
&= 4 - N_s - 2N_V - \sum_a N_a d_a + \sum_{V_i} (d_i - 4) \\
r(\gamma) &\geq 4 - N_s - 2N_V + \sum_{a,b}^{a\leq b} I_{ab} (r_a + r_b) + \sum_{V_i} (D_i - 4) = \\
&= 4 - N_s - 2N_V - \sum_a N_a r_a + \sum_{V_i} (r_i - 4)
\end{aligned}
\tag{8.25}
$$

Here we have introduced the degrees associated with a vertex, V_i,

$$
\left.\begin{matrix} d_i \\ r_i \end{matrix}\right\} = D_i + \sum_a c_{ai} \left\{\begin{matrix} d_a \\ r_a \end{matrix}\right.
\tag{8.26}
$$

The bounds (8.25) combined with a power counting theorem [III.14] which covers also the IR-behaviour will tell which subtractions have to be performed in order to obtain a finite integral. In fact, this task is simplified by [III.13] where for ordinary field theories a subtraction scheme has been constructed so that only the applicability to our more general case has to be checked. We note that formal integration of I_γ (8.11) over the loop momenta k would yield just the Θ-structure of supersymmetric Green's functions, i.e. the structure which can be shown to follow from supersymmetry covariance. Our aim in the next section will be to implement a subtraction scheme preserving it.

We conclude this section with an interesting application of the power counting estimates (8.25), namely the non-renormalization theorem for chiral vertices [II.4, III.23,8,10]. We consider Green's functions of elementary fields, in which case all vertices correspond to the interaction Lagrangian, its UV dimension being bounded by 4. For graphs or subgraphs γ with only chiral and antichiral legs the estimate of their UV degree of divergence takes the simpler form

$$d_\omega(\gamma) \leq 4 - N_s - \sum_a N_a d_a + \frac{1}{2}|\omega| \tag{8.27}$$

and is bounded by (see (8.12))

$$d_\Omega(\gamma) \leq 3 - \sum_a N_a d_a \tag{8.28}$$

if only chiral or only antichiral legs are present, and by

$$d_\Omega(\gamma) \leq 2 - \sum_a N_a d_a \tag{8.29}$$

in the mixed case. In the theories expounded in chapter II the UV dimension of the chiral fields A is 1 - for this choice is the first of inequalities (8.18) fulfilled and is the dimension of the various Lagrangians bounded by 4. Thus the only potentially divergent graphs are those contributing to the vertex functions $\Gamma_{A\bar{A}}$ $(d_\Omega(\gamma) \leq 0)$, Γ_{AA} $(d_\Omega(\gamma) \leq 1)$, Γ_{AAA} $(d_\Omega(\gamma) \leq 0)$ together with their complex conjugates. We already see that only logarithmic divergences are present - also for Γ_{AA} due to Lorentz invariance. But it is possible to show that in fact only graphs of the type $\Gamma_{A\bar{A}}$ are superficially divergent. The argument is the following. Let γ be a 1PI graph with only chiral external legs, e.g. a graph contributing to Γ_{AA} or Γ_{AAA}. γ has n_s, $\bar{n}_s$ and n_v vertices of the chiral, antichiral and vector type, respectively, and let us set to zero all external momenta p. Before the $\tilde{\Theta}$-integrations over the vertices are performed, the integrand I_γ (at p = o) is a function of the differences $\tilde{\Theta}_{ij}$, i,j labelling all vertices and legs of γ. Next, all $\bar{\Theta}$'s have to be integrated, only Θ's being involved in external legs. There are $\bar{n}_s + n_v$ Θ-integrations, but only $\bar{n}_s + n_v - 1$ variables $\bar{\Theta}_{ij}$ are at our disposal, so that the result is zero. Having shown in this way that the integrand vanishes at zero external momenta we conclude that its effective degree of divergence is lowered by one, in fact by two due to Lorentz invariance. This demonstrates that the graphs of the type Γ_{AA}, Γ_{AAA} (and conjugates) are effectively convergent. Since these graphs contribute to the renormalization of the masses and self-coupling constants of the chiral fields, we conclude that these parameters are not subjected to infinite renormalizations. This is the content of the non-renormalization theorem. (Finite renormalizations may very well occur for specific normalization conditions.)

Section 9. The subtraction scheme *

In order to extend the BPHZ (Bogoliubov, Parasiuk, Hepp, Zimmermann) renormalization scheme [III.4,13] to the class of theories described above [III.11] we have to define the forest formula of Zimmermann [III.4] in superspace, with the help of a suitable subtraction operator. Let us first associate with each graph or subgraph γ the UV and IR subtraction degrees

$$\delta(\gamma) = 4 - N_S(\gamma) - 2N_V(\gamma) + \sum_{a,b}^{a\leq b} I_{ab}(\gamma)(d_a + d_b) + \sum_{V_i\epsilon\gamma} (D_i + \delta_i - d_i - 4)$$

$$= 4 - N_S(\gamma) - 2N_V(\gamma) - \sum_a N_a(\gamma)d_a + \sum_{V_i\epsilon\gamma} (\delta_i - 4)$$

$$\rho(\gamma) = 4 - N_S(\gamma) - 2N_V(\gamma) + \sum_{a,b}^{a\leq b} I_{ab}(\gamma)(r_a + r_b) + \sum_{V_i\epsilon\gamma} (D_i + \rho_i - r_i - 4)$$

$$= 4 - N_S(\gamma) - 2N_V(\gamma) - \sum_a N_a(\gamma)r_a + \sum_{V_i\epsilon\gamma} (\rho_i - 4) \qquad (9.1)$$

and

$$\delta^{(\omega)}(\delta) = \delta(\gamma) + \frac{|\omega|}{2} \quad , \quad \rho^{(\omega)}(\gamma) = \rho(\gamma) + \frac{|\omega|}{2} . \qquad (9.2)$$

One notes that these expressions are obtained from the upper and lower bounds (8.25) by substituting for the vertex degrees d_i and r_i the numbers δ_i and ρ_i. The latter are subjected to the restrictions

$$\delta_i \geq d_i \quad , \quad 0 \leq \rho_i \leq \min(r_i, \delta_i) , \qquad (9.3)$$

moreover $\rho_i \geq 4$ if the vertex V_i is integrated.

With the help of these subtraction degrees one could now readily proceed, define a subtraction operator and show that it renders finite either diagrams not containing divergent subdiagrams or "nested" ones: where any divergent subdiagram is properly contained in another one

* This section is rather technical. Its main results are summarized at the end.

or completely disjoint. But matters are complicated by the fact that divergent subdiagrams may "overlap": their intersection is non-empty, but they are also not properly contained in each other. Example: Consider in pure SYM

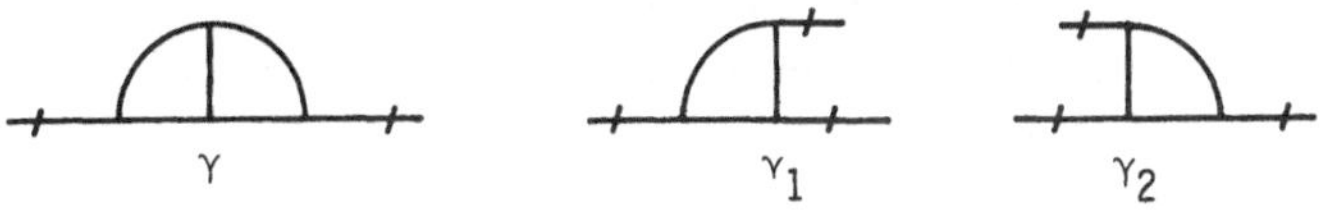

Fig. 9.1

$$\gamma_i \subset \gamma \quad i = 1,2 \qquad \gamma_1 \cap \gamma_2 \neq \emptyset \, .$$

All diagrams have

$$\delta_\Omega = \rho_\Omega = 4 - 2N_V + \frac{1}{2} 4\,(N_V - 1) = 2$$

i.e. require subtractions. These overlaps are disentangled by the forest formula and enforce as prerequisite a precise definition of momentum flow in a graph as well as the notion of reduced graph. Standard routing of momentum in a one-particle-irreducible diagram γ and its subdiagrams: let $p_1 \ldots p_n$ be a basis for the external momenta of γ, $k_1 \ldots k_m$ a basis for the integration momenta and k_i associated with the loop C_i in γ. Then the momentum $\ell_{ab\nu}$ flowing through $L_{ab\nu}$, the ν-th line connecting vertex V_b to V_a, is written

$$\ell_{ab\nu} = q_{ab\nu}(p) + k_{ab\nu}(k) \tag{9.4}$$

where $$k_{ab\nu} = \sum_i \epsilon_{ab\nu i} k_i$$

$$\varepsilon_{ab\nu i} = \begin{cases} 1 & \text{if } L_{ab\nu} \in C_i \quad \text{(including orientation)} \\ -1 & \text{if } L_{ba\nu} \in C_i \quad \text{(including orientation)} \\ 0 & \text{otherwise} \end{cases}$$

and $g_{ab}\,(p)$ is determined by "Kirchhoff's laws" of circuit theory:

$$\sum_{b,\nu} q_{ab\nu} + q_a = 0 \qquad \forall\, V_a \in \mathcal{V}(\gamma)$$
$$\sum_{L_{ab\nu}\in C} r_{ab\nu}\, q_{ab\nu} = 0 \qquad \forall \text{ loops } C \text{ of } \gamma \tag{9.5}$$

($\mathcal{V}(\gamma)$ denotes the set of vertices of γ). The "resistances" $r_{ab\nu} = r_{ba\nu}$ may be chosen in any convenient manner (including $r_{ab\nu} = 0$ or ∞), except that closed loops of zero-resistance or infinite resistance lines are not permitted.

In addition to (9.4), we shall also need an expression for the momentum flow through $L_{ab\nu}$ considered as a line of $\lambda \subset \gamma$. If $p^\lambda = p_1^\lambda \dots p_{n(\lambda)}^\lambda$ is a basis for the external momenta of λ we write

$$\ell_{ab}^\lambda = q_{ab\nu}^\lambda(p^\lambda) + k_{ab\nu}^\lambda(k) \tag{9.6}$$

where $g_{ab\nu}^\lambda$ is determined by "Kirchhoff's laws" restricted to λ

$$\sum_{b,\nu} q_{ab}^\lambda + q_a^\lambda = 0 \qquad \forall\, V_a \in \mathcal{V}(\lambda) \tag{9.7}$$

$$\sum_{L_{ab\nu}\in C} r_{ab\nu}\, q_{ab\nu}^\lambda = 0 \qquad \forall \text{ loops } C \text{ of } \lambda \tag{9.8}$$

and $k_{ab\nu}^\lambda$ is determined by the identity

$$q_{ab\nu}(p) + k_{ab\nu}(k) = q_{ab\nu}^\lambda(p^\lambda(p,k)) + k_{ab\nu}^\lambda \tag{9.9}$$

with $p^\lambda(p,k)$ determined by momentum conservation at the vertices of λ. The $k^\lambda_{ab\nu}$ defined in this way is in fact a function only of k. (All momentum assignments leading to this latter consequence are called "admissible" momentum flows.)

We now turn to the definition of a reduced graph. Let $\lambda_j \subset \lambda$ be a set of disjoint subgraphs of γ. We define the integrand

$$I_{\bar\gamma} \equiv I_{\gamma/\lambda_1\dots\lambda_c}\,(p^\gamma, k^{\bar\gamma}, s^{\bar\gamma}, \Theta^\gamma, \Theta^\lambda) \tag{9.10}$$

corresponding to the <u>reduced graph</u> $\bar\gamma = \gamma/\lambda_1\dots\lambda_c$ in the following way: the subgraphs λ_i are contracted to vertices $\bar V_i$; we have the usual Feynman rules, stated above, for all (internal and external) lines and vertices of $\bar{\gamma}$ which have nothing in common with any λ_i; the Θ's being internal to γ but external to some λ_j appear now with their new name $\Theta^{\lambda_j}_{ext}$ at their old place (namely in propagators of $\bar{\gamma}$ having $\bar{V}_j$ as an endpoint); at vertices $\bar{V}_i$ of $\bar{\gamma}$ there appears for each external line of γ carrying Θ^{γ}_{ext}, p^{γ} which is also an external line of some λ_j carrying $\Theta^{\lambda_j}_{ext}$, a delta function (8.2) $\delta(p^{\gamma},\Theta^{\gamma}_{ext},\Theta^{\lambda_j}_{ext})$.

We then have

$$I_{\gamma}(p^{\gamma},k^{\gamma},s^{\gamma},\Theta^{\gamma}) = \int d\Theta^{\lambda}_{ext}\, I_{\bar{\gamma}}(p^{\gamma},k^{\bar{\gamma}},s^{\bar{\gamma}},\Theta^{\gamma},\Theta^{\lambda}_{ext}) \times$$
$$\times\, S_{\gamma} \prod_{i=1}^{c} I_{\lambda_i}(p^{\lambda_i},k^{\lambda_i},s^{\lambda_i},\Theta^{\lambda_i}) \tag{9.11}$$

where the integration runs over all Θ^{λ}_{ext} of $\lambda_1\cup\ldots\cup\lambda_c$ and S_{γ} is the substitution operator [III.4] defined by

$$(S_{\gamma}f)(p^{\gamma},k^{\gamma},s^{\gamma}) = f(p^{\lambda}(p^{\gamma},k^{\gamma}),k^{\lambda}(k^{\gamma}),s^{\gamma}) \text{ and} \tag{9.12}$$

$S_{\gamma}\Theta^{\lambda_i} = \Theta^{\gamma}$ if Θ^{λ_i} is not one of the $\Theta^{\lambda_i}_{ext}$; then $S_{\gamma}\Theta^{\lambda_i}_{ext} = \Theta^{\lambda_i}_{ext}$

if $\lambda\subset\gamma$; an admissible flow of momentum is assumed.

Inserting into (9.11) the Θ-expansions (8.11) for I_{γ}, I_{λ_i} we obtain, with an automatic rearrangement of the exponential factors,

$$I^{(\omega)}_{\gamma}(p^{\gamma},k^{\gamma},s^{\gamma}) = \sum_{(\omega_1),\ldots,(\omega_c)} I^{(\bar{\omega})}_{\bar{\gamma}}(p^{\gamma},k^{\bar{\gamma}},s^{\bar{\gamma}})S_{\gamma} \prod_{i=1}^{c} I^{(\omega_i)}_{\lambda_i}(p^{\lambda_i},k^{\lambda_i},s^{\lambda_i}) \tag{9.13}$$

where

$$I^{(\omega)}_{\bar{\gamma}} \equiv I^{(\omega)(\omega_1)\ldots(\omega_c)}_{\gamma/\lambda_1\ldots\lambda_c}(p^{\gamma},k^{\bar{\gamma}},s^{\bar{\gamma}})$$

is the (ω) - component of the Θ^{γ}-expansion of

$$\int d\Theta^{\lambda}_{ext}\, I_{\bar{\gamma}}(p^{\gamma},k^{\bar{\gamma}},s^{\bar{\gamma}},\Theta^{\gamma},\Theta^{\lambda}_{ext})S_{\gamma} \prod_{i=1}^{c}\left[(\Theta^{\lambda_i}_{..})^{(\omega_i)} e^{E(p^{\lambda_i},\Theta^{\lambda_i})}\right] \tag{9.14}$$

that is the (ω) - component of the reduced graph $\bar{\gamma}$, where with each vertex $\bar{V}_i$ we have associated the factor

$$(\Theta^{\lambda_i})^{(\omega_i)}\, e^{E(p^{\lambda_i},\Theta^{\lambda_i})}$$

One has the following useful relations between the degrees of a graph γ, of disjoint subgraphs γ_i and of the reduced graph $\bar{\gamma} = \gamma/\gamma_1\ldots\gamma_c$ (see eq. (9.13)):

$$\chi^{(\omega)}(\gamma) = \chi^{(\bar{\omega})}(\bar{\gamma}) + \sum_{i=1}^{c} \chi^{(\omega_i)}(\gamma_i) \tag{9.15}$$

for any choice of the (ω_i)'s where χ = d or r and where

$$\chi^{(\bar{\omega})}(\bar{\gamma}) \equiv \chi^{(\omega)(\omega_1)\ldots(\omega_c)}(\gamma/\gamma_1\ldots\gamma_c);\ \chi = d \text{ or } r \tag{9.16}$$

is the UV or IR degree of divergence of the reduced graph, whose integrand is given by (9.14).

Let us consider now a connected graph Γ and its integrand $I_\Gamma(p,k,s,\Theta)$ defined in the preceding section. Its renormalized integrand is given by the forest formula

$$R_{\Gamma_\varepsilon}(p,k,s,\Theta) = \sum_{U\varepsilon F_\Gamma} \prod_{\gamma\varepsilon U} (-T_\gamma) I_\Gamma(p,k,s,\Theta) \tag{9.17}$$

where the sum runs over all forests U of Γ (i.e. all sets of nonoverlapping, one-particle irreducible (1PI) subgraphs $\gamma\subset\Gamma$). The ε subscript reminds one of the ε dependence of the propagators. The subtraction operators T_γ are defined as follows (notation $y^\gamma = (p^\gamma,k^\gamma,s^\gamma)$)

i) if γ_1 and $\gamma_2\varepsilon U$, $\gamma_1\supset\gamma_2$, then $(-T_{\gamma_1})$ is to be placed to the left of $(-T_{\gamma_2})$,

ii) if $\Gamma\varepsilon U$, then

$$T_\Gamma T_{\gamma_1} \ldots T_{\gamma_n} I_\Gamma(y,\Theta) = \tau_\Gamma T_{\gamma_1} \ldots T_{\gamma_n} I_\Gamma(y,\Theta) \tag{9.18}$$

iii) if $\gamma,\gamma_1,\ldots,\gamma_c\varepsilon U$ and $\{\gamma_1,\ldots,\gamma_c\}$ are maximal subgraphs of γ (i.e. if $\gamma_i\subset\lambda\subset\gamma$ then $\lambda = \gamma_i$ or $\lambda = \gamma$) then

$$T_{\gamma_1}\cdots T_{\gamma_c}\begin{pmatrix}\Pi & T\\ \lambda^1_i c\gamma_1 & \lambda^1_i\end{pmatrix}\cdots\begin{pmatrix}\Pi & T\\ \lambda^c_i c\gamma_c & \lambda^c_i\end{pmatrix} I_\gamma(y^\gamma,\Theta^\gamma)$$

$$= \int d\Theta^{\gamma_1}\ldots d\Theta^{\gamma_c}\, I_{\gamma/\gamma_1\ldots\gamma_c}(y^\gamma,\Theta^\gamma,\Theta^{\gamma_i})\; S_\gamma \quad \times \tag{9.19}$$

$$\times \prod_{j=1}^{c}\left[\tau_{\gamma_j}\begin{pmatrix}\Pi & T\\ \lambda^j_i \subsetneq \gamma & \lambda^j_i\end{pmatrix} I_{\gamma_j}(y^{\gamma_j},\Theta^{\gamma_j})\right]$$

where here and in what follows it is understood that we are integrating over the external Θ's: $\left\{\Theta^{\gamma_1}_{ext},\ldots,\Theta^{\gamma_c}_{ext}\right\}$ (see (9.10), (9.11) for the definitions of reduced graphs $\gamma/\gamma_1\ldots\gamma_c$ and associated integrands),

iv) the operator τ_γ was defined in [III.15]

$$\tau_\gamma = \tau^0_\gamma$$

$$\tau^{|\omega|}_\gamma f(p^\gamma,k^\gamma,s^\gamma) = \tag{9.20}$$

$$= \left[t^{\rho^{(\omega)}(\gamma)-1}_{p^\gamma,(s^\gamma-1)} + t^{\delta^{(\omega)}(\gamma)}_{p^\gamma,s^\gamma} - t^{\rho^{(\omega)}(\gamma)-1}_{p^\gamma,(s^\gamma-1)}\, t^{\delta^{(\omega)}(\gamma)}_{p^\gamma,s^\gamma}\right] f(p^\gamma,k^\gamma,s^\gamma)$$

where $t^n_x f(x)$ is the Taylor expansion of f around x = 0 up to and including degree n, and $\rho^{(\omega)}(\gamma)$ and $\delta^{(\omega)}(\gamma)$ are given by (9.1) and (9.2), further

$$\tau^{|\omega|}_\gamma \Theta^\gamma f(y^\gamma,\Theta^\gamma) = \Theta^\gamma \tau^{|\omega|+1/2}_\gamma f(y^\gamma,\Theta^\gamma) \tag{9.21}$$

$\tau^{|\omega|}_\gamma$ commutes with Θ^λ if λ and γ are disjoint;

v) if T_γ and T'_γ are different subtraction operators for the same subgraph γ, the operator $aT_\gamma+bT'_\gamma$ is defined as above, but with τ_γ replaced by $a\tau_\gamma+b\tau'_\gamma$.

These rules ensure that each component $I^{(\omega)}_\gamma$ of each integrand I_γ (see eqs. (8.11) and (9.2)) will be subtracted with the subtraction operator (9.20), which is the one given by Lowenstein [III.13] for theories containing massless propagators.

An essential feature of the forest formula (9.17) and of the subtraction operators T_γ is that supersymmetry is preserved. In other words, if the unsubtracted integrand I_Γ is supersymmetric, i.e. it possesses the structure (8.7), the renormalized integrand R_Γ is also supersymmetric, since the T_γ defined above commute with any homogeneous function $E(p^\gamma,\Theta^\gamma)$ of degree 0. Chirality properties, as expressed by the detailed structures (8.8 - 8.10) are preserved in the same way.

We turn now to the task of proving that, under suitable conditions to be stated later on,

$$\int dk R_{\Gamma_\varepsilon}(p,k,s,\Theta) \tag{9.22}$$

exists, in the limit $\varepsilon \to 0^+$, as a Lorentz covariant tempered distribution in p, or, in the case that Γ is one-particle-irreducible, as a complex-valued function in p for p (Euclidean) non-exceptional. We shall do this by reducing the problem to a form for which the proof of Lowenstein and Speer [III.13,3] directly applies.

Let us write the forest formula (9.17) for the (ω)-component of the graph Γ (see (8.11)), using the definitions above. We get first a recursive form for the integrand:

a) $$R^{(\omega)}_{\Gamma_\varepsilon}(y) = \sum_{U\varepsilon F_\Gamma} R^{(\omega)}_{\Gamma_\varepsilon}(U) \tag{9.23}$$

b) $$R^{(\omega)}_{\Gamma_\varepsilon}(U) = \begin{cases} (-\tau_\Gamma^{|\omega|})\, X_\Gamma^{(\omega)}(y) & \text{if } \Gamma\varepsilon U \\ X_\Gamma^{(\omega)}(y) & \text{if } \Gamma\not\varepsilon U \end{cases} \tag{9.24}$$

c) $$X_\gamma^{(\omega_\gamma)}(y^\gamma) = \sum_{(\omega_{\gamma_1})\ldots(\omega_{\gamma_c})} I_{\gamma/\gamma_1\ldots\gamma_c}^{(\omega_\gamma)(\omega_{\gamma_1})\ldots(\omega_{\gamma_c})}(y^\gamma)S_\gamma \quad \times$$

$$\times \quad \prod_{j=1}^{c}\left[-\tau_{\gamma_j}^{|\omega_j|} X_{\gamma_j}^{(\omega_{\gamma_j})}(y^{\gamma_j})\right] \tag{9.25}$$

if $\gamma_1,\ldots,\gamma_c$ are all the maximal subgraphs of γ which belong to U, and

d) $$X_\lambda^{(\omega_\lambda)}(y^\lambda) = I_\lambda^{(\omega_\lambda)}(y^\lambda) \tag{9.26}$$

if λ is a minimal element of U. Applying the recurrence formula (9.24) again and again from the minimal elements of U to Γ itself, one obtains the following closed form, an extension of that given by Zimmermann [III.4]. Let

$$U = \{\Gamma,\gamma_1,\ldots,\gamma_n\} \qquad \text{or} \qquad U = \{\gamma_1,\ldots,\gamma_n\}$$

$$(\bar\omega_U) = ((\omega_1),\ldots,(\omega_n))$$

then

$$X_\Gamma^{(\omega)}(y) = \sum_{(\bar\omega_U)} \prod_{k=1}^{n}\left(-\tau_{\gamma_k}^{|\omega_k|} S_{\gamma_k}\right) I_\Gamma^{(\omega)(\bar\omega_U)}(U) \tag{9.27}$$

where (with $\gamma_0 = \Gamma$)

a) $$I_\Gamma^{(\omega)(\bar\omega_U)}(U) = \prod_{k=0}^{n} I_{\bar\gamma_k}^{(\omega)(\bar\omega_n)}(y^{\gamma_k})$$

b) $$I_{\bar\gamma_k(U)}^{(\omega)(\bar\omega_U)}(y^{\gamma_k}) = I_{\gamma_k/\gamma_{k_1}\cdots\gamma_{k_c}}^{(\omega_k)(\omega_{k_1})\ldots(\omega_{k_c})}(y^{\gamma_k})$$

if $\{\gamma_{k_i},\ i = 1,\ldots,c\}$ are all maximal subgraphs of γ_k,

c) $$I_{\gamma_k(U)}^{(\omega)(\bar\omega_U)}(y^{\gamma_k}) = I_{\gamma_k}^{(\omega_k)}(y^{\gamma_k})$$

if γ_k is a minimal element of U.

Thus we see from equation (9.23), 9.24) and (9.27) that the renormalized (ω)-component of the integrand R_{Γ_ε} is a sum of terms, each of which has the usual forest structure, the recurrence flowing now from Θ-component graph to Θ-component graph. We note, moreover, that the notion of "complete forest" and of "sum over complete forests" [III.4], which we shall not define here but which are crucial for the convergence proof, depend only on topological considerations about graphs, and so remain valid in our case. From these two remarks it follows that the convergence of (9.22) is guaranteed if the following four criteria of Lowenstein [III.13] are fulfilled.

$$\text{(C1)} \qquad \rho^{(\omega)}(\gamma) \leq \delta^{(\omega)}(\gamma) + 1 \qquad (9.28)$$

for any 1P1 sugraph $\gamma \subset \Gamma$ and any (ω).

If $\gamma, \gamma_1, \ldots, \gamma_c$ are 1P1 subgraphs of Γ, if $\{\gamma_1, \ldots, \gamma_c\} \subset \gamma$ and if $\gamma_i \cap \gamma_j = \emptyset$ for $i \neq j$, then

$$\text{(C2)} \qquad \delta^{(\omega)}(\gamma) \geq d^{(\bar{\omega})}(\bar{\gamma}) + \sum_{k=1}^{c} \delta^{(\omega_k)}(\gamma_k) \qquad (9.29)$$

$$\text{(C3)} \qquad \rho^{(\omega)}(\gamma) \leq r^{(\bar{\omega})}(\bar{\gamma}) + \sum_{k=1}^{c} \rho^{(\omega_k)}(\gamma_k) \qquad (9.30)$$

for any possible choice of $((\omega), (\omega_1), \ldots, (\omega_c))$; the UV and IR degrees $d^{(\bar{\omega})}(\bar{\gamma})$ and $r^{(\bar{\omega})}(\bar{\gamma})$ of the reduced graph $\gamma/\gamma_1 \ldots \gamma_c$ were given by (9.15). Before stating the last criterion, let us give one more definition: if Γ is a connected graph, the augmented graph $\hat{\Gamma}$ is obtained from Γ by drawing special lines (q-lines) which carry the external momenta q(p), connected to the external points of Γ and which meet at a new internal vertex V_0; to each q-line there is assigned a scalar massive propagator

$$[q^2 - \mu^2 + i\varepsilon(\bar{q}^2 + \mu^2)]^{-\nu}$$

with ν so large that one can choose $\rho^{(\omega)}(\gamma)$ and $\delta^{(\omega)}(\gamma)$ negative (and consistent with criteria (C1-3) for any subgraph γ of Γ which contains V_0. Note that $\hat{\Gamma}$ and Γ have the same Θ-structure. The last criterion is then

$$\text{(C4)} \qquad E^{(\bar{\omega})}(\Lambda) \equiv r^{(\bar{\omega})}(\Lambda) + \sum_{k=0}^{c} \max(0, \rho^{(\omega_k)}(\gamma_k)) > 0 \qquad (9.31)$$

with $\Lambda \equiv \hat{\Gamma}/\gamma_0\gamma_1\cdots\gamma_c$ for any set $\{\gamma_0,\gamma_1,\ldots,\gamma_c\}$ of disjoint, non-trivial, one-particle irreducible subgraphs of $\hat{\Gamma}$, and any choice of $((\omega), (\omega_0),\ldots,(\omega_c)) = (\bar{\omega})$.

We now want to show that the criteria (C1-4) are fulfilled, under the following assumptions.

i) The UV and IR dimensions of the superfields Φ_a are restricted by

$$r_a > 0 \tag{9.32a}$$

$$r_a \geq d_a \tag{9.32b}$$

$$D_{ab} \equiv -d_a - d_b + d_{ab} + 4 \leq 0 \tag{9.32c}$$

$$R_{ab} \equiv -r_a - r_b + r_{ab} + 4 \geq 0 \tag{9.32d}$$

where d_{ab}, r_{ab} were defined in equations (8.17).

ii) The UV and IR degrees for vertices V_i associated with superfield monomials Q_i (eq. (8.4), (8.26)) satisfy the constraints

$$\rho_i \leq \delta_i \quad , \quad \delta_i \geq d_i \quad , \quad r_i \geq \rho_i \geq 0 \quad , \tag{9.33}$$

but $r_i \geq \rho_i \geq 4$ if V_i is associated with an integrated monomial

$$\int d^4 x \, Q_i(x,\Theta,\bar{\Theta}) \; . \tag{9.34}$$

Note that additional constraints, which will be explained later, are needed if massless vector superfields with IR dimension $r_a = 0$ are present. In any case one has to make sure, that the effective subtraction degrees $\rho_\omega(\gamma)$ (9.2) are integers. This is not automatically the case for all assignments satisfying i) and ii).

Inequality (9.28) for criterion (C1) follows simply from $r_a \geq d_a$ and $\rho_i \leq \delta_i$. One even sees that one can allow for one vertex having $\rho_i > \delta_i$, provided that $\rho_i - \delta_i \leq 1$. In order to prove inequality (9.29) for criterion (C2) one writes the UV subtraction degree in the form

$$\delta^{(\omega)}(\gamma) = d^{(\omega)}(\gamma) + \sum_{V_i\varepsilon\gamma}(\delta_i - d_i) - \sum_{a,b}^{a<b} I_{ab}(\gamma)D_{ab} . \tag{9.35}$$

Using the relation (9.15) one obtains

$$\delta^{(\omega)}(\gamma) - \sum_{k=1}^{c}\delta^{(\omega_k)}(\gamma_k) - d^{(\bar{\omega})}(\bar{\gamma}) =$$

$$= \sum_{V_i\varepsilon\bar{\gamma}}(\delta_i - d_i) - \sum_{a,b}^{a<b} I_{ab}(\bar{\gamma})D_{ab} \geq 0 .$$

Inequality (9.30) for criterion (C3) is proved in the same way.

We turn now to the criterion (C4). Let us suppose that the subgraph γ_0 contains the vertex V_0, and let us call $Q(\gamma)$, $N_Q(\gamma)$ the number of internal, resp. external, q-lines of graph γ. Clearly (Γ being a connected graph)

$$r^{(\omega)}(\hat{\Gamma}) = \frac{|\omega|}{2} + 4 + \sum_{a\leq b} I_{ab}(\Gamma)(r_{ab} + 4)$$
$$+ \sum_{V_i\varepsilon\Gamma}(D_i - 4) + 4(Q(\hat{\Gamma}) - 1) \tag{9.36}$$

and

$$r^{(\omega_0)}(\gamma_0) = \frac{|\omega_0|}{2} + 4 - (N_s + 2N_v)(\gamma_0)$$
$$+ \sum_{a\leq b} I_{ab}(\gamma_0)(r_{ab} + 4) + \sum_{V_i\varepsilon\gamma_0}(D_i - 4) \tag{9.37}$$
$$+ 4(Q(\gamma_0) - 1)$$

Recall that $r^{(\omega_k)}(\gamma_k)$ for $k = 1,\ldots,c$ are given by equations (8.16) and (8.24). Using equation (9.15) for the degree of the reduced graph Λ, together with the relation

$$Q(\hat{\Gamma}) - Q(\gamma_0) = Q(\Lambda) + \sum_{k=1}^{c} N_Q(\gamma_k) ,$$

$Q(\Lambda)$ = number of non-integrated vertices of Γ which remain in Λ, one obtains the expression

$$r^{(\bar{\omega})}(\Lambda) = (N_s+2N_v)(\gamma_0) + \sum_{k=1}^{c} [(N_s+2N_v+4N_Q)(\gamma_k) - \frac{|\omega_k|}{2} - 4]$$

$$+ \sum_{a\leq b} I_{ab}(\Lambda)(r_{ab}+4) + \sum_{V_i\varepsilon\Gamma\cap\Lambda}^{ext} D_i + \sum_{V_i\varepsilon\Gamma\cap\Lambda}^{int} (D_i-4) \tag{9.38}$$

$$+ (1/2)(|\omega|-|\omega_0|)$$

where we have split the sum over vertices into a sum over non-integrated ("ext") and a sum over integrated ("int") ones. From inequality (9.32d) and the topological relation

$$\sum_{b}^{b\neq a} I_{ab}(\Lambda) + 2I_{aa}(\Lambda) = \sum_{V_i\varepsilon\Gamma\cap\Lambda} C_{a_i} + \sum_{k=0}^{c} N_a(\gamma_k)$$

one gets the lower bound

$$r^{(\bar{\omega})}(\Lambda) \geq N_r(\gamma_0) + \sum_{k=1}^{c} [N_r(\gamma_k) + 4N_Q(\gamma_k) - \frac{|\omega_k|}{2} - 4]$$

$$+ \sum_{V_i\varepsilon\Gamma\cap\Lambda}^{ext} r_i + \sum_{V_i\varepsilon\Gamma\cap\Lambda}^{int} (r_i-4) + (1/2)(|\omega|-|\omega_0|) \ , \tag{9.39}$$

where we have used the abbreviation

$$N_r(\gamma) = (N_s + 2N_v + \sum_a N_a r_a)(\gamma) \ . \tag{9.40}$$

Now from

$$\max(0,\rho^{(\omega_k)}(\gamma_k)) \geq \rho^{(\omega_k)}(\gamma_k) \quad , \quad k = 1,\ldots,c$$

$$\max(0,\rho^{(\omega_0)}(\gamma_0)) = 0 \quad ,$$

since γ_0 contains the vertex V_0 and from

$$N_Q(\gamma_k) = \text{number of non-integrated vertices of } \gamma_k$$

one obtains for the expression (9.31)

$$E^{(\bar{\omega})}(\Lambda) \geq N_r(\gamma_0) + (1/2)(|\omega| - |\omega_0|) + \sum_{V_i \varepsilon \Gamma \cap \Lambda}^{ext} r_i$$

$$+ \sum_{V_i \varepsilon \Gamma \cap \Lambda}^{int} (r_i - 4) \tag{9.41}$$

$$+ \sum_{k=1}^{c} \left[\sum_{V_i \varepsilon \gamma_k}^{ext} \rho_i + \sum_{V_i \varepsilon \gamma_k}^{int} (\rho_i - 4) \right] .$$

A careful counting of the Θ's in the relations (9.11), (9.13), and (9.14) applied to $\{\Gamma, \gamma_0\}$ shows that

$$(1/2)(|\omega| - |\omega_0|) \geq (N_s + 2N_v)(\gamma_0) \tag{9.42}$$

This yields the estimate

$$E^{(\bar{\omega})}(\Lambda) \geq \sum_a N_a(\gamma_0) r_a + \sum_{V_i \varepsilon \Gamma \cap \Lambda}^{ext} r_i$$

$$+ \sum_{V_i \varepsilon \Gamma \cap \Lambda}^{int} (r_i - 4) \tag{9.43}$$

$$+ \sum_{k=1}^{c} \left[\sum_{V_i \varepsilon \gamma_k}^{ext} \rho_i + \sum_{V_i \varepsilon \gamma_k}^{int} (\rho_i - 4) \right]$$

which is strictly positive if all of the conditions (9.32), (9.33), and (9.34) are met.

However one of these conditions, namely inquality (9.32a) is violated in the case of supersymmetric gauge theories (Ch.II). Indeed, there occur massless vector superfields with dimensions $d = r = 0$; so there are choices of the subgraphs $\{\gamma_0, \gamma_1, \ldots, \gamma_c\}$ such that the right hand side of (9.43) is zero. For example such a divergent graph in SQED (even in the $\alpha = 1$ Feynman gauge) is given in figure 9.2

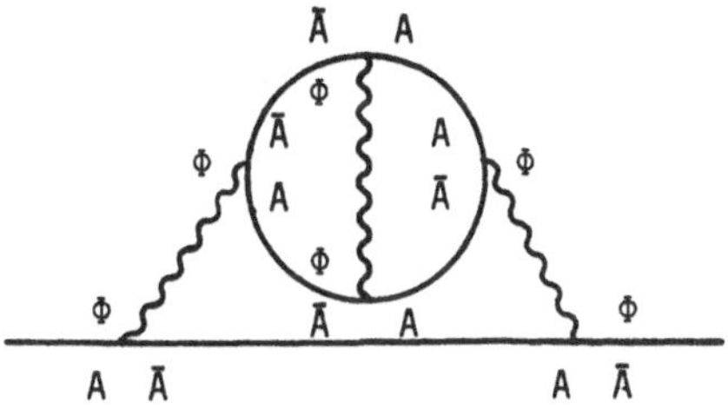

Fig. 9.2: An example of an IR divergent diagram

Looking more closely into the origin of this IR divergence, one clearly finds its cause to be one specific Θ-component only (see figure 9.3). This indicates that we can still refine the estimate (9.30) by exhibiting Θ-components.

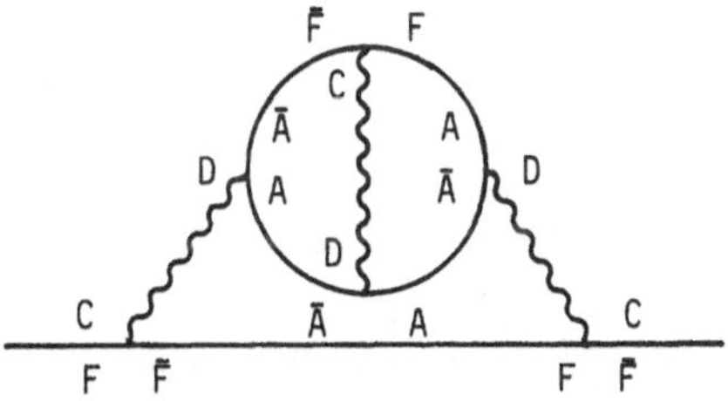

Figure 9.3: An example of an IR divergent diagram in terms of components.

In fact, if we associate with each amputated external line a component $\Phi^{(\omega_a)}$, due to chiral (vectorial) δ-functions it will contribute to the Θ-power of the diagram according to

$$|\omega| = |\omega_{a_1}| + \ldots + |\omega_{a_n}| = \sum_{(\omega_s)} N_s^{(\omega_s)} (2 - |\omega_a|) \tag{9.44}$$

$$+ \sum_{(\omega_v)} N_v^{(\omega_v)} (4 - |\omega_v|)$$

where $N_s^{(\omega_s)}$, $N_V^{(\omega_V)}$ count components of chiral, resp. vectorial superfields. Therefore, we find instead of (9.42) an equality

$$1/2(|\omega| - |\omega_0|) = -(N_s + 2N_V)(\gamma_0) + (1/2) \sum_{a,(\omega_a)} N_a^{(\omega_a)}(\gamma_0)\,\omega_a + (1/2) \sum_{V_i \in \Lambda \cup \gamma_1 \cup \ldots \cup \gamma_c} \left| \omega_i^{V_i} \right| \tag{9.45}$$

and obtain the improved bound

$$E^{(\bar{\omega})}(\Lambda) \geq \sum_{a,(\omega_a)} N_a^{(\omega_a)} r_a^{(\omega_a)} + \sum_{V_i \in \Lambda \cap \Gamma}^{ext} r_i^{(\omega_i^{V_i})} + \sum_{V_i \in \Lambda \cap \Gamma}^{int} (r_i^{(\omega_i^{V_i})} - 4) + \sum_{k=1}^{c} \sum_{V_i \in \gamma_k}^{ext} \rho_i^{(\omega_i^{V_i})} + \left[\sum_{V_i \in \gamma_k}^{int} (\rho_i^{(\omega_i^{V_i})} - 4) \right] \tag{9.46}$$

which is strictly positive if, instead of (9.32a) we have $r_{a_0}^{(\omega_a)} > 0$. This condition is violated in the case of SQED only by the Θ^0-component, C, of the massless vector superfield whose presence, as the example shows, indeed causes IR divergence.

This result, that for massless vector superfields one will not have IR-convergence in general, forces one to a case by case study which we shall present in chapter V. It should also be noted, that Lowenstein's criteria (C1) ... (C4) (9.28) ... (9.31) are immediately applicable for any supersymmetric theory formulated entirely in components (and not in superspace). By doing so one simply does not "see" that the subtractions maintain manifest supersymmetry. In the example of pure SYM we shall proceed in this way after having rendered massive the field C (cf. below chapter V).

Let us summarize the results of this section.

Starting with subtraction degrees (9.1)-(9.3) for diagrams in superspace we have set up a systematic procedure to render them finite (forest formula (9.17)) provided the conditions (9.32) on the fields and the conditions (9.33) on the vertices in the theory are satisfied. The subtractions maintain supersymmetry, i.e. starting from a naively supersymmetric theory one will not violate this supersymmetric structure by the subtractions. In the case that massless vector-superfields are present, IR-convergence (of Green's functions, i.e. off-shell) is not guaranteed by the above procedure (condition (9.32a) is violated).

Section 10. Normal products

In the preceding section we have described a recipe to produce finite Feynman diagrams in superspace for a large class of theories, restricted by the conditions (9.32) on fields and by conditions (9.33) on vertices. We therefore have a tool for the construction of Green's functions, namely the modified Gell-Mann-Low formula (7.1), not only of elementary fields but also of composite operators [III.16]. Via the reduction formalism (when applicable) we have thus defined operators for the theories we started with. Since, besides the S-matrix, composite operators - essentially currents and charges - are the primary object of interest in any theory, it is important to collect all relevant information on them. This is the aim of the present section.

Let us first of all note a simple, but important fact which follows immediately from the form of the subtraction operator (9.20)

$$(1-\tau_{p,s-1}^{\rho^{(\omega)}-1}(\gamma))(1-\tau_{p,s}^{\delta^{(\omega)}}(\gamma)) \equiv 1 - \tau_{\gamma}^{\omega} \tag{10.1}$$

($\tau_{x_1x_2}^{d}$ was the Taylor series about $x_1 = x_2 = 0$ up to and including order d for $d \geq 0$; $\tau_{x_1x_2}^{d} \equiv 0$ for $d < 0$): the normalization conditions for vertex functions.

If for some ω, $\rho^{(\omega)}(\gamma) \leq 0$ and $\delta^{(\omega)}(\gamma) \geq 0$ then

$$(1 - \tau_{\gamma}^{\omega})I_{\gamma}^{\omega}(p,k,s)\Big|_{\substack{p=0\\ s=0}} = 0 \tag{10.2}$$

i.e. only the trivial graphs (no loops) will contribute to the ω-component of this vertex function at p = 0, s = 0.

If for some ω, $\rho^{(\omega)}(\gamma) \geq 1$, then

$$(1 - \tau_{\gamma}^{\omega}) I_{\gamma}^{\omega}(p,k,s)\Big|_{\substack{p=0 \\ s=1}} = 0 \tag{10.3}$$

i.e. only the trivial graphs will contribute to the ω-component of this vertex function at $p = 0$, $s = 1$.

In order to deal with the above mentioned operators and their Green's functions we introduce some additional notation. Let Q_i $i = 1,\ldots,n$ be a set of monomials in the free fields of the theory (8.4).

$$Q_i(x_i,\Theta_i,\bar{\Theta}_i) = (s-1)^{a_i} \left(\frac{\partial}{\partial x}\right)^{(\mu_i)} (\tilde{D})^{\nu_i} [\prod_a \Phi_a^{c_{ai}}(x_i,\tilde{\Theta}_i)] \tag{10.4}$$

i.e. with naive UV-(IR-) dimensions (8.23) (8.24)

$$d_i = a_i + |\mu_i| + \frac{1}{2}|\nu_i| + \sum_a c_{ai} d_a$$

$$r_i = a_i + |\mu_i| + \frac{1}{2}|\nu_i| + \sum_a c_{ai} r_a \tag{10.5}$$

and let Q_i now occur in a diagram as a vertex. Above we have associated subtraction degrees $\delta_i \geq d_i$, $\rho_i \leq r_i$ with the vertex Q_i and we have shown that such a diagram will be finite after subtractions via (9.17) provided $0 \leq \rho_i \leq \delta_i$ (and (9.32) for the fields being true).

The subtraction degrees were (9.1) (9.2)

$$\delta^{(\omega)}(\gamma) = \delta(\gamma) + \frac{|\omega|}{2}$$

$$\rho^{(\omega)}(\gamma) = \rho(\gamma) + \frac{|\omega|}{2} \tag{10.6}$$

$$\delta(\gamma) = 4 - N_s(\gamma) - 2N_v(\gamma) - \sum_a d_a N_a(\gamma) + \sum_{i\in\mathfrak{V}(\gamma)} (\delta_i - 4)$$

$$\rho(\gamma) = 4 - N_s(\gamma) - 2N_v(\gamma) - \sum_a r_a N_a(\gamma) + \sum_{i\in\mathfrak{V}(\gamma)} (\rho_i - 4) \tag{10.7}$$

It is then natural to define as Green's functions of the operator Q_i just those given by the modified Gell-Mann-Low formula *

$$< T(Q_i(z_i)\Phi_i\ldots\Phi_m) > = R \frac{<TQ_i(z_i)\Phi_i\ldots\Phi_m\, e^{i\int L_{int}} >_{(0)}}{< e^{i\int L_{int}} >_{(0)}} \tag{10.8}$$

$(z_i \equiv (x_i,\Theta_i,\bar{\Theta}_i))$, where the right-hand-side is to be expanded into a sum of diagrams according to Wick's theorem and R is the subtraction procedure described in section 9 with Q_i being taken into account. I.e. one finds just all diagrams having Q_i as a vertex. We shall call Q_i an insertion (or normal product) [III.16] and write its general Green's functions more precisely than in (10.8) as $<T\, N_{\delta_i}^{\rho_i}\, [Q_i(z_i)]\, X>$ in order to record the power counting with which it contributes to the subtraction degrees. In the language of generating functionals (cf. App B) we write: $N_{\delta_i}^{\rho_i}\,[Q_i]\cdot Z$, $N_{\delta_i}^{\rho_i}\,[Q_i]\cdot Z_c$, $N_{\delta_i}^{\rho_i}\,[Q_i]\cdot\Gamma$ for general connected -, one-particle-irreducible Green's functionals, respectively. They generate all Green's functions having the vertex Q_i as special vertex. If we consider Green's functions of several insertions, our recipe of section 9 works as well and the notation is accordingly:

$$< T\prod_i N_{\delta_i}^{\rho_i}\, Q_i(z_i)\, X > \qquad \text{for Greens functions,}$$

$$\prod_i N_{\delta_i}^{\rho_i}[Q_i(z_i)]\cdot Z\,, \qquad \prod_i N_{\delta_i}^{\rho_i}[Q_i(z_i)]\cdot Z_c\,,$$

$$\prod_i N_{\delta_i}^{\rho_i}[Q_i(z_i)]\cdot\Gamma \qquad \text{for the functionals.}$$

If $\Delta_i = \int dx\, Q_i(x,\Theta,\bar{\Theta})$ is an x-integrated monomial, we have convergence for the corresponding normal product if in addition to the above conditions $r_i \geq \rho_i \geq 4$.

The more precise notation is analogously $< T\, N_{\delta_i}^{\rho_i}\, [\Delta_i]\, X>$ for Green's

---

* It is a remarkable consistency check of this approach that for minimal subtraction $\delta_i = d_i$, (massive theory) the above definition coincides with the one where one fixes external lines of a diagram to form the vertex Q_i and renders finite the new diagram by taking into account the divergences of the additional loops one has created in the fusion process. The latter is the original definition of normal products [III.16].

functions

$$N_{\delta_i}^{\rho_i}[\Delta_i] \cdot Z, \quad N_{\delta_i}^{\rho_i}[\Delta_i] \cdot Z_c, \quad N_{\delta_i}^{\rho_i}[\Delta_i] \cdot \Gamma$$

for the functionals. $N_\delta^\rho\,[\Delta]$ (or Δ_δ^ρ) will be called integrated normal product, (integrated) insertion or differential vertex operation [III.17]. The reason for this last name will become clear in section 12.

The handling of normal products is governed by only a few calculational rules which we now present.

(1) $\underline{\Delta_\delta^\rho \cdot \Gamma = \Delta + O(\hbar\Delta)}$ (10.9)

The functional of vertex functions all having the special vertex Δ, $\Delta\cdot\Gamma$, is given by the sum of two terms: first of all by its trivial part - the vertex Δ, this is the local contribution to $\Delta\cdot\Gamma$, second diagrams having the vertex Δ and at least one loop ($\hbar$) - this is the non-local part of the functional $\Delta\cdot\Gamma$. Diagrammatically this equation reads:

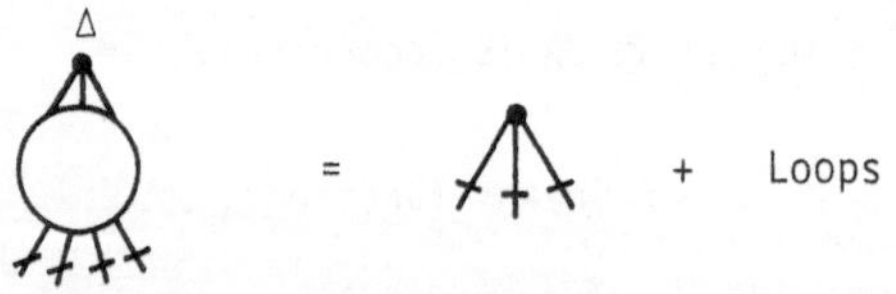

(10.10)

Although the equation (10.9) is extremely simple it will often help us to reduce true quantum problems to classical ones (cf. chapters IV,V).

(2) derivative rule

$$\partial_\mu^x N_\delta^\rho[Q(x,\theta,\bar\theta)] \cdot \Gamma = N_{\delta+1}^{\rho+1}[\partial_\mu^x Q(x,\theta,\bar\theta)] \cdot \Gamma \tag{10.11}$$

$$\tilde\theta\, N_\delta^\rho[Q(z)] \cdot \Gamma = N_{\delta-\frac{1}{2}}^{\rho-\frac{1}{2}}\,[\tilde\theta Q(z)] \cdot \Gamma \quad (\tilde\theta{:}\theta \text{ or } \bar\theta) \tag{10.12}$$

$$\tilde D_z N_\delta^\rho[Q(z)] \cdot \Gamma = N_{\delta+\frac{1}{2}}^{\rho+\frac{1}{2}}[\tilde D_z Q(z)] \cdot \Gamma \quad (\tilde D{:}D_\alpha \text{ or } \bar D_{\dot\alpha}) \tag{10.13}$$

This rule is proved by expanding both sides of the equations into diagrams and then the diagrams into forests. The equations follow due to the corresponding commutation property of $p_\mu,\tilde\theta,\tilde D$ with the subtraction operator τ^{ω}_{γ} according to which then the degrees change: e.g.

$$p^{\gamma}_{\mu}\tau^{\delta}_{p\gamma} = \gamma^{\delta+1}_{p\gamma}\; p^{\gamma}_{\mu} \; .$$

(3) Zimmermann identities

Starting from any monomial Q_i (10.4) one may construct, according to the above, different normal products out of it by assigning different subtraction degrees to it. Zimmermann identities tell how these different normal products are related to each other.

To begin with let us assume a massive theory to be given and treat just one insertion differently with respect to degree. Then the corresponding Zimmermann identity reads [III.16,17]:

$$N_{\delta}[Q] \cdot \Gamma = N_{\varphi}[Q] \cdot \Gamma + \sum_i r_{Qi} N_{\varphi}[Q_i] \cdot \Gamma \tag{10.14}$$

where by assumption: $\varphi > \delta \geq d(Q)$ (10.15)

$$r_{Qi} = O(\hbar) \tag{10.16}$$

and the Q_i span a basis of all (classical) monomials with $\delta + 1 \leq d(Q_i) \leq \varphi$ and equality of all other quantum numbers referring to symmetries which are naively maintained by the subtraction scheme. Amongst these symmetries is at least Lorentz invariance and (if present classically) supersymmetry; also all conserved discrete symmetries are operative.

Let us give a concrete example. The Wess-Zumino model (3.26) is supersymmetric and parity invariant. According to section 9 we may therefore construct finite diagrams and thus finite Green's functions with the subtraction rules given there, since the conditions (9.32) (9.33) are satisfied. Supersymmetry and parity are conserved in the subtraction process. If we want to relate now $N_2[A^2]$, the minimally subtracted normal product, to $N_3[A^2]$, the once oversubtracted one, we may write according to (10.14)

$$mN_2[A^2] \cdot \Gamma = mN_3[A^2] \cdot \Gamma + N_3[r_1 A\bar{D}\bar{D}\bar{A} + r_2 A^3 + r_3 \bar{D}\bar{D}(\bar{A}^2)] \cdot \Gamma \tag{10.17}$$

because the terms $A\bar{D}\bar{D}\bar{A}$, A^3, $\bar{D}\bar{D}(\bar{A}^2)$ are a basis for all classical monomials which are chiral and of naive dimension 3. Due to conservation of parity we also have

$$mN_2[\bar{A}^2]\cdot\Gamma = mN_3[\bar{A}^2]\cdot\Gamma + N_3[r_1\bar{A}DDA + r_2\bar{A}^3 + r_3DD(A^2)]\cdot\Gamma \quad (10.18)$$

with the same (real) coefficients r_i. Multiplying (10.17) by DD, (10.18) by $\bar{D}\bar{D}$, using (10.13) adding and integrating over x-space we have therefore

$$\begin{aligned} mN_3[\int dSA^2 + \int d\bar{S}\bar{A}^2]\cdot\Gamma &= mN_4[\int dSA^2 + \int d\bar{S}\bar{A}^2]\cdot\Gamma \\ &+ N_4[2\, r_1\int dVA\bar{A} + r_2(\int dSA^3 + \int d\bar{S}\bar{A}^3)]\cdot\Gamma \end{aligned} \quad (10.19)$$

The r_3-term dropped out since DD $\bar{D}\bar{D}(\bar{A}^2)$ = $[DD,\bar{D}\bar{D}]$ $(\bar{A}^2)$ = total divergence (same for conjugate).

The coefficients r_i are obviously of order $\hbar$ - they come from loop diagrams - since in the tree approximation there are no subtractions i.e. the N-symbol is ineffective. The best way to calculate them is via normalization conditions:

$$\begin{aligned} mN_3[\int dSA^2 + \int d\bar{S}\bar{A}^2]\cdot\Gamma_{A\bar{A}}\big|_{p=0} &= N_4[m(\int dSA^2 + \int d\bar{S}\bar{A}^2) \\ &+ 2\, r_1\int dVA\bar{A} + r_2(\int dSA^3 + \int d\bar{S}\bar{A}^3)]\cdot\Gamma_{A\bar{A}}\big|_{p=0} \;. \end{aligned} \quad (10.20)$$

Due to the subtractions:

$$\begin{aligned} &\text{left-hand-side : } \delta_\Omega = 4-2\cdot2 + (3-4) + \tfrac{1}{2}\cdot 0 = -1 \\ &\text{right-hand-side: } \delta_\Omega = 4-2\cdot2 + (4-4) + \tfrac{1}{2}\cdot 0 = \;\;0 \end{aligned} \quad (10.21)$$

on the right-hand-side there survives only the trivial contribution i.e.:

$$mN_3[\int dSA^2 + \int d\bar{S}\bar{A}^2]\cdot\Gamma_{A\bar{A}}\big|_{p=0} = 2\, r_1 \quad (10.22)$$

For instance in the one-loop approximation one has two contributions from the diagram

$$N_3[\int dSA^2 + \int d\bar{S}\bar{A}^2]$$

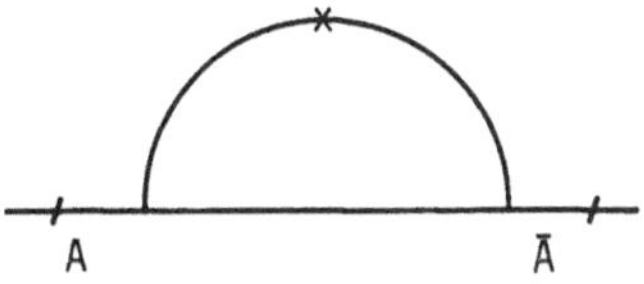

which are of the form (up to a numerical factor)

$$\begin{aligned} 2\, r_1 &\sim \frac{1}{(2\pi)^4} \left(\frac{ig}{48}\right)^2 \int dk \, \frac{m^2 \bar{D}\bar{D}_1 \delta_{\bar{S}}(1,2) \bar{D}\bar{D}_1 \delta_{\bar{S}}(1,2)}{(k^2 - m^2 + i\varepsilon)^3} \\ &= \frac{1}{(2\pi)^4} \left(\frac{ig}{48}\right)^2 \int dk \, \frac{m^2}{(k^2 - m^2 + i\varepsilon)^3} \neq 0 \end{aligned} \tag{10.23}$$

(since $\bar{D}\bar{D}_1(k)\ \delta_{\bar{S}}(1,2;k) = 1$).

Similarly, r_2 is given e.g. by the test

$$mN_3[\int dSA^2 + \int d\bar{S}\bar{A}^2] \cdot \Gamma_{A_1 A_2 A_3}\Big|_{p=0} = 6\, r_2 \delta_S(1,2)\delta_S(1,3) \tag{10.24}$$

since again the l.h.s. is not subtracted whereas the r.h.s. is

$$\begin{aligned} \text{l.h.s.} :\ \delta_\Omega &= 4-2\cdot 3 + (3-4) + \tfrac{1}{2}\cdot 2\cdot 2 = -1 \\ \text{r.h.s.} :\ \delta_\Omega &= 4-2\cdot 3 + (4-4) + \tfrac{1}{2}\cdot 2\cdot 2 = \ \ 0 \end{aligned} \tag{10.25}$$

Looking into the one-loop contributions we find as possible diagrams

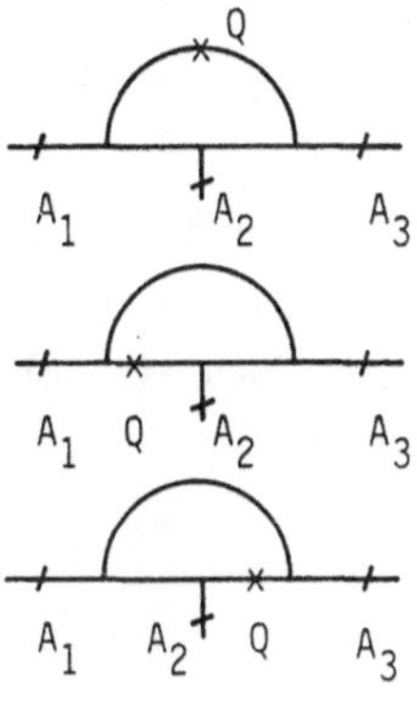

$$Q \equiv N_3[\int dSA^2 + \int d\bar{S}\bar{A}^2]$$

which all vanish at $p = 0$ since they contain the square of a chiral (resp. antichiral) propagator. Hence

$$r_2 = O(\hbar^2) \tag{10.26}$$

(In fact we shall prove below that $r_2 = 0$ to all orders.)

The next case which we treat is a theory containing massless fields for which an auxiliary mass term has been used. Then the need may arise to compare a normal product $N[(s-1)^a Q]$ with $(s-1)^a N[Q]$ - for instance in checking massless limits via $s \to 1$. Let us assume subtractions degrees δ, ρ for Q satisfying $d(Q) \leq \delta$, $r(Q) \geq \rho, \delta \geq \rho \geq 0$. Then the relevant <u>Zimmermann identity</u> reads:

$$(s-1)^a N^{\rho}_{\delta}[Q] \cdot \Gamma = N^{\rho+a}_{\delta+a}[(s-1)^a Q] \cdot \Gamma + \sum_i r_{Qi} N^{\rho+a}_{\delta+a}[Q_i] \cdot \Gamma \tag{10.27}$$

again $$r_{Qi} = o(\hbar) \tag{10.28}$$

and the characterization of the Q_i relative to the Q is analogous to the previous case ($d(Q_i) \leq \delta + a$, $r(Q_i) \geq \rho + a$; equal quantum numbers for naively maintained symmetries). Since now $(s-1)^a Q$ is not an over-subtracted quantity it may appear again amongst the Q_i (and may be eliminated there by repeated use of (10.27)).

Let us as an example again consider the Wess-Zumino model (3.26), now in its massless limit. The criteria (9.32) (9.33) are satisfied with an auxiliary mass term

$$\Gamma_m = \frac{1}{8} m (s-1)(\int dSA^2 + \int d\bar{S}\bar{A}^2) \tag{10.29}$$

and s participating in the subtractions as dictated by the forest-formula (9.17). Hence we have finite diagrams, i.e. finite Green's functions (at non-exceptional momenta). The Zimmermann identity (10.27) corresponding to (10.17) reads

$$N_3^3\,[m(s-1)A^2]\cdot\Gamma = m(s-1)N_2^2\,[A^2]\cdot\Gamma + N_3^3\,[qm(s-1)A^2 + rA\bar{D}\bar{D}\bar{A} + tA^3 + n\bar{D}\bar{D}\bar{A}^2]\cdot\Gamma + v\Box A + xm(s-1)\,\bar{D}\bar{D}\bar{A} + ym^2(s-1)^2A \qquad (10.30)$$

Similarly to the massive case the coefficients are best calculated from normalization conditions.

The last case we shall present is the Zimmermann identity

$$N_\delta^\rho\,[Q]\cdot\Gamma = N_\varphi^\sigma\,[Q]\cdot\Gamma + \sum_i r_{Qi}N_\delta^\rho\,[Q_i]\cdot\Gamma \qquad (10.31)$$

Here

$$\delta \geq d(Q) \qquad \rho \leq r(Q) \qquad\qquad \rho \leq \sigma \leq \varphi \leq \delta$$
$$\varphi \geq d(Q) \qquad \sigma \leq r(Q) \qquad (10.32)$$

In other words: the common basis for normal products with mixed degrees is spanned by the products with smallest IR-degrees and the highest UV-degrees involved.

An explicit example will be given below in connection with the O'Raifeartaigh model.

For the proof of these (and related) Zimmermann identities we refer to the literature (for ordinary component theories [III.16,18], in superspace [III.11]). We should like to make only one comment. The main property to be proved is that the correction terms between two differently subtracted normal products is again a normal product. This requires regrouping the two forests involved and showing that the remainder is again a forest for certain insertions restricted only by power counting and the naive invariances. The hard part of the proof is thus the existence of equations (10.14) (10.27) (10.31), not the actual determination of the coefficients involved (as could have been suggested by our presentation).

Section 11. The action principle

Under this heading run essentially three theorems which describe the variation of (1) a parameter, (2) a field, (3) an external field of the theory. As will be seen in the subsequent chapters they govern all studies of parametric dependences and symmetries, hence are truely fundamental. Their most detailed proofs have been given in the context of the BPHZ-renormalization scheme, but they hold in an analogous fashion in all other schemes developed thus far and are therefore considered as general theorems valid in perturbative quantum field theory.

In order to formulate the principle we have to specify somewhat the action on which we base our theory. We write

$$\Gamma_{eff} = \Gamma_o + \Gamma_{int} \tag{11.1}$$

where Γ_o is the free part of the theory which determines the free propagators; Γ_{int} is the tree approximation interaction plus all sorts of counterterms (depending on $\hbar$) which may be needed in higher orders for ensuring normalization conditions or symmetries. It is Γ_{int} which appears in the Gell-Mann-Low formula (7.1). Since the notion "effective action" has different meaning to different people let us emphasize the difference between Γ_{eff} (11.1) and the full functional

$$\Gamma = \Gamma^{(o)} + \hbar\Gamma^{(1)} + \hbar^2\Gamma^{(2)} + \ldots \tag{11.2}$$

of one-particle-irreducible Green's functions (vertex functions): although they coincide in the tree approximation (zeroth order in $\hbar$)

$$\Gamma^{(o)} = \Gamma^{(o)}_{eff} = \text{classical action} \tag{11.3}$$

they differ in all higher orders $n \gtrsim 1$ in $\hbar$:

$$\Gamma^{(n)} = \Gamma^{(n)}_{eff} + \Gamma'^{(n)} \tag{11.4}$$

Here $\Gamma^{(n)}_{eff}$ collects all <u>trivial</u> graphs (point vertices) at the considered order in $\hbar$, it is the local contribution to Γ at this order, whereas $\Gamma'^{(n)}$ stands for all genuine loop graphs of this order: i.e. $\Gamma'^{(n)}$ has at least one loop (and $\hbar$-dependent vertices such that the total

$\hbar$-power equals n) and is the non-local part of the functional Γ. (Cf. the analogous discussion following (10.9)). This distinction will be crucial in the treatment of symmetries.

Let us now state the theorems.

(1) Action principle - variation of a parameter

$$\frac{\partial\Gamma}{\partial\lambda} = [\frac{\partial\Gamma_{eff}}{\partial\lambda}]_4^4 \cdot \Gamma \tag{11.5}$$

for any parameter λ of Γ_{eff}.

(2) Action principle - variation of a propagating field

(a) in ordinary space

$$\frac{\delta\Gamma}{\delta\Phi} =[\frac{\delta\Gamma_{eff}}{\delta\Phi}]_{4-d(\Phi)}^{4-r(\Phi)} \cdot \Gamma \tag{11.6}$$

("linear field equation")

$$\Phi' \frac{\delta\Gamma}{\delta\Phi} = [\Phi' \frac{\delta\Gamma_{eff}}{\delta\Phi}]_{4-d(\Phi)+d(\Phi')}^{4-r(\Phi)+r(\Phi')} \cdot \Gamma \tag{11.7}$$

("bilinear field equation")

(b) in superspace *

$$\frac{\delta\Gamma}{\delta A} = [\frac{\delta\Gamma_{eff}}{\delta A}]_{3-d(A)}^{3-r(A)} \cdot \Gamma \tag{11.8}$$

$$A' \frac{\delta\Gamma}{\delta A} = [A' \frac{\delta\Gamma_{eff}}{\delta A}]_{3-d(A)+d(A')}^{3-r(A)+r(A')} \cdot \Gamma \tag{11.9}$$

for chiral fields A,A'

$$\frac{\delta\Gamma}{\delta\Phi} = [\frac{\delta\Gamma_{eff}}{\delta\Phi}]_{2-d(\Phi)}^{2-r(\Phi)} \cdot \Gamma \tag{11.10}$$

---

* The difference in power counting assignment between ordinary space and superspace arises from the different integration measures dx versus dS = dxDD, dV = dxDD$\bar{D}\bar{D}$ which are "integrated away" with the corresponding δ-functions.

$$\Phi' \frac{\delta\Gamma}{\delta\Phi} = [\Phi' \frac{\delta\Gamma_{eff}}{\delta\Phi}]^{2-r(\Phi) + r(\Phi')}_{2-d(\Phi) + d(\Phi')} \cdot \Gamma \tag{11.11}$$

for vector fields Φ , Φ'

(3) Action principle - variation of an external field

Suppose in Γ_{eff} there is a non-linear field monomial Q coupled to an external field η, then

$$\int dz \frac{\delta\Gamma}{\delta\eta(z)} \frac{\delta\Gamma}{\delta\Phi(z)} = \Delta^{\rho}_{\delta} \cdot \Gamma \tag{11.12}$$

with

$$\begin{aligned} \delta &= 4-d(\Phi) + 4-d(\eta) = 4-d(\Phi) + d(Q) \\ \rho &= 4-r(\Phi) + 4-r(\eta) = 4-r(\Phi) + r(Q) \end{aligned} \tag{11.13}$$

in ordinary space (dz = dx)

$$\begin{aligned} \delta &= 3-d(\Phi) + d(Q) \\ \rho &= 3-r(\Phi) + r(Q) \end{aligned} \tag{11.14}$$

in superspace, for dz = dS, $d\bar{S}$

$$\begin{aligned} \delta &= 2-d(\Phi) + d(Q) \\ \rho &= 2-r(\Phi) + r(Q) \end{aligned} \tag{11.15}$$

for dz = dV

Let us note that (2) controls linear and inhomogeneous field variations, hence is suitable for deriving Ward identities of (spontaneously broken) linear symmetries, whereas (3) is of a very general nature and may be used to control non-linear symmetries. It will serve us below for BRS-invariance.

For the general proofs of the theorems we refer to the literature (ordinary space [III.19,20,18] superspace [III.11]), but we shall illustrate them by treating a simple example.

Let us consider the Wess-Zumino model (3.26) with free action

$$\Gamma_0 = \frac{1}{16} \int dV\bar{A}A + \frac{m}{8} (\int dSA^2 + \int d\bar{S}\bar{A}^2) \tag{11.16}$$

and interaction given by

$$\Gamma_{int} = \frac{z-1}{16} \int dV\bar{A}A - \frac{a}{8} (\int dSA^2 + \int d\bar{S}\bar{A}^2) + \frac{\lambda}{48} (\int dSA^3 + \int d\bar{S}\bar{A}^3) \tag{11.17}$$

We then define

$$\Gamma_{eff} = \Gamma_0 + \Gamma_{int} \tag{11.18}$$

and Green's functions by *

$$\langle TA(1)\ldots A(n)\bar{A}(1)\ldots\bar{A}(m)\} = R \langle TA\ldots\bar{A}\ldots e^{i\Gamma_{int}}\rangle_{(o)} \tag{11.19}$$

(cf. (7.1) and (9.17) for the subtractions R). Since the criteria (9.32) (9.33) are fulfilled with $\delta_{\omega(\gamma)} = 4-2N_s + \frac{|\omega|}{2}$ we know that (11.19) yields convergent Green's functions (from convergent diagrams) and furthermore that supersymmetry and parity are maintained. The coefficients z,a,λ may be fixed by normalization conditions:

$$\Gamma_{F\bar{F}} = 1 \qquad \Gamma_{AF} = m \qquad \Gamma_{AAF} = g \tag{11.20}$$

either at p = 0, then they have their tree approximation value

$$z = 1 \qquad a = 0 \qquad \lambda = g \tag{11.21}$$

or at $p^2 = \mu^2$, a normalization point, then they are formal power series in $\hbar$ and functions of m,g,μ^2:

---

* Vacuum-to-vacuum diagrams are suppressed.

$$z = z\left(\frac{m^2}{\mu^2}, g\right) = 1 + O(\hbar) \qquad a = a(m,\mu^2,g) = O(\hbar)$$
$$\lambda = \lambda\left(\frac{m^2}{\mu^2}, g\right) = g + O(\hbar) \tag{11.22}$$

Due to the subtractions and (11.17) we have also $\langle F \rangle = 0$.

Let us now derive (11.5) for the parameters g, μ^2, m of the model. to begin with we re-interprete first the Gell-Mann-Low formula (11.19) by writing [III.21]

$$\begin{aligned} G^{(n,m)} &\equiv \langle TA(1)\ldots A(n)\bar{A}(1)\ldots\bar{A}(m)\rangle \\ &= e^{i(z-1)\Delta_1 - ai\Delta_2 + \lambda i\Delta_3}\, G^{(n,m)}_{(0)} \end{aligned} \tag{11.23}$$

here

$$\begin{aligned} \Delta_1 &\equiv N_4\left[\tfrac{1}{16}\int dV\bar{A}A\right] \\ \Delta_2 &\equiv N_4\left[\tfrac{1}{8}\left(\int dSA^2 + \int d\bar{S}\bar{A}^2\right)\right] \\ \Delta_3 &\equiv N_4\left[\tfrac{1}{48}\left(\int dSA^3 + \int d\bar{S}\bar{A}^3\right)\right] \end{aligned} \tag{11.24}$$

i.e. the full Green's function $G^{(n,m)}$ is understood as being obtained from the free one, $G^{(n,m)}_{(0)}$, by inserting into the latter the normal products $\Delta_1, \Delta_2, \Delta_3$. (I.e. we have used the definition

$$\Delta\cdot G^{(n,m)} = \langle T\Delta A^{(1)}\ldots A^{(n)}\bar{A}^{(1)}\ldots\bar{A}(m)\rangle \tag{11.25}$$

and then repeatedly applied the Gell-Mann-Low formula.) Differentiating now (11.23) with respect to g we obtain

$$\begin{aligned} \frac{\partial}{\partial g} G^{(n,m)} &= i\,\frac{\partial\Gamma_{int}}{\partial g}\, e^{i\Gamma_{int}} G^{(n,m)}_{(0)} \\ &= i\,\frac{\partial\Gamma_{int}}{\partial g}\cdot G^{(n,m)} \end{aligned} \tag{11.26}$$

(an equality which one may also check explicitly by going through the series expansion of the exponential).

On the functional Z (11.26) reads

$$\frac{\partial}{\partial g} Z = [i \frac{\partial \Gamma_{int}}{\partial g}]_4 \cdot Z \tag{11.27}$$

and hence on Γ

$$\frac{\partial}{\partial g} \Gamma = [\frac{\partial \Gamma_{int}}{\partial g}]_4 \cdot \Gamma \tag{11.28}$$

Since Γ_0 is independent of g, one may also write $\frac{\partial \Gamma_{int}}{\partial g} = \frac{\partial \Gamma_{eff}}{\partial g}$ and thus one has finally (11.5).

The same procedure goes through for differentiation with respect to μ^2. If we now differentiate with respect to m we have to consider not only the analogous contribution to the above, namely when $\partial/\partial m$ acts on the vertices, but also when it acts on $G_0^{(n,m)}$, the free lines. With the free propagators (3.18) (3.19) one verifies that *

$$\frac{\partial}{\partial m} \underset{\Theta_1 \quad k \quad \Theta_2}{\longrightarrow} = \frac{-4_i \delta_s(1,2)(k^2+m^2)}{(k^2-m^2+i\varepsilon)^2} = \underset{\Theta_1 \qquad \Theta_2}{\overset{i\Delta_2}{-\!\!\times\!\!-}} \tag{11.29}$$

analogously for $\frac{\partial}{\partial m} \underset{\bar\Theta_1 \quad k \quad \bar\Theta_2}{\longrightarrow}$ and also for

$$\frac{\partial}{\partial m} \underset{\Theta_1 \quad k \quad \bar\Theta_2}{\longrightarrow} = \frac{2im\bar D\bar D\delta_{\bar s}(1,2)}{(k^2-m^2+i\varepsilon)^2} = \underset{\Theta_1 \qquad \bar\Theta_2}{\overset{i\Delta_2}{-\!\!\times\!\!-}} \tag{11.30}$$

I.e. on each line $\partial/\partial m$ equals the insertion $i\Delta_2$ (in accordance with its power counting $\delta = 4$). Since furthermore

$$\partial_m \Gamma_{(0)} = \Delta_2 \tag{11.31}$$

(once we assign $\delta = 4$ to $\Gamma_{(0)}$) we finally have

$$\partial_m Z = i\Delta_2 \cdot Z = \partial_m [i\Gamma_{int}]_4 \cdot Z \tag{11.32}$$

and thus on Γ (11.5)

---

* ε-contributions in the numerators are neglected in view of the eventual limit $\varepsilon \to 0$.

$$\partial_m \Gamma = [\partial_m \Gamma_{eff}]_4 \cdot \Gamma \qquad (11.33)$$

This concludes the derivation of (11.5).

Let us now give a derivation of (11.8).

The aim is to prove that

$$\frac{\delta\Gamma}{\delta A} = [\frac{\delta\Gamma_{eff}}{\delta A}]_2 \cdot \Gamma \qquad (11.34)$$

with (cf. (11.18))

$$\frac{\delta\Gamma_{eff}}{\delta A} = \frac{z}{16} \bar{D}\bar{D}\bar{A} + \frac{m-a}{4} A + \frac{\lambda}{16} A^2 \qquad (11.35)$$

Since Γ is the vertex functional and a linear term inserted always gives rise to one-particle-reducible graphs, (11.34) is more precisely to be written as

$$\frac{\delta\Gamma}{\delta A} = \frac{z}{16} \bar{D}\bar{D}\bar{A} + \frac{m-a}{4} A + N_2 [\frac{\lambda}{16} A^2] \cdot \Gamma \qquad (11.36)$$

Legendre transformation yields

$$- J_A = \frac{z}{16} \bar{D}\bar{D} \frac{\delta Z_c}{\delta J_{\bar{A}}} + \frac{m-a}{4} \frac{\delta Z_c}{\delta J_A} + N_2 [\frac{\lambda}{16} A^2] \cdot Z_c \qquad (11.37)$$

Going over to $Z = e^{iZ_c}$ we find

$$-iJ_A Z = \frac{z}{16} \bar{D}\bar{D} \frac{\delta Z}{\delta J_{\bar{A}}} + \frac{m-a}{4} \frac{\delta Z}{\delta J_A} + N_2[i \frac{\lambda}{16} A^2] \cdot Z \qquad (11.38)$$

or

$$\frac{z}{16} \bar{D}\bar{D} \frac{\delta Z}{i\delta J_{\bar{A}}(z)} + \frac{m-a}{4} \frac{\delta Z}{i\delta J_A(z)} = -J_A(z)Z - N_2 [\frac{\lambda}{16} A^2] \cdot Z \qquad (11.39)$$

i.e. for a Green's function with n chiral, m antichiral external legs

$$\langle T((\frac{z}{16}\bar{D}\bar{D}\bar{A} + \frac{m-a}{4}A)(z)A(1)\dots A(n)\bar{A}(1)\dots\bar{A}(m))\rangle$$

$$= + \sum_{k=1}^{n} i\delta_s(z,z_k) \langle T(A(1)\dots\check{A}(k)\dots A(n)\bar{A}(1)\dots\bar{A}(m)\rangle$$

$$+ \langle T(N_2 [-\frac{\lambda}{16} A^2](z)A(1)\dots A(n)\bar{A}(1)\dots\bar{A}(m)\rangle \tag{11.40}$$

($\check{A}(k)$ means omission of the leg $A(k)$.)

Equation (11.40) is the suitable form for a proof starting from the Gell-Mann-Low formula.

Let us consider a Green's function with n chiral, m anti-chiral legs and one leg whose chirality we do not yet specify ($\tilde{z}$). Then the graphical alternatives are

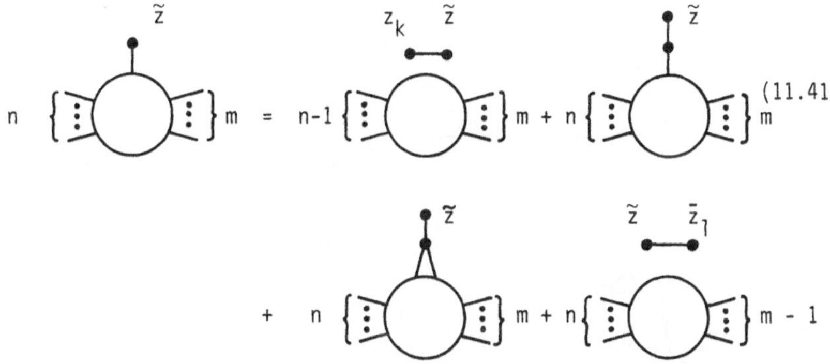

(11.41)

i.e. $\tilde{z}$ is connected immediately with another external line ("disconnected" part), $\tilde{z}$ enters the diagram via a bilinear or via a trilinear vertex of Γ_{int}.

Combining now two such Green's functions:

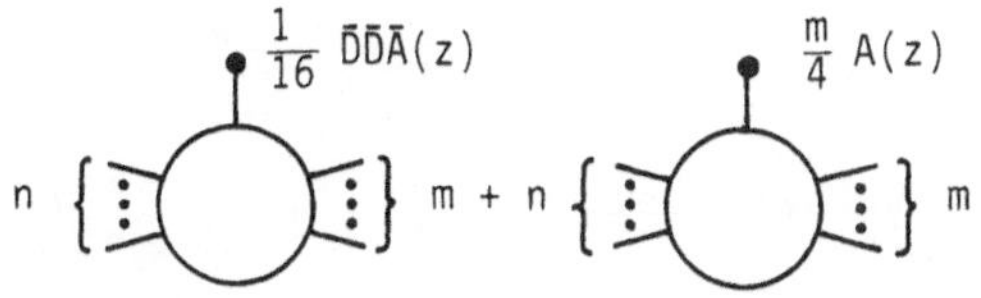

we see that the disconnected diagrams yield

$$\sum_{k=1}^{n} <T \; (\tfrac{1}{16} \, \bar{D}\bar{D}\bar{A}(z) + \tfrac{m}{4} \, A(z))A(z_k)>_{(o)} <T(A(1)..\overset{v}{A}(k)..A(n)\bar{A}(1)..\bar{A}(m)>$$

$$+ \sum_{k} <T \; (\tfrac{1}{16} \, \bar{D}\bar{D}\bar{A}(z) + \tfrac{m}{4} \, A(z))\bar{A}(z_k)>_{(o)} <T(A(1)..A(n)\bar{A}(1)..\overset{v}{\tilde{A}}(k)..\bar{A}(m))>$$

$$= \sum_{k=1}^{n} i \, \delta_s(z,z_k)(T(A(1)..\overset{v}{A}(k)..A(n)\bar{A}(1)..\bar{A}(m))> + 0 \tag{11.42}$$

due to (3.14) (differentiated with respect to $J_A, J_{\bar{A}}$, resp.).

The bilinear vertex diagrams contribute

$$<T \; (- \tfrac{(z-1)}{16} \, \bar{D}\bar{D}\bar{A} \; (z) - \tfrac{-a}{4} \, A(z)) \; X> \tag{11.43}$$

whereas the trilinear interaction vertices contribute $<T(- \tfrac{\lambda}{16} \, A^2(z)>$, as can be seen from

$$(\tfrac{1}{16} \, \bar{D}\bar{D}\bar{A}(z) + \overline{\tfrac{m}{4} \, A(z))\int dy \; \tfrac{i\lambda}{48}} \cdot 3A(y)A^2(y) = - \tfrac{\lambda}{16} \, A^2(z)$$

$$\tag{11.44}$$

$$(\tfrac{1}{16} \, \bar{D}\bar{D}\bar{A}(z) + \overline{\tfrac{m}{4} \, A(z))\int d\bar{y} \; \tfrac{i\lambda}{48}} \cdot 3\bar{A}(y)\bar{A}^2(y) = 0$$

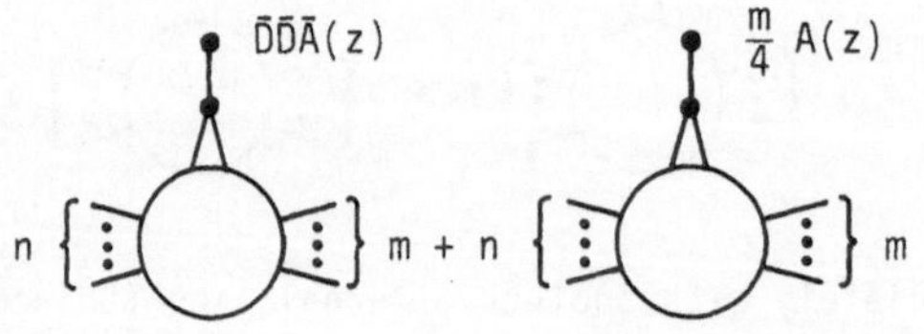

The power counting assignment follows from the fact that amputating an external line attached to a vertex V of degree 4 raises the degree by two $(N_s \rightarrow N_s - 1)$

$$\delta_\omega(\gamma) = 4 - 2N_s(\gamma) + \sum_{i \in \mathcal{V}} (\delta_i - 4) + \frac{|\omega|}{2} \tag{11.45}$$

hence if for any diagram γ containing V the total counting must not change one has to lower the counting of the vertex V (from 4 to 2).

Collecting the contributions (11.42)-(11.44) we obtain (11.40); going backwards to (11.36) we have proved (11.34) i.e. the action principle (11.8) for the Wess-Zumino model.

The name "linear field equation" for (11.8) is obvious from (11.40): amputating the lines $A(1)...A(n)\bar{A}(1)...\bar{A}(m)$ one obtains the operatorequivalent of the classical field equation:

$$(\frac{1}{16}\bar{D}\bar{D}\bar{A} + \frac{m}{4}A)_{op} = (-\frac{z-1}{16}\bar{D}\bar{D}\bar{A} + \frac{a}{4}A)_{op} - \frac{\lambda}{16}N_2[A^2]_{op} \tag{11.46}$$

For the proof of (11.9) (e.g. with $A' \equiv A$) one proceeds exactly as for the previous case, i.e. one writes first all graphical possibilities of connecting the vertex $A(\frac{1}{16}\bar{D}\bar{D}\,\bar{A} + \frac{m}{4}A)$ with the remainder of the diagram and then goes through all consequences arising from use of the free propagator for the line carrying $\frac{1}{16}\bar{D}\bar{D}\,\bar{A} + \frac{m}{4}A$. The result, (11.9), we shall not prove in detail.

Section 12. Symmetric operators

Suppose we have proved for a theory that a (rigid) Ward-identity holds:

$$W_x\Gamma = 0 \qquad x \in \text{symmetry algebra} \tag{12.1}$$

and let λ be a parameter of Γ_{eff}. Then the action principle (11.5) defines via

$$\frac{\delta\Gamma}{\delta\lambda} = N_4^4\left[\frac{\delta\Gamma_{eff}}{\delta\lambda}\right]\cdot\Gamma \equiv [\Delta_\lambda]_4^4\cdot\Gamma \tag{12.2}$$

a unique insertion Δ_λ. Now act on (12.1) with $\partial/\partial\lambda$ and on (12.2) with W_x and form the difference:

$$\left[\frac{\delta}{\delta\lambda}, W_x\right]\cdot\Gamma = -W_x(\Delta_\lambda\cdot\Gamma) \tag{12.3}$$

If $\partial/\partial\lambda$ commutes with W_x we have

$$0 = W_x(\Delta_\lambda \cdot \Gamma) = W_x \Delta_\lambda + O(\hbar\Delta_\lambda) \tag{12.4}$$

(here we have used (10.9)) which expresses nothing but the invariance of the insertion Δ_λ under this rigid symmetry transformation W_x. We have thus found a criterion for the construction of symmetric operators: within a symmetric theory (12.1) any differential operator $\partial/\partial\lambda$ which commutes with the Ward-identity operator generates a symmetric insertion Δ_λ.

The simplest examples are provided by unbroken symmetries where the corresponding W_x does not depend on a parameter, e.g. in section 11 (Wess-Zumino model) derivatives with respect to m and g generate symmetric insertions whose tree approximation are just the mass, respectively the interaction terms. It is natural to ask, whether the third invariant - the kinetic term - can also be associated with a symmetric differential operator. The answer is yes: for all linear homogeneous Ward-identity operators of the type $W \equiv \int dx\, \Phi' \,\delta/\delta\Phi$, the leg counting operator

$$\mathcal{N} \equiv \int dx (\Phi' \frac{\delta}{\delta\Phi'} + \Phi \frac{\delta}{\delta\Phi}) \tag{12.5}$$

is symmetric (commutes with W). In the example (use (11.9) and (11.24))

$$\mathcal{N}\Gamma \equiv (\int dSA \frac{\delta}{\delta A} + d\bar{S}\bar{A} \frac{\delta}{\delta \bar{A}})\Gamma = [\mathcal{N}\Gamma_{eff}]^4_4 \cdot \Gamma \tag{12.6}$$

$$= [2\,\Delta_1 + 2\,\Delta_2 + 3\,\Delta_3]^4_4 \cdot \Gamma$$

In fact there is a one-to-one relation between symmetric insertions and symmetric differential operators.

In the case of spontaneously broken symmetries W depends on parameters and the definition of symmetric insertions is then greatly simplified by utilizing differential operators appropriately symmetrized by making their commutators with W vanishing (cf. below the O'Raifeartaigh model).

The above treatment is appropriate for rigid symmetries linear in Γ. The Γ-nonlinear BRS-symmetry (5.84) (5.85) enforces a slight modification.

We shall call an insertion Δ BRS-symmetric if

$$\mathcal{S}(\Gamma_\varepsilon^{(o)}) = O(\varepsilon^2) \tag{12.7}$$

$$\text{for } \Gamma_\varepsilon^{(o)} = \Gamma^{(o)} + \varepsilon\Delta \tag{12.8}$$

and $\mathcal{S}(\Gamma^{(o)}) = 0$.

According to (5.84) (α-gauge) we have

$$\mathcal{S}(\Gamma_\varepsilon^{(o)}) = \mathcal{S}(\Gamma^{(o)}) + \varepsilon\Big\{\frac{\delta\Delta\Gamma^{(o)}}{\delta\eta}\cdot\frac{\delta\Gamma^{(o)}}{\delta\Phi} + \frac{\delta\Gamma^{(o)}}{\delta\eta}\cdot\frac{\delta\Delta\Gamma^{(o)}}{\delta\Phi}$$

$$+\Phi\left(\frac{\delta}{\delta c_-}+\frac{\delta}{\delta c_-}\right)\Delta\Gamma^{(o)}+\left(\frac{\delta\Delta\Gamma^{(o)}}{\delta\sigma}\cdot\frac{\delta\Gamma^{(o)}}{\delta c_+}+\frac{\delta\Gamma^{(o)}}{\delta\sigma}\cdot\frac{\delta\Delta\Gamma^{(o)}}{\delta c_+}+\frac{\delta\Delta\Gamma^{(o)}}{\delta Y}\cdot\frac{\delta\Gamma^{(o)}}{\delta A}+\frac{\delta\Gamma^{(o)}}{\delta Y}\cdot\frac{\delta\Delta\Gamma^{(o)}}{\delta A}\right.$$

$$\left.+\text{ conj.}\right)\Big\} + O(\varepsilon^2)$$

$$= (\Gamma^{(o)} + \varepsilon\Big\{(\mathcal{B}_{\bar{\Gamma}^{(o)}} + \Phi(\frac{\delta}{\delta c_-} + \frac{\delta}{\delta\bar{c}_-}))\Delta\Gamma^{(o)}\Big\} + O(\varepsilon^2) \tag{12.9}$$

i.e. Δ is symmetric (in the α-gauge formulation) if

$$(\mathcal{b} + \mathrm{Tr}\int dV\ \Phi(\frac{\delta}{\delta c_-} + \frac{\delta}{\delta\bar{c}_-}))\ \Delta\Gamma^{(o)} = 0 \tag{12.10}$$

A somewhat stronger notion is also useful: we shall call Δ $\mathcal{b}$-symmetric if already

$$\mathcal{b}\Delta\Gamma^{(o)} = 0 \tag{12.11}$$

i.e. if an insertion is symmetric and does not contain the ghost field c_- it is $\mathcal{b}$-symmetric.

As far as methods are concerned let us remark that the use of an infinitesimal parameter ε was the means to linearize the action of $\mathcal{s}$ on the complete functional $\Gamma_\varepsilon^{(0)}$. It is no surprise then that we arrive at the operator $B_{\bar{\Gamma}(0)} \equiv \mathcal{b}$ as the one defining symmetry of an insertion. In fact, in more complicated cases (like the one treated in section 5.4) it is the above technique and the question "How does an insertion transform?" which leads to the correct $\mathcal{b}$-operator. I.e. it is not necessarily the requirement: "linearity" which defines that transformation law.

If Δ is a BRS symmetric insertion generated by a differential operator

$$\Delta \cdot \Gamma^{(0)} = \frac{\partial}{\partial\lambda}\Gamma^{(0)} \tag{12.12}$$

then (12.10) yields the definition of a BRS-symmetric differential operator

$$(\mathcal{b} + \mathrm{Tr}\int dV\Phi\,(\frac{\delta}{\delta c_-} + \frac{\delta}{\delta \bar{c}_-}))\,\frac{\partial}{\delta\lambda}\Gamma^{(0)} = 0 \tag{12.13}$$

and (12.11) the definition of a $\mathcal{b}$-symmetric differential operator:

$$\mathcal{b}\,\frac{\partial}{\partial\lambda}\Gamma^{(0)} = 0 \tag{12.14}$$

For the extension to higher orders one simply drops the superscript (o) and thus has as definitions for BRS-symmetric insertions:

$$(\mathcal{B}_{\bar{\Gamma}} + \mathrm{Tr}\int dV\Phi\,(\frac{\delta}{\delta c_-} + \frac{\delta}{\delta \bar{c}_-}))\Delta \cdot \Gamma = 0 \tag{12.15}$$

$$(\mathcal{B}_{\bar{\Gamma}} + \mathrm{Tr}\int dV\Phi\,(\frac{\delta}{\delta c_-} + \frac{\delta}{\delta \bar{c}_-}))\,\frac{\partial}{\partial\lambda}\Gamma = 0$$

and for $\mathcal{B}_{\bar{\Gamma}}$-symmetric:

$$\mathcal{B}_{\bar{\Gamma}}\,(\Delta \cdot \Gamma) = 0 \tag{12.16}$$

$$\mathcal{B}_{\bar{\Gamma}}\,(\frac{\partial}{\partial\lambda}\Gamma) = 0$$

For the B-gauge formulation let us only give the definitions of symmetry for insertions $\acute{\Delta}$ not depending on the field B.

Writing the equivalent of (12.9) and acting on it with $\delta/\delta B$ one finds that Δ has to satisfy the respective ghost equation of motion (cf. (5.89) (5.90)):

$$\frac{\delta\Delta}{\delta c_-} + \frac{1}{8}\bar{D}\bar{D}DD\frac{\delta\Delta}{\delta\rho} = 0 \qquad (\alpha,B)\text{-gauge} \qquad (12.17)$$

$$\frac{\delta\Delta}{\delta c_-} - \frac{1}{128}\bar{D}\bar{D}\frac{\delta\Delta}{\delta\rho} = 0 \qquad (B)\text{-gauge} \qquad (12.18)$$

and then Δ will be (BRS-) symmetric if it satisfies

$$\mathfrak{b}\Delta = 0 \qquad (12.19)$$

(i.e. B independence of Δ enforces the validity of the ghost equations and makes the notion of BRS-invariance coincident with that of $\mathfrak{b}$-invariance).

The extension to higher orders is obtained by replacing Δ by $\Delta\cdot\Gamma$ and $\mathfrak{b}$ by $\mathfrak{B}_{\bar{\Gamma}}$

As the final example of symmetric operator we should like to discuss operators invariant with respect to abelian gauge symmetry. In the α-gauge formulation we have derived the Ward-identities in SQED (4.69)

$$\begin{aligned} w_\Lambda\Gamma &= \frac{1}{8\alpha}(\Box + \alpha M^2)\,\bar{D}\bar{D}\Phi \\ w_{\bar{\Lambda}}\Gamma &= \frac{1}{8\alpha}(\Box + \alpha M^2)\,DD\Phi \end{aligned} \qquad (12.20)$$

The physical Hilbert space $\mathcal{H}_p$ in SQED is defined by that subspace of the entire indefinite metric Fock - space $\mathcal{H}$ whose states $|\text{phys}\rangle$ satisfy

$$(DD\Phi)^-\,|\text{phys}\rangle = 0 = (\bar{D}\bar{D}\Phi)^-\,|\text{phys}\rangle \qquad (12.21)$$

We therefore call on operator Q gauge invariant, if

(1) it commutes with $DD\Phi$ in $\mathcal{H}$:

$$[Q,DD\Phi] = 0 = [Q,\bar{D}\bar{D}\Phi] \tag{12.22}$$

and (2) its matrix elements in $\mathcal{H}_p$ are α-independent.

The requirements (12.21), (12.22) make sense since $DD\Phi$, $\bar{D}\bar{D}\Phi$ are free fields once the Ward-identities (12.20) are satisfied (cf. the discussion in section 4.). A sufficient condition on the gauge variation of an operator Q still guaranteeing (12.22) is the following:

$$w_\Lambda(1)Q(2) = P(2)\,(\Box + \alpha M^2)\delta_s(1,2) \tag{12.23}$$

(P denotes an operator) [III.22]. If (12.23) is satisfied with P = 0 we call the corresponding Q's symmetric operators. Examples will be given below in chapter V.

CHAPTER IV

RENORMALIZATION : HARD ANOMALIES

In the present chapter we answer the question which symmetries (out of the class of superconformal and gauge symmetries) can be established to all orders of perturbation theory, at least in the asymptotic (deep Euclidean) region. I.e. we are searching for hard anomalies. It will turn out that supersymmetry, rigid gauge invariance and R-symmetry are anomaly free, whereas the conformal symmetries, abelian gauge- and BRS-invariance have in general anomalies which are essentially supersymmetric extensions of the known anomalies in ordinary (non-supersymmetric) component theories. Although N = 1 supersymmetry links certain anomalies it does not prevent them as one might have guessed from the very special character which anomalies have as far as their symmetry properties are concerned.

The main tool of our study is the action principle established in the preceding chapter in combination with the algebraic technique of Becchi - Rouet - Stora (BRS) [IV.1, II.12]. Independent of any regularization and regardless of any specific subtraction scheme this technique permits to determine all possible anomalies of a symmetry. Anomalies are thereby understood as peculiar non-trivial solutions of a system of consistency conditions provided by the algebra of the Ward-identity operators. Once they are algebraically characterized in this way, usually a one-loop calculation suffices to determine their presence in a model - and a theorem has to guarantee absence to all orders (if feasible).

As concrete illustrations of the general theorems we shall treat the extension to higher order of the models presented in chapter II, in the tree approximation.

Section 13. Rigid symmetries

13.1 Consistency conditions, the algebraic technique

Suppose a set of (linear) field transformation laws is given

$$i\,[Q_x,\Phi]_\pm = \delta_x\Phi \qquad (13.1)$$

and the charges Q_x satisfy an algebra

$$[Q_x,\, Q_y]_\pm = i\, f_{xyz}\, Q_z \qquad (13.2)$$

Here Φ may be a superfield or an ordinary one; the algebra may be a Lie algebra with structure constants f_{xyz} or e.g. the supersymmetry algebra (1.4) - (1.6). We may then translate the transformations into the language of functional differential operators

$$W_x \equiv -i \int dz\, \delta_x\Phi \frac{\delta}{\delta\Phi} \qquad (13.3)$$

acting on Γ, the vertex functional, and having the same algebra as the charges

$$[W_x,\, W_y]_\pm = i\, f_{xyz}\, W_z \qquad (13.4)$$

The aim is now to construct order by order in the loop expansion the functional

$$\Gamma = \Gamma^{(0)} + \hbar\Gamma^{(1)} + \ldots \qquad (13.5)$$

with $\Gamma^{(0)}$ = classical action, such that

$$W_x\Gamma = 0 \qquad (13.6)$$

i.e. the Ward-identity holds for the respective symmetry. If we succeed in devicing such a Γ to all orders for the given set of W_x's we have perturbatively implemented the given symmetry to all orders.

In order to illustrate the systematic construction of such a Γ we look first of all into the one-loop approximation. Suppose

$$\tilde{\Gamma} = \tilde{\Gamma}^{(0)} + \hbar\, \tilde{\Gamma}^{(1)} \tag{13.7}$$

is the vertex functional as calculated up to one loop from a given $\Gamma^{(0)}$ which is symmetric:

$$W_x\, \Gamma^{(0)} = 0 \tag{13.8}$$

(Everything is finite due to our subtraction procedure.)

The action principle, section 11.2, combined with Zimmermann identities, section 10.3, tells us that

$$W_x \tilde{\Gamma} = \hbar \Delta_x \cdot \tilde{\Gamma} = \hbar \Delta_x + o(\hbar^2) \tag{13.9}$$

(the right-hand-side being of order $\hbar$ is a consequence of (13.8), also (10.9) has been used).

Applying W_y on (13.9), combining it with the equation for order $W_x W_y$ and using it also for W_z we find that the algebra (13.4) constitutes constraints on the possible insertions Δ_x [IV.1, IV.2]

$$W_x\, \Delta_y \pm W_y\, \Delta_x = i\, f_{xyz}\, \Delta_z + O(\hbar^2) \tag{13.10}$$

i.e. we have a <u>classical</u> equation to solve, namely finding the general solution Δ_x for (13.10).

Let us assume for the moment that for any local insertion Δ_x which solves (13.10) there exists a local Δ with

$$\Delta_x = W_x\, \Delta \tag{13.11}$$

then we can define $\Gamma^{(o)}_{eff} = \Gamma^{(o)} - \hbar\Delta$ i.e. have

$$\Gamma = \tilde{\Gamma} - \hbar\Delta \tag{13.12}$$

as our final vertex functional of order $\hbar$ and convince ourselves that indeed

$$W_x\Gamma = W_x\tilde{\Gamma} - \hbar\, W_x\Delta = O(\hbar^2) \tag{13.13}$$

By induction one can proceed to all orders. But the crucial assumption is (13.11): there may be solutions Δ_x of (13.10) which are not variations; there may be Δ_x's of the form (13.11) but having a Δ with power-counting degrees not permitted for a term of Γ_{eff} (i.e. either $\delta>4$ or $\rho<4$). In all of these cases we call such a Δ_x an anomaly for the corresponding Ward-identity (ultraviolet anomaly if Δ_x is not a variation or variation of a Δ with $\delta>4$; infrared anomaly if it is a variation but $\rho(\Delta)<4$). If furthermore an explicit computation shows in a model, that the anomaly has nonvanishing coefficient then the symmetry in question can never be restored in the ordinary loop expansion (of a renormalizable theory), i.e. there is no permissible choice of counterterms in Γ_{eff} which could compensate the corresponding breaking of symmetry.

13.2 Symmetry breaking

The case of spontaneous symmetry breaking is covered by the preceding subsection: shifts by constant amounts in certain fields do not alter the algebra of the Ward-identity operators. But due to the shifts the dimensional assignment of the insertions Δ_x changes and infrared anomalies may be picked up. Those have then to be treated in a specific way (cf. chapter V).

Explicit symmetry breaking can be reduced to the one of spontaneous with the help of a shifted external field $\hat{\eta}$ which transforms suitably (R. Stora in [R.14]). Suppose that on the classical level one has added to a symmetric action $\Gamma^{(o)}$ a breaking term B:

$$\Gamma = \Gamma^{(o)} + B \tag{13.14}$$

$$W_x\, \Gamma^{(o)} = 0 \qquad W_x\, \Gamma = W_x\, B \neq 0 \tag{13.15}$$

Then this action has to be converted into one which is invariant under the original transformation + such of an external field $\hat{\eta}$:

$$\Gamma^{(\eta)} = \Gamma^{(o)} + \int dz\hat{\eta}b \tag{13.16}$$

$$\hat{\eta} = \eta + \eta_0 \qquad \int dz \eta_0 b \equiv B$$

$$W_x^{(\eta)} = W_x + i\int dz \delta_x \hat{\eta} \frac{\delta}{\delta \hat{\eta}} \tag{13.17}$$

$$\delta_x \hat{\eta}\Big|_{\eta=0} = \beta$$

$$W_x^{(\eta)} \Gamma^{(\eta)} = 0$$

$$W_x^{\text{homog}} \Gamma^{(\eta)} - i\int dz_1 \delta_x \hat{\eta}(1) \frac{\delta}{\delta \hat{\eta}(1)} \int dz_2 \hat{\eta}(2) b(2) = 0 \tag{13.18}$$

$$\text{at} \quad \eta = 0 : \quad W_x \Gamma = -i\int dz \beta b = W_x B \tag{13.19}$$

The renormalization program for (13.18) is then exactly one of the preceding subsection and (13.19) renormalized will yield the renormalization of the breaking B [IV.1,3]. (For explicit examples cf. chapter V.)

13.3 Supersymmetry

Let there be given a set of superfields. They may be chiral, antichiral or vectorsuperfields, they may be propagating or external fields. Their F resp. D components may be shifted. Supersymmetry transformations are then (on functionals) generated by Ward-identity operators W_α ($\bar{W}_{\dot\alpha}$) which have a homogeneous as well as a shift part:

$$W_\alpha \equiv -i\int dz \left(\delta_\alpha \Phi_k \frac{\delta}{\delta \hat{\Phi}_k} + \delta_\alpha v_k \frac{\delta}{\delta \hat{\Phi}_k} = -i\int dz\, \delta_\alpha \hat{\Phi}_k \frac{\delta}{\delta \hat{\Phi}_k}\right. \tag{13.20}$$

$$\hat{\Phi}_k = \Phi_k + v_k \tag{13.21}$$

$$v_k = \theta^2 f_k \qquad \text{for shift in a chiral field}$$

$$v_k = \frac{1}{4}\theta^2 \bar{\theta}^2 d_k \qquad \text{for shift in a vector field}$$

They satisfy the supersymmetry algebra (1.4 - 1.6)

$$\{W_\alpha, \bar{W}_{\dot\alpha}\} = 2\sigma^\mu_{\alpha\dot\alpha} W^P_\mu$$
$$\{W_\alpha, W_\beta\} = 0 = \{\bar{W}_{\dot\alpha}, \bar{W}_{\dot\beta}\} \tag{13.22}$$

where

$$W^p_\mu = -i\int dz\,\partial_\mu\hat{\Phi}_k \frac{\delta}{\delta\hat{\Phi}_k}$$

is the translation Ward-identity operator.

If the classical theory we start with is strictly supersymmetric, i.e. $v_k = 0$ in (13.20), then the supersymmetry Ward-identities

$$W_\alpha\Gamma = 0 \qquad \bar{W}_{\dot\alpha}\Gamma = 0 \tag{13.23}$$

follow from the subtraction rules defined in Section 9, which preserve supersymmetry. For this one has to use an explicitly supersymmetric Γ_{eff} (expressed in terms of superfields), the terms of which have UV- and IR-dimension $d \leq 4$, $r \geq 4$, with associated subtraction degrees $\delta = \rho = 4$, and no massless field of dimension zero must be present.

In the general case of spontaneous or explicit breakdown ($v_k \neq 0$) in (13.20)) we have to work in field components and to consider the Ward-identity operators (13.20) in their component form. Suppose that finite diagrams, hence finite vertex functions are constructed according to some subtraction scheme, e.g. the BPHZ procedure [III.4,13] . We start with a Γ_{eff} (with subtraction degrees $\delta = \rho = 4$, and no massless field of dimension zero present) which coincides at $\hbar = 0$ with the classical action $\Gamma^{(o)}$ fulfilling the supersymmetric Ward-identities. Then, in higher orders, the action principle (11.6, 11.7) yields (at $s = 1$ if needed)

$$\begin{aligned} W_\alpha\Gamma &= \Delta_\alpha\cdot\Gamma = \Delta_\alpha + O(\hbar\Delta) \\ \bar{W}_{\dot\alpha}\Gamma &= \bar{\Delta}_{\dot\alpha}\cdot\Gamma = \bar{\Delta}_{\dot\alpha} + O(\hbar\bar{\Delta}) \end{aligned} \tag{13.24}$$

where use has been made of (10.9) in the right-hand-side. The power counting assignment of Δ_α is determined from the dimensions of W_α : its homogeneous part raises dimensions at most by 1/2 its inhomogeneous parts lowers at most by 3/2, hence

$$\delta(\Delta_\alpha) = \frac{9}{2} \qquad \rho(\Delta_\alpha) = \frac{5}{2} \tag{13.25}$$

i.e. in Δ_α all integrated monomials with one spinor index are permitted whose UV-dimension is smaller than or equal to 9/2, whose IR-dimension is greater than or equal to 5/2. Now, the algebra (13.22) together with (13.24) and translation invariance

$$W^p_\mu \Gamma = 0 \tag{13.26}$$

implies consistency conditions on the possible insertions Δ_α:

$$\begin{aligned} &W_\alpha \bar{\Delta}_{\dot\alpha} + \bar{W}_{\dot\alpha} \Delta_\alpha = 0 \ (\hbar\Delta) \\ &W_\alpha \Delta_\beta + W_\beta \Delta_\alpha = 0 \ (\hbar\Delta) \\ &\bar{W}_{\dot\alpha} \bar{\Delta}_{\dot\beta} + \bar{W}_{\dot\beta} \bar{\Delta}_{\dot\alpha} = 0 \ (\hbar\Delta) \end{aligned} \tag{13.27}$$

As explained in sect. 13.1 the quantum problem of satisfying a Ward-identity has been reduced to a classical problem. The solution of it is provided in the following theorem [IV.4].

<u>Theorem 13.3</u> Let

$$\begin{aligned} \Delta_\alpha &= \int dx \ \mathcal{P}^1_\alpha(x) \\ \bar{\Delta}_{\dot\alpha} &= \int dx \ \mathcal{P}^2_\alpha(x) \end{aligned} \tag{13.28}$$

be polynomials in the fields and let (13.28) satisfy the consistency conditions (13.27). Then the general solution Δ of

$$\begin{aligned} W_\alpha \Delta &= \Delta_\alpha \\ \bar{W}_\alpha \Delta &= \bar{\Delta}_{\dot\alpha} \end{aligned} \tag{13.29}$$

is given by

$$\Delta = \Delta_{sym} + \hat{\Delta} \tag{13.30}$$

where

$$W_\alpha \Delta_{sym} = \bar{W}_{\dot\alpha} \Delta_{sym} = 0 \tag{13.31}$$

and $\hat{\Delta}$ is a <u>local</u>, translation invariant functional of UV-dimension ≤ 4.

The IR-dimension $r(\Delta_\alpha)$ being $\geq 5/2$ (see (13.25)), $\hat{\Delta}$ can have terms of IR-dimension ≥ 2. Let us first suppose that $r(\hat{\Delta}) \geq 4$. Then, by the process described in Section 13.1 we may absorb $\hat{\Delta}$ into Γ_{eff} order by order in $\hbar$ and arrive at the desired Ward-identities (13.23). If on the other hand $\hat{\Delta}$ contains terms of IR-dimension less than 4, those cannot be absorbed and remain in the r.h.s. as <u>infrared anomalies</u>. We shall discuss such cases in Chapter V. We note that these anomalies are soft: they have UV-dimension less than 9/2, since $d \leq r$ for any insertion, hence $d(\Delta_\alpha) \leq r(\Delta_\alpha) < 9/2$ for $\Delta_\alpha = W_\alpha \hat{\Delta}$ with $r(\hat{\Delta}) < 4$. We can therefore write the result as

$$W_\alpha \Gamma \sim 0 \quad , \quad \bar{W}_{\dot{\alpha}} \Gamma \sim 0 \tag{13.32}$$

where $\sim$ means "up to terms of UV-dimension less than that of the other terms of the equations" - here 9/2. These terms are negligible in the deep Euclidean region of momentum space.

We shall encounter situations where supersymmetry is softly broken already in the tree approximation, e.g. by non-supersymmetric mass terms:

$$W_\alpha \Gamma^{(o)} = \text{mass contributions} \tag{13.33}$$

Our arguments apply in this case too, namely for the hard terms, and still lead to the softly broken Ward-identities (13.32) in higher orders.

Let us go through the list of our examples in order to familiarize ourselves with the scope of these statements.

13.3.1 Wess - Zumino model

The massive case has been treated in section 11 as example for the action principle (cf. (11,16)). So let us look at the massless limit achieved via the auxiliary mass term [I.6]

$$\Gamma_{M(s-1)} = \frac{1}{8} M(s-1) \left(\int dSA^2 + \int d\bar{S}\bar{A}^2\right). \tag{13.34}$$

Assigning r = d = 1 to the fields A and $\bar{A}$ all criteria (9.32) for the fields are satisfied, i.e. the Gell-Mann-Low formula (7.1) (9.17) yields finite Green's functions for non-exceptional momenta, if we use as subtraction degrees just the standard ones (9.1) (9.2)

$$\begin{aligned} \delta^{(\omega)}(\gamma) &= 4 - 2N_s(\gamma) + \sum_{i\in\mathcal{V}} (\delta_i - 4) + \tfrac{1}{2}|\omega| \\ \rho^{(\omega)}(\gamma) &= 4 - 2N_s(\gamma) + \sum_{i\in\mathcal{V}} (\rho_i - 4) + \tfrac{1}{2}|\omega| \end{aligned} \tag{13.35}$$

Choosing s = 1 at the end of any calculation is taking the massless limit. Since the subtraction procedure (section 9) maintains supersymmetry we need not rely on theorem 13.3 but know already from the supersymmetric form of the action principle (section 11.2) that we have strict supersymmetry:

$$W_\alpha \Gamma = 0 \qquad \bar{W}_{\dot{\alpha}} \Gamma = 0 \tag{13.36}$$

(Of course, with another subtraction scheme or in components one might use theorem 13.3; one then had to show that the soft anomalies are absent.)

13.3.2 O'Raifeartaigh model

The normalization conditions (3.45) - (3.49) and the symmetry requirements: spontaneously broken supersymmetry (3.41), R-invariance (3.42), parity (3.43), I-invariance (3.44), fixed uniquely the action in the tree approximation:

$$\begin{aligned} \Gamma^{(o)} = \frac{1}{16} \int dV\, \bar{A}_k A_k &+ \int dS \left(\frac{m}{4} A_1 A_2 + \frac{g}{32}(A_0 + f\Theta^2) A_1^{\,2}\right) \\ &+ \int d\bar{S} \left(\frac{m}{4} \bar{A}_1 \bar{A}_2 + \frac{g}{32}(\bar{A}_0 + f\bar{\Theta}^2) A_1^{\,2}\right) \end{aligned} \tag{13.37}$$

$(f = -\frac{4\xi}{g})$.

For the extension to higher orders one first has to check the degree requirements for convergence and one finds from propagators and vertices of the quadratic field part in $\Gamma^{(0)}$ that the assignment - in components:

$$d(A_k) = 1 \;, \quad d(\psi_k) = 3/2 \quad , \quad d(F_k) = 2$$

$$r(A_0) = r(A_2) = 1 \quad , \quad r(A_2) = 2 \tag{13.38}$$

$$r(\psi_k) = r(A_k) + 1/2 \;, \quad r(F_k) = r(A_k) + 1$$

together with the subtraction degrees:

$$\begin{aligned} \delta(\gamma) &= 4 - \sum_k d_k N_k + \sum_{i\in \mathcal{V}} (\delta_i - 4) \\ \rho(\gamma) &= 4 - \sum_k r_k N_k + \sum_{i\in \mathcal{V}} (\rho_i - 4) \end{aligned} \tag{13.39}$$

leads to finite diagrams (i.e. the criteria (9.32) are satisfied and the vertices satisfying (9.33) are permitted). Here we have added to (13.37) an auxiliary mass term for the multiplet A_0:

$$\Gamma_{M(s-1)} = \frac{M(s-1)}{8} \left(\int dSA_0^2 + \overline{\int dSA_0^2}\right) \tag{13.40}$$

(cf. [IV.5] for more details).

Let us now apply theorem 13.3 for the supersymmetry Ward-identity (3.41)

$$\begin{aligned} W_\alpha \Gamma &= [W_\alpha \hat{\Delta}]^{5/2}_{9/2} \cdot \Gamma \\ \bar{W}_{\dot\alpha} \Gamma &= [\bar{W}_{\dot\alpha} \hat{\Delta}]^{5/2}_{9/2} \cdot \Gamma \end{aligned} \tag{13.41}$$

and let us look at those terms of $\hat{\Delta}$ which are naively R-invariant.* Amongst those there is the monomial $\int dx\bar{A}_0 A_0$ (A_0 denotes the first component of the superfield A_0). This term has IR-dimensions 2, its supersymmetry variation $\int dx\psi_{o\alpha}\bar{A}_0$ has IR-dimension 5/2. Therefore $W_\alpha \int dx A_0\bar{A}_0$ is permitted in Δ_α, but $\int dx A_0\bar{A}_0$ cannot be absorbed in Γ_{eff} because of its IR-dimension: inserted into any (otherwise finite diagram) it causes IR-divergence

---

* It will be shown in subsection 13.5 that only R-invariant terms have to be considered.

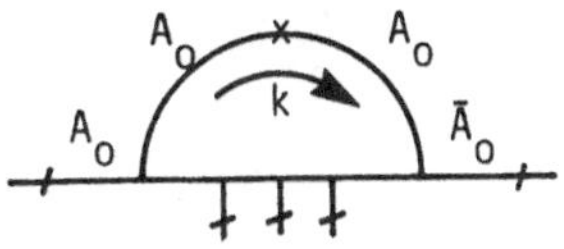

$\sim \int dk \, \frac{1}{k^2} \cdot \frac{1}{k^2} \, f(p,k)$

Fig. 13.3.2

with $f(p,k) \neq 0$ at $k = 0$. If now by explicit calculations the coefficient of $W_\alpha \int dx A_0 \bar{A}_0$ in (13.41) turns out to be non-zero, we have what we have called above an IR-anomaly: the supersymmetry Ward-identity cannot be proved to be strictly true (in the $\hbar$-expansion). This coefficient has been calculated [IV.5], it is proportional to $g^2 \xi^2/m^2 \neq 0$ (already in one loop). We shall in chapter V show how to deal further with this anomaly.

13.3.3 SQED

In the version with massive vector field (4.66) (4.69) one just performs UV-subtractions according to

$$\delta^{(\omega)}(\gamma) = 4 - 2N_s - 2N_v + \sum_{i \in \mathcal{V}} (\delta_i - 4) + \frac{|\omega|}{2} \tag{13.42}$$

with the (8.12, 8.13) for $|\omega|$, and has convergence and manifest supersymmetry.

But for the case of massless vector field we have pointed out in section 9 (cf. Fig. 9.2) that there are IR-diverent diagrams even in the gauge $\alpha = 1$, (4.32). A possible way out is (actually as in ordinary QED) just not to rely on the auxiliary mass term and subtractions with respect to $s - 1$ but instead to show that due to the specific couplings the UV-subtractions do not lead to IR-divergences. Technically easiest is to introduce formally the auxiliary mass:

$$\Gamma_{M(s-1)} = \frac{1}{16} M^2 (s-1)^2 \int dV \, \phi^2 \tag{13.43}$$

and to choose as IR-subtraction degree just the upper bound (9.28) for it:

$$\rho^{(\omega)}(\gamma) = \delta^{(\omega)}(\gamma) + 1 \tag{13.44}$$

Due to the specific form of the subtraction operator (9.20) one finds

$$\tau_{\gamma}^{|\omega|} = \tau_{p,s-1}^{\rho^{(\omega)}-1}(\gamma) + \tau_{p,s}^{\delta^{(\omega)}}(\gamma) - \tau_{p,s-1}^{\rho^{(\omega)}-1}\,\tau_{p,s}^{\delta^{(\omega)}}$$

$$= \tau_{p,s-1}^{\delta^{(\omega)}}(\gamma) + \tau_{p,s}^{\delta^{(\omega)}}(\gamma) - \tau_{p,s-1}^{\delta^{(\omega)}}\,\gamma_{p,s}^{\delta^{(\omega)}}$$

$$= \tau_{p,s}^{\delta^{(\omega)}}(\gamma) \tag{13.45}$$

i.e. at the end of the calculation, where one puts s = 1 one has effectively just performed the UV-subtractions. In this form one can now go through the criteria (9.28) - (9.31) and indeed prove convergence for the vertices of the tree action (4.67) and the gauge $\alpha = 1$. Supersymmetry is then again manifest, theorem 13.3 is not needed within the subtraction scheme of section 9 [IV.6].

13.3.4 S'QED

The tree action

$$\Gamma^{(0)} = \frac{1}{16}\int dV(\Phi(\square + M^2(s-1)^2)\Phi + \bar{A}_+ e^{g\hat{\Phi}} A_+ + A_- e^{-g\hat{\Phi}} \bar{A}_-)$$

$$- \frac{1}{4} m\,(\int dS\, A_+A_- + \overline{\int dS\, \bar{A}_-\bar{A}_+}) \tag{13.46}$$

$$\hat{\Phi} \equiv \Phi + \frac{1}{4}\,\theta^2\,\bar{\theta}^2\, v,$$

where we have already added an auxiliary mass term, tells one the degrees and dimensions to be assigned and one finds [IV.7] that the dimensions

	$A_\pm$	$\psi_\pm$	$F_\pm$	C	χ	M	v_μ	λ	D
d	1	3/2	2	0	1/2	1	1	3/2	2
r	2	3/2	2	0	1/2	1	1	3/2	2

Table 13.3.4

together with the subtraction degrees

$$\delta(\gamma) = 4 - \sum_a d_a N_a (\gamma)$$
$$\rho(\gamma) = 5 - \sum_a r_a N_a (\gamma) \qquad (13.47)$$

lead to convergent diagrams for the vertices of the tree action.*

Using now theorem 13.3 for $\Gamma_{int} = \Gamma_{int}^{(o)}$ it assures, that

$$W_\alpha \Gamma = \Delta_\alpha \cdot \Gamma = [W_\alpha \Delta]^{5/2}_{9/2} \cdot \Gamma \qquad (13.48)$$

but of course not, that Δ is made up from the vertices of $\Gamma_{int}^{(o)}$. In fact, if one wants to exclude infrared anomalies one has to go through the entire system of Ward-identities which serves to define the model. Supersymmetry alone is not sufficient to restrict appropriately Δ. We shall present this analysis below.

13.3.5 <u>SYM</u>

Due to the self-interaction of the vectorsuperfields the convergence problem is even more involved than in SQED or S'QED. Since for many considerations, in particular for the question whether in the Slavnov identity (hard) anomalies show up, masses are irrelevant, one may put mass terms for all fields irrespective of symmetries. One is then only interested in asymptotic supersymmetry:

$$W_\alpha \Gamma \sim 0 \qquad \bar{W}_{\dot\alpha} \Gamma \sim 0 \qquad (13.49)$$

and this is guaranteed by theorem 13.3.

---

* This non-supersymmetric assignment could also be used to construct SQED along the lines of S'QED. It would then provide a subtraction scheme using s-1 non-trivially, but also enforcing a Γ_{eff} which were not naively supersymmetric.

13.4 Rigid gauge invariance

Suppose a (semi-) simple Lie group G (e.g. SU(n)) is given, also a set of fields which transform under the adjoint representation and another set of fields which transform under a unitary transformation generated by Hermitean matrices T^i (i = 1,..., number of generators). We may then put the first set into matrices τ_i (generating the fundamental representation of G), write the second as vector and have a transformation law (5.19) (5.19a):*

$$\Phi \equiv \Phi^i \tau^i \qquad (\Phi^i = \mathrm{Tr}\, \tau^i \Phi) \tag{13.50}$$

$$A \equiv (A_a) \tag{13.51}$$

$$\delta_{rig}\, \Phi = i[\Phi,\omega] \qquad \omega \equiv \omega^i \tau^i \tag{13.52}$$

$$\delta_{rig}\, A = -i\, \tilde{\omega}\, A \qquad \tilde{\omega} \equiv \omega^i T^i \tag{13.53}$$

$$\delta_{rig}\, \bar{A} = i\, \bar{A}\, \tilde{\omega} \tag{15.54}$$

(ω^i real constant parameters)

The Ward-identity operator

$$W_\omega = -i \int dz\, \delta_{rig.}\varphi \frac{\delta}{\delta\varphi} \tag{13.55}$$

(sum over the fields φ, appropriate integration measure dz understood) satisfy then the Lie algebra (13.4) with f_{xyz} being the structure constants of the Lie algebra of G.

The main theorem governing the renormalization of models with the above symmetry has been proved by BRS [IV.1].

Theorem 13.4

There are no anomalies possible for rigid transformations belonging to a semi-simple Lie group G. I.e.

---

* For the subsequent considerations it is irrelevant whether the fields Φ,A are superfields or ordinary ones.

$$W_\omega \Gamma = 0 \tag{13.56}$$

In the case of spontaneous breaking there may arise infrared anomalies i.e. the Ward-identity of the spontaneously broken symmetry can only be proved up to a term with too low infrared dimension:

$$W_\omega \Gamma = [W_\omega \Delta] \cdot \Gamma \tag{13.57}$$

$\delta(\Delta) \leq 3$, $\rho(\Delta) = 2$ or 3; hence Δ cannot be absorbed in Γ_{eff} [IV.8].

Explicit breaking reveals the content of Symanzik's theorem [IV.3]: breaking terms of ultraviolet dimension less than four only yield breakings of the same dimension.

Their algebraic properties are explained in [R.14] (R. Stora).

13.5 R-invariance

The transformations

$$i[R,\varphi] = \delta_R\varphi \equiv i(n + \Theta\partial_\Theta + \bar{\Theta}\partial_{\bar{\Theta}})\varphi \tag{13.58}$$

where n is a real number, the "R-weight" of φ, have a non-trivial commutator with supersymmetry

$$[R,Q_\alpha] = Q_\alpha \qquad [R,\bar{Q}_{\dot{\alpha}}] = -\bar{Q}_{\dot{\alpha}} \tag{13.59}$$

and appear even on the right-hand side of the superconformal algebra (cf. section 2). Indeed, the latter closes on real superfields Φ with $n(\Phi) = 0$, on chiral superfields A with $n(A) = -2/3\, d(A)$, where d(A) is the dilatational weight of A.

Acting with

$$W_R \equiv -i\int dz\, \delta_R\varphi \frac{\delta}{\delta\varphi} \tag{13.60}$$

on Γ we obtain via the action principle (section 11.2)

$$W_R \Gamma = \Delta_R \cdot \Gamma = \Delta_R + O(\hbar\, \Delta_R)$$

$$\delta(\Delta_R) = \rho(\Delta_R) = 4 \tag{13.61}$$

(we have also used (10.9)).

Now we observe that

$$\Delta_R^{(i)} = \frac{1}{n_i} W_R\, \Delta^{(i)} \tag{13.62}$$

for any non-invariant integrated insertion $\Delta_R^{(i)}$ (n_i: R-weight of $\Delta^{(i)}$), thus we have

$$W_R \Gamma = \Delta_R^{inv} + \sum_i \frac{a_i}{n_i} W_R\, \Delta^{(i)} + O(\hbar\, \Delta_R) \tag{13.63}$$

Testing with respect to hard invariants one finds zero for their coefficients, so it remains to absorb all absorbable term $\Delta^{(i)}$. Assuming - which is the case for all our specific examples - that the tree action has only soft R-breakings, we may absorb at least the terms Δ^i of UV-dimension 4. Soft breaking terms (i.e. terms with UV-dimension strictly less than 4) may not be absorbable because they may cause conflict with normalization conditions. We formulate this result as a theorem.

Theorem 13.5

The R-symmetry is at most softly broken. I.e.

$$W_R \Gamma \sim 0 \tag{13.64}$$

To illustrate the theorem in concrete cases let us look into our standard examples.

13.5.1 Wess-Zumino model

For the massive version (11.18) and the action principle (11.9) yield explicitly

$$W_R\Gamma = [W_R\,\Gamma_{eff}]_4 \cdot \Gamma$$

$$= [-\frac{a}{8}(2n+2)\,(\int dSA^2 - \int d\overline{SA^2}) \qquad (13.65)$$

$$+ \frac{\lambda}{48}(3n+2)\,(\int dSA^3 - \int d\overline{SA^3})]_4 \cdot \Gamma$$

i.e. the superconformal choice for the weight

$$n = -\frac{2}{3}\,d(A) = -\frac{2}{3}$$

makes disappear the second term, whereas the first one can be reduced to a soft term + corrections by combining the Zimmermann identities (10.17) and (10.18):

$$W_R\Gamma = -\frac{a}{12}[\int dSA^2 - \int d\overline{SA^2}]_3 \cdot \Gamma$$
$$+ \frac{a}{12m} r_2 [\int dSA^3 - d\overline{SA^3}]_4 \cdot \Gamma \qquad (13.66)$$

(the r_1 terms combine in $W_R\,\Gamma$ to total divergences and drop out in the integration). In order to determine r_2 we test with respect to $\frac{\delta^3}{\delta A(1)\delta A(2)\delta A(3)}$ and obtain

$$-2\,\Gamma_{123} + \sum_{i=1}^{3}\frac{\partial}{\partial\Theta_i}(\Theta_i\Gamma_{123}) = -\frac{a}{12}[\int d\overline{SA^2} - \int d\overline{SA^2}]_3 \cdot \Gamma$$
$$+ \frac{ar_2}{2m}\,\delta_S(1,2)\delta_S(1,3) + O(\hbar r_2) \qquad (13.67)$$

Recalling that the higher Θ-components of Γ_{123} in momentum space are of the form $\Theta_{12}^2\Theta_{13}^2\,\gamma(p_2,p_3)$ (cf. (8.11, 8.13))we find by differentiating with respect to $(\Theta/\partial\Theta_2)^2\,(\partial/\partial\Theta_3)^2$ zero on the left-hand side:

$$0 = -\frac{a}{12}\left(\frac{\partial}{\partial\Theta_2}\right)^2\left(\frac{\partial}{\partial\Theta_3}\right)^2[\int dSA^2 - \int d\overline{SA^2}]_3 \cdot \Gamma_{AAA}$$
$$+ 8\,\frac{a}{m}\,r_2 + O(\hbar r_2)\ . \qquad (13.68)$$

The insertion on the right-hand side is soft, hence does not contribute at large momenta, hence $r_2 = 0$. This argument via asymptotic behaviour [IV.9] shows that the effect of the mass is negligible in the asymptotic region and that there is no hard anomaly and this is the statement of (13.64). In fact this can independently be checked by formulating with the help of the auxiliary mass immediately the massless theory [I.6]. There the vanishing of the coefficient of the hard insertion is an immediate consequence of the normalization conditions.

For the massless Wess-Zumino model we therefore have strict (superconformal) R-invariance:

$$W_R \Gamma \Big|_{s=1} = 0 \tag{13.69}$$

13.5.2 O'Raifeartaigh model

The general argument above for establishing R-invariance is needed, the consistency conditions amongst supersymmetry and R:

$$W_R \Delta_\alpha - W_\alpha \Delta_R = \Delta_\alpha + O(\hbar\Delta) \tag{13.70}$$

restrict then Δ_α to an insertion with a definite R-weight:

$$W_R \Delta_\alpha = \Delta_\alpha + O(\hbar\Delta) \tag{13.71}$$

hence only those have to be considered in the solution of (13.27). Only in the R-invariant sector IR-anomalies are possible, since Δ_R was a $\delta = \rho = 4$ insertion.

To avoid confusion we state again the result:

$$W_R^{(n)} \Gamma \Big|_{s=1} = 0 \tag{13.72}$$

for $n(A_0) = n(A_2) = -2$, $n(A_1) = 0$. For superconformal weights we have

$$W_R \Gamma \Big|_{s=1} \sim 0 \tag{13.72a}$$

13.5.3. SQED

On the basis of a Γ_{eff} which is just made up from the terms occurring in the tree approximation it is clear that for

$n(A_+) = n(A_-) = -1$ one has strict R-invariance to all orders

$$W_R^{(n=-1)} \Gamma = 0 \,, \tag{13.73}$$

since Γ_{eff} is naively invariant. For other choices of weights the matter mass term is not R-invariant, has to be expanded in a basis of N_4 products and that will be determined by supersymmetry and gauge invariance (cf. next subsection). It will turn out that the correction terms have to be gauge invariant i.e. are just the tree approximation terms which are R-invariant and then do not contribute:

$$W_R^{(n)} \Gamma = -\frac{n+1}{2} (m+a) \, [\int dSA_+A_- - \int d\bar{S}\bar{A}_-\bar{A}_+]_3 \cdot \Gamma \tag{13.73a}$$

The subtleties really involved in this procedure will be pointed out below.

13.5.4 S'QED

Although R-invariance (with weights $n(A_+) = n(A_-) = -1$) is an accidental symmetry in the tree approximation it plays a vital role for the construction of higher orders. Since it is intricately related to the other symmetries we shall discuss it below in the next section where abelian gauge invariance is considered.

13.5.5 SYM

R-invariance with conformal weights is a very valuable tool in order to restrict the counter-terms which are needed for establishing BRS-invariance. One puts general mass terms to render the theory completely massive and then proves asymptotic R-invariance along the lines of the main section.

Section 14: Abelian gauge invariance

14.1 SQED

We noted already in section 13.3.3 that for the massless vector case finite diagrams can be constructed with

$$\Gamma_{eff} = \int dSL + \int d\overline{SL}$$

$$L = L_1 + \frac{1}{\alpha} L_2 + (m+a) L_3 + L_4$$

$$L_1 = - \frac{1}{2\cdot 128} N_3 [\overline{DDD}\Phi\ \overline{DDD}\Phi],\quad L_2 = - \frac{1}{2} \frac{1}{128} N_3 [\overline{DDDD}\Phi\ \overline{DD}\Phi]$$

$$L_3 = - \frac{1}{4} N_3 [A_+A_-] \qquad L_4 = \frac{1}{2} \overline{DD} I_5$$

$$I_5 = \frac{1}{16} N_2 (\bar{A}_+ e^{g\Phi} A_+ + A_- e^{-g\Phi} \bar{A}_-) \tag{14.1}$$

Introducing

$$\Delta_k = \int dSL_k + \int d\bar{S}L_k \tag{14.2}$$

we may write Γ_{eff} as the sum

$$\Gamma_{eff} = \Delta_1 + \frac{1}{\alpha} \Delta_2 + (m+a)\Delta_3 + \Delta_4 \tag{14.3}$$

But convergence was only ensured for $\alpha = 1$ and having as interaction vertices just Δ_3 and Δ_4, Δ_3 being the oversubtracted mass term. Let us recall that we used as subtraction degrees

$$\delta^{(\omega)}(\gamma) = 4 - 2N_2(\gamma) - 2N_v(\gamma) + \sum_{i\in\mho} (\delta_i - 4) + \frac{|\omega|}{2}$$

$$\rho^{(\omega)}(\gamma) = \delta^{(\omega)}(\gamma) + 1 \tag{14.4}$$

(range (8.12) for ω) and assigned in the check of the criteria the infrared dimension 3/2 to the superfields $A_\pm$, 0 to Φ.

The supersymmetric and (up to Δ_2) gauge invariant form of Γ_{eff} implies via the action principle (11.8), (11.9) that supersymmetry is maintained

$$W_\alpha \Gamma = 0 \qquad \bar{W}_{\dot\alpha} \Gamma = 0 \tag{14.5}$$

and also that the gauge Ward-identities hold in their naive form

$$\begin{aligned} w_\Lambda \Gamma &\equiv (\overline{DD}\frac{\delta}{\delta\Phi} - g\, A_+ \frac{\delta}{\delta A_+} + g\, A_- \frac{\delta}{\delta A_-})\Gamma = \frac{1}{8}\,\Box\,\overline{DD}\,\Phi \\ w_{\bar\Lambda} \Gamma &\equiv (DD\frac{\delta}{\delta\bar\Phi} - g\, \bar{A}_+ \frac{\delta}{\delta \bar{A}_+} + g\, \bar{A}_- \frac{\delta}{\delta \bar{A}_-})\Gamma = \frac{1}{8}\,\Box\,DD\,\bar\Phi \end{aligned} \tag{14.6}$$

But what is needed for deriving the Ward-identity of R-symmetry and also e.g. for the Callan-Symanzik-equation is slightly more than the above. We need a Zimmermann identity relating the <u>over</u>subtracted mass term Δ_3 to its minimally subtracted normal product

$$\Delta_m := -\frac{1}{4}\, N_3[\int dSA_+A_- + \int d\bar{S}\bar{A}_+\bar{A}_-] \equiv \int dSL_m + \int d\bar{S}\bar{L}_m \tag{14.7}$$

and this does not exist! Indeed with its counting $\delta = 3$ we find for its diagrams:

$$\delta^{(\omega)}(\gamma) = 4 - 2N_s - 2N_v + (3-4) + \frac{|\omega|}{2} \tag{14.8}$$

for instance with $N_s = 0$, $N_v = 2$ (in Fig. (14.1)

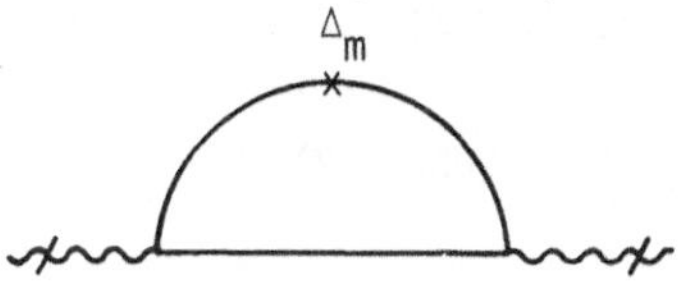

$$\delta^{(\omega)}(\gamma) = -1 + \frac{|\omega|}{2}$$

Fig. 14.1

therefore its $\omega = 0$ - the $\Delta_m \cdot \Gamma_{DD}$ - component is not subtracted and will thus cause IR-divergence, e.g. in a diagram like Fig. 14.2

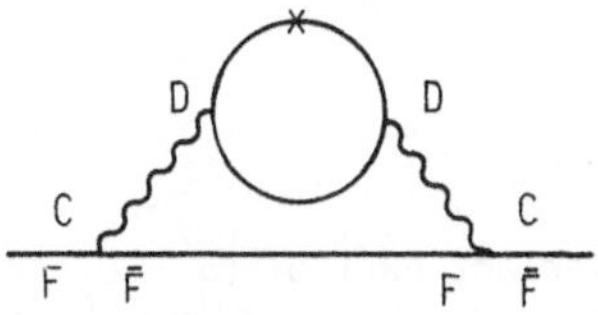

Fig. 14.2

(similar to Fig. 9.3). Now a IR-divergence of an integrated insertion is best studied by looking into diagrams containing its non-integrated counterpart. So one might try to establish a Zimmermann identity between $N_3\ [A_+A_-]$ and $N_2\ [A_+A_-]$; both being formally gauge invariant one expands in a basis of gauge invariant, chiral insertions of dimension 3. A member of this basis is L_1 and again one has trouble: it does not exist either. In the diagram of Fig. 14.3 one has also

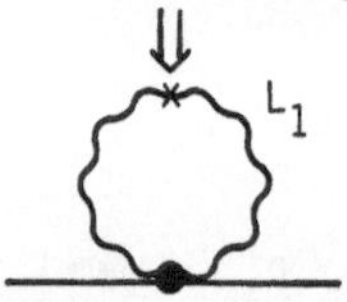

$$\delta^{(\omega)} = -1 + \frac{|\omega|}{2}$$

Fig. 14.3

and $\Omega = 2$. Hence the Ω-component is subtracted (at $p = 0$) and the subtraction term contains $1/(k^2)^2$ i.e. is IR-divergent. By explicit calculation one finds that the unsubtracted Ω-component does not vanish, has no factor k and thus the problem is real.

To overcome this difficulty [IV.6] one first goes over from L_1 to $L_1 + L_2$ in the basis: an insertion which is not gauge invariant in the sense of section 12 (cf. (12.23)):

$$\begin{aligned} w_\Lambda(1)\ (L_1 + L_2)\ (2)\cdot\Gamma &= \frac{1}{16}\ \Box\,\delta_S(1,2)\ \overline{DD}\ \Phi(2) \\ w_{\bar{\Lambda}}(1)\ (L_1 + L_2)\ (2)\cdot\Gamma &= \frac{-1}{256}\ \overline{DD}\ \delta_{\bar{S}}\ (1,2)\ \overline{DD}DD\Phi\ (2) \end{aligned} \tag{14.9}$$

but exists. Then one defines an insertion without any subtraction

$$L_0 = \overline{DD}\,\mathcal{L}_0 = -\frac{1}{256}\ N_0[\overline{DD}DD\Phi\ \overline{DD}\Phi],\ \mathcal{L}_0 = -\frac{1}{256}\ N_0[DD\Phi\ \overline{DD}\Phi] \tag{14.10}$$

and chooses as basis elements of chiral, gauge invariant N_3-normal products:

$$L_{kin} = L_1 + L_2 - L_0, L_3, L_4, \overline{DDL}_m \tag{14.11}$$

L_{kin} is gauge invariant because the non-invariant part of $L_1 + L_2$ is exactly compensated by L_0 (their variation is linear in the quantized field, hence there is no collision with renormalization) and it exists provided $L_1 + L_2$ and L_0 exist. This is shown in [IV.6].

We now have a Zimmermann identity

$$(m+a)L_m = (m+a)L_3 + r_k L_{kin} + r_g L_4 + r_m \overline{DDL}_m \tag{14.12}$$

and also its conjugate

$$(m+a)\bar{L}_m = (m+a)\bar{L}_3 + r_k \bar{L}_{kin} + r_g \bar{L}_4 + r_m DDL_m \tag{14.13}$$

with real coefficients r_k, r_g, r_m, which are, of course, finite. If we now try to integrate (14.12) with the chiral measure we obtain on the left-hand side still a divergent quantity, but on the right-hand side too: since

$$\int dS(L_1 + L_2) = \frac{1}{32} N_4[\int dV \ \Phi \Box \Phi] \tag{14.14}$$

which exists, the divergence must come from the integrated L_0 part. Hence

$$\Delta_m^{*} = \Delta_m + \frac{r_k}{m+a} N_0[\int dSL_0 + \int d\overline{SL}_0] \tag{14.15}$$

exists. This insertion will show up below in the Callan-Symanzik equation.

In order to establish the R-Ward-identity (13.73a) one needs the difference of the appropriately integrated Zimmermann identities (14.12, 14.13). And in this difference all correction terms - including those coming from L_0 - compensate:

$$N_3[\int dSL_m] - N_3[\int d\overline{SL}_m] = N_4[\int dSL_2] - N_4[\int d\overline{SL}_2] \tag{14.16}$$

hence (13.73a).

Let us conclude this discussion of the global aspects (we have not yet introduced currents) with a remark on the α-independence of Green's functions. Since we have convergence only for $\alpha = 1$ it obviously makes no sense to consider $\partial/\partial\alpha$ in the theory with massless vector field.

In the completely massive theory $\partial/\partial\alpha$ is the variation of a parameter governed by the action principle and this we shall still present.

We enlarge Γ_{eff} (13.74) to

$$\Gamma_{eff} = z_1\Delta_1 + \frac{1}{\alpha}\Delta_2 + (m+a)\,\Delta_3 + z_4\Delta_4 + M^2\Delta_5$$
$$\Delta_5 = \int dSL_5 + \int d\overline{SL}_5 \qquad L_5 = \frac{1}{32} N_3[\overline{DD}\,\Phi^2] \tag{14.17}$$

and use intermediate normalization conditions for the vector mass (normalization at zero momentum).

The action principle (11.5) on the functional Z yields for the variation of α:

$$\frac{\partial Z}{\partial\alpha} = [\frac{\partial\Gamma_{eff}}{\partial\alpha}]_4 \cdot Z$$
$$= [\partial_\alpha z_1\Delta_1 - \frac{1}{\alpha^2}\Delta_2 + \partial_\alpha a\Delta_3 + \partial_\alpha z_4\Delta_4]Z \tag{14.18}$$

With the help of the Zimmermann identity

$$\mathcal{L}_0 = \mathcal{L}_2 + r\,\mathcal{L}_4 + r_m(\mathcal{L}_m + \mathcal{L}_m) \tag{14.19}$$

where $\mathcal{L}_2 = -\frac{1}{2\alpha}\frac{1}{128} N_2\,[DD\Phi\;\overline{DD\Phi}]$, $\mathcal{L}_4 = \frac{1}{2} I_5$, $\mathcal{L}_m = -\frac{1}{4} N_2[A_+A_-]$

which after dV-integration yields

$$\Delta_0 = \Delta_2 + r\,\Delta_4 \tag{14.20}$$

(14.18) becomes

$$\partial_\alpha Z = [\partial_\alpha z_1\Delta_1 - \frac{1}{\alpha^2}\Delta_0 + \partial_\alpha a\Delta_3 + (\partial_\alpha z_4 + \frac{r}{\alpha^2})\Delta_4] \cdot Z \tag{14.21}$$

Choosing now

$$z_1 = 1 \qquad \partial_\alpha a = 0 \qquad z_4 = -\int_1^\alpha d\beta\,\frac{r}{\beta^2} + 1 \tag{14.21}$$

we obtain

$$\partial_\alpha Z = -\frac{1}{\alpha^2}\Delta_0 \cdot Z \tag{14.23}$$

This shows that the physical S-matrix is α-independent since the α-variation is an unphysical insertion, Δ_0, vanishing between physical states.

For later use we note a generalization: the α-independence of physical matrix elements of an operator Q is conveniently expressed by

$$\partial_\alpha(Q\ Z) = -\frac{1}{\alpha^2}\Delta_0\cdot Q\cdot Z \tag{14.24}$$

14.2 S'QED

Above in section 13.3.4 we have already given the subtraction degrees which produce finite diagrams. We have now to construct a Γ_{eff} such that to all orders the symmetries of the model are established.

Although abelian symmetries are seemingly "simpler" than non-abelian ones they are definitely less restrictive than those as far as their cohomology properties are concerned: consistency conditions of the type

$$w(1)\ Q(2) - w(2)\ Q(1) = 0 \tag{14.25}$$

for an abelian (local) Ward-identity operator w do in general not imply that Q is a variation. This is the case for non-abelian symmetries (based on a semi-simple group) (cf. sections 13.4 and 15). Hence in determining Γ_{eff} one has to restrict it suitably from the very beginning and then to show that the restricted Γ_{eff} is nevertheless rich enough to prove the desired symmetries. "Rich enough" means that it provides all the counter-terms needed for the symmetries which are not naively maintained by the subtraction scheme. At this point R-invariance becomes useful: it is homogeneous in the fields i.e.

$$W_R\Gamma = [W_R \Gamma_{eff}]_4^4 \cdot \Gamma \tag{14.26}$$

and we may and will choose a Γ_{eff} which is naively R-invariant (weights: $n(A_+) = n(A_-) = -1$, $n(\Phi) = 0$).

Similarly, amongst the gauge symmetries connected with the superfield

$$\Lambda = A + \Theta\psi + \Theta^2 F \tag{14.27}$$

the symmetry generated by A(x) is homogeneous with respect to the IR-, UV-dimensions of the fields and thus Γ_{eff} may be chosen naively A (and $\bar{A}$) invariant. (The variation of the breaking term which contains s-1, cf. (13.46), is linear in the quantized field, hence not dangerous.) As far as the other symmetries are concerned subtraction degrees differ for different fields hence non-naive use of the action principle has to be made. As example consider the gauge variation

$$\delta_\psi \psi_+ = -\, ig\, A_+ \qquad \delta_\psi \chi = i\psi$$
$$(r(\psi_+) = 3/2,\ r(A_+) = 2,\ r(\chi) = 1/2)) \tag{14.28}$$

Supersymmetry is anyway inhomogeneous.

As far as discrete symmetries are concerned one can only rely on CP since parity was broken "by hand" in the tree approximation, D being pseudo-scalar.

The remaining task is now to write down the most general Γ_{eff} satisfying CP, R-invariance, A-($\bar{A}$-) invariance and to check that the Ward-identities

$$W_\alpha \Gamma = 0\ , \qquad \bar{W}_{\dot\alpha}\Gamma = 0 \tag{14.29}$$

$$w_F\Gamma = i\Box \bar{M}\ , \quad w_\psi \Gamma = -\frac{i}{2}\Box(\lambda - i\partial\!\!\!/\bar{\chi}) \tag{14.30}$$

can be satisfied at s=1. At our disposal for restricting the insertions Δ_α, $\bar{\Delta}_{\dot\alpha}$, Δ_ψ, Δ_F we have the algebraic relations:

$$
\begin{aligned}
&[W_R, W_A] = 0 \\
&[W_R, W_\psi] = - W_\psi \\
&[W_R, W_F] = - 2W_F
\end{aligned}
\qquad (14.31)
$$

the algebra of supersymmetry (cf. (13.28) and its solution) and also

$$
\begin{aligned}
&[W_\alpha, W_\Lambda] = iW_{\delta_\alpha \Lambda} \\
&[\bar{W}_{\dot{\alpha}}, W_\Lambda] = iW_{\bar{\delta}_{\dot{\alpha}} \Lambda}
\end{aligned}
\qquad (14.32)
$$

for the gauge transformation Λ; $\delta_\alpha \Lambda$ denotes gauge transformations with supersymmetry transformed parameters.

It has been shown in [IV.7] that these restrictions are strong enough to make all insertions Δ_x occurring in the right-hand side of the Ward-identity $W_x \Gamma$ to variations and that the coefficients of all IR-anomaly candidates vanish. Thus, there remains an entire massless vector superfield to all orders and supersymmetry breaking only affects the mass splitting in the matter multiplets.

Section 15. Non-Abelian gauge invariance

15.1 Statement of the problem

In Section 5 we have introduced non-Abelian gauge transformations in the context of supersymmetric theories, studied the most general transformation law and converted gauge transformations into Becchi-Rouet-Stora (BRS) transformations. We were led to define the theory by the Slavnov identity (5.85) - we shall work now in the (α, B) gauge defined by the condition (5.86) - and found that due to the vanishing dimension of the superfield Φ, already in the tree approximation the theory admits the replacement (5.115)

$$
\Phi_i \rightarrow \Phi_i = \mathcal{F}_i(\Phi) = \Phi_i \sum_{k \geq 2} \sum_{\omega=0}^{\Omega(k)} c_{\omega k} \, s^{\omega}_{i(i_1 \ldots i_k)} \Phi_{i_1} \ldots \Phi_{i_k} \qquad (15.1)
$$

with arbitrary coefficients $c_{\omega k}$. We had to fix the latter in terms of infinitely many parameters $a_{\omega k}$ by the normalization conditions (5.182). We have also shown that the $a_{\omega k}$'s are gauge parameters like α, i.e. the physical quantities do not depend on them provided the enlarged Slavnov identity (5.165)

$$\begin{aligned}
\mathcal{S}(\Gamma) \equiv \mathrm{Tr} \int dV \frac{\delta\Gamma}{\delta\rho}\frac{\delta\Gamma}{\delta\Phi} \\
+ \int dS \, \{\mathrm{Tr}\, [B \frac{\delta\Gamma}{\delta c_-} + \frac{\delta\Gamma}{\delta\sigma}\frac{\delta\Gamma}{\delta c_+}] + \frac{\delta\Gamma}{\delta Y}\frac{\delta\Gamma}{\delta A}\} \\
+ \int d\bar{S} \, \{\mathrm{Tr}\, [\bar{B} \frac{\delta\Gamma}{\delta\bar{c}_-} + \frac{\delta\Gamma}{\delta\bar{\sigma}}\frac{\delta\Gamma}{\delta\bar{c}_+}] + \frac{\delta\Gamma}{\delta\bar{Y}}\frac{\delta\Gamma}{\delta\bar{A}}\} \\
+ \chi\partial_\alpha\Gamma + \sum_{\omega,k} x_{\omega k}\, \partial_{a_{\omega k}} \Gamma = 0
\end{aligned} \tag{15.2}$$

holds, where χ, $x_{\omega k}$ are anticommuting parameters, the (α,B)-gauge condition (5.86) being modified into

$$\frac{\delta\Gamma}{\delta B} = \alpha\, \overline{DD}\, \bar{B} + \frac{1}{8}\, \overline{DD}DD\, \Phi + \frac{1}{2}\, \chi\overline{DD}\, \bar{c}_- \tag{15.3}$$

To simplify the problem we have required rigid gauge invariance (cf. the last remark in Section 5).

$$W_\omega \Gamma \equiv - i\int dz\delta_{rig}\, \varphi \frac{\delta\Gamma}{\delta\varphi} = 0 \tag{15.4}$$

with $\delta_{rig}\varphi = i[\varphi,\omega]$ (15.5)

for $\varphi = \Phi, c_\pm, \rho, \eta, \sigma, A, Y$

$$\varphi \equiv \varphi^i\tau^i \qquad \omega = \omega^i\tau^i$$

τ^i: generators of the fundamental representation; but with

$$\begin{aligned}
\delta_{rig} A = -i\tilde{\omega}A \qquad \tilde{\omega} \equiv \tilde{\omega}^i T^i \\
\delta_{rig} Y = iY\tilde{\omega}
\end{aligned} \tag{15.6}$$

if the matter (A_a) transforms under a unitary transformation generated by T^i.

As a consequence the coefficients $s^{\omega}_{i(i_1...i_k)}$ in (15.3) are invariant tensors under the rigid group and completely symmetric in the indices $(i_1...i_k)$. We have first written the classical solution of (15.2) at $x_{\omega k} = a_{\omega k} = 0$ (5.78, 5.81, 5.73, 4.35, 5.20, 5.172)

$$\Gamma_s^{(o)}(\Phi, c_+, A, \rho, \sigma, Y, c_-, B) = \bar{\Gamma}_s^{(o)}(\Phi, c_+, A, \eta, \sigma, Y)$$

$$+ \operatorname{Tr} \int dV \, [B \, DD\Phi + \bar{B}\, \overline{DD}\Phi + \alpha B\bar{B} + \frac{1}{2}\chi(c_-\bar{B} + \bar{c}_- B)] \tag{15.7}$$

$$\bar{\Gamma}_s^{(o)} = -\frac{1}{128 g^2} \operatorname{Tr} \int dS \, [\bar{D}\bar{D}(e^{-\Phi} D e^{\Phi})]^2 + \frac{1}{16}\int dV \, \bar{A} e^{\Phi} A$$

$$+ \int dS \, [m_{ab} A_a A_b + h_{abc} A_a A_b A_c] + \text{c.c.}$$

$$+ \operatorname{Tr} \{\int dV \eta \, Q_s(\Phi, c_+) - \int dS \, \sigma \, c_+ c_+ - \int d\bar{S} \, \bar{\sigma} \, \bar{c}_+ \bar{c}_+\}$$

$$- \int dS \, Y \, \tilde{c}_+ A - \int d\bar{S} \, \bar{A} \, \tilde{\bar{c}}_+ \bar{Y} \tag{15.8}$$

where

$$\psi = \psi_i \tau^i \, , \; \tilde{\psi} = \psi_i T^i \quad \text{for} \quad \psi = \Phi, c_+, c_-, B, \rho, \sigma$$

$$\eta = \rho - \frac{1}{8}(DD \, c_- + \overline{DD} \, \bar{c}_-) \tag{15.9}$$

and

$$Q_s(\Phi, c_+) = c_+ - \bar{c}_+ + \frac{1}{2}[\Phi, c_+ + \bar{c}_+] + \frac{1}{12}[\Phi, [\Phi, c_+ - \bar{c}_+]]$$

$$+ \dots \tag{15.10}$$

is the special ($a_{\omega k} = 0$) BRS-transformation (5.65) of Φ, more simply written in the exponential form as

$$s e^{\Phi} = e^{\Phi} c_+ - \bar{c}_+ e^{\Phi} \; . \tag{15.11}$$

We then found the general classical solution (5.177), rigid invariance (15.4) taken into account: it has the form (15.7) but with

$\bar{\Gamma}_s^{(o)}$ (15.8) replaced by - we add here the contributions of the matter fields which were not considered in (5.177) -

$$\bar{\Gamma}^{(o)}(\Phi,c_+,A,\eta,\sigma,Y,p_k,z_k) = \bar{\Gamma}_s^{(o)}(\hat{\Phi},\hat{c}_+,\hat{A},\hat{\eta},\hat{\sigma},\hat{Y})$$

$$+ \sum_k z_k \{ \mathrm{Tr}\int dV\, \eta\, G_k(\Phi,p)$$

$$+ \int dS\, [h_k^c(p)\mathrm{Tr}(\sigma c_+) + h_k^A(p)\, YA] + \text{c.c.}\} \tag{15.12}$$

where $\hat{\Phi}$ is given by (15.1) with $c_{\omega k}$ an arbitrary function of α and the $a_{\omega k}$'s and, with the notation

$$(p_k) = (\alpha, a_{\omega k}) \quad , \quad z_k = (\chi, x_{\omega k}) \tag{15.13}$$

$$\begin{aligned}
\hat{c}_+ &= t_c(p)c_+ & \hat{A} &= t_A(p)A \\
\hat{\sigma} &= \frac{1}{t_c(p)}\sigma & \hat{Y} &= \frac{1}{t_A(p)}Y \\
h_k^c &= -\frac{1}{t_c}\partial_{p_k} t_c & h_k^A &= \frac{1}{t_A}\partial_{p_k} t_A \\
\hat{\eta} &= \frac{\delta}{\delta\hat{\Phi}} \mathrm{Tr}\int dV\, \eta\, \mathcal{F}^{-1}(\hat{\Phi},p)\Big|_{\hat{\Phi} = \mathcal{F}(\Phi,p)} \\
G_k &= -\frac{\partial}{\partial p_k} \mathcal{F}^{-1}(\hat{\Phi},p)\Big|_{\hat{\Phi} = \mathcal{F}(\Phi,p)}
\end{aligned} \tag{15.14}$$

$t_c(p)$ and $t_A(p)$ being arbitrary functions of the gauge parameters. The normalization conditions (5.180-5.183) completed with appropriate conditions for the matter fields fix the solution (15.12) with

$$\begin{aligned}
t_c(p) &= t_A(p) = 1 \\
c_{\omega k}(p) &= c_{\omega k}\,(a_{\omega' k'};\ k' \leq k) \\
&= \text{solution of the eqs. (5.140)}
\end{aligned} \tag{15.15}$$

At the very beginning we had of course required the supersymmetry Ward-identity to hold

$$W_\alpha \Gamma = 0 \qquad \bar{W}_{\dot\alpha} \Gamma = 0 \tag{15.16}$$

and also found that superconformal R-invariance is maintained up to soft breakings:

$$W_R \Gamma \sim 0 \tag{15.17}$$

The aim is now to generalize - if possible - the above Ward-identities to all orders. The missing one, after sections 13.3,13.4,13.5, is the Slavnov-identity (15.2) . In ordinary component theories the Slavnov-identity may be ruined in higher orders [II.12] by an anomaly which is hard, not the variation of a local insertion of UV-dimension smaller than or euqals to four, but passing through the consistency conditions following from the BRS-symmetry. The subject of the present section is to perform the corresponding analysis for the supersymmetric case i.e. to find out, whether the Slavnov identity (15.2) will be broken by a hard anomaly. For these considerations masses are irrelevant and, since masslessness of the superfield Φ causes problems with the IR-convergence (see end of sect. 9), we add mass terms to (15.12) such that all fields become massive and thus try to prove Ward identities up to soft breakings:

$$W_\alpha \Gamma \sim 0 \qquad \bar{W}_{\dot\alpha} \Gamma \sim 0 \tag{15.18}$$

$$W_R \Gamma \sim 0 \tag{15.19}$$

$$W_\omega \Gamma \sim 0 \tag{15.20}$$

$$\mathcal{S}(\Gamma) \sim 0 \tag{15.21}$$

With the theorems of Sections 13.3, 13.4, 13.5 the Ward-identities (15.18-15.20) are proved. Our remaining task is to check whether a suitable choice of hard counterterms enables us to prove the Slavnov identity (15.21) at each order of perturbation theory.

15.2 The consistency condition

In order to check, whether one can prove the Slavnov identity (15.21,15.2), we follow rather closely the discussion of the tree approximation (Sections 5.2, 5.3 and 5.5). Differentiation of (15.2) with respect to B and use of the gauge condition (15.3) yields the ghost equation (5.170) (from now on every equation holds up to soft contributions)

$$\mathcal{G}\Gamma \equiv [\frac{\delta}{\delta c_-} + \frac{1}{8}\overline{DD}\,DD\frac{\delta}{\delta\rho}]\,\Gamma \sim -\frac{1}{2}\chi\,\overline{DD}\,\bar{B}\,. \tag{15.22}$$

As a consequence the looked for solution Γ can depend on c_- only via the combination η (15.9). The introduction of $\bar{\Gamma}$ defined by (cf. (15.12))

$$\begin{aligned}\Gamma(\Phi,c_+,A,\eta,\sigma,Y,c_-,B) &= \bar{\Gamma}(\Phi,c_+,A,\eta,\sigma,Y)\\ &+ \mathrm{Tr}\int dV\,[BDD\,\Phi + \bar{B}\,\overline{DD}\,\bar{\Phi} + \alpha\,B\bar{B} + \frac{1}{2}\chi(c_-\bar{B} + \bar{c}_-B)]\end{aligned} \tag{15.23}$$

then permits to rewrite the Slavnov identity in the form (5.171):

$$\begin{aligned}\mathcal{S}(\Gamma) \sim \mathcal{B}(\bar{\Gamma}) &\equiv \mathrm{Tr}\int dV\,\frac{\delta\bar{\Gamma}}{\delta\eta}\frac{\delta\bar{\Gamma}}{\delta\Phi}\\ &+ \int dS\,[\mathrm{Tr}\,\frac{\delta\bar{\Gamma}}{\delta\sigma}\frac{\delta\bar{\Gamma}}{\delta c_+} + \frac{\delta\bar{\Gamma}}{\delta Y}\frac{\delta\bar{\Gamma}}{\delta A} \quad + \text{terms }(\bar{\sigma},\bar{c}_+,\bar{Y},\bar{A})\\ &+ \sum_k z_k\partial_{p_k}\bar{\Gamma} \sim 0\end{aligned} \tag{15.24}$$

Let us define the $\bar{\Gamma}$-depending linear operator

$$\begin{aligned}\mathcal{B}_{\bar{\Gamma}} &\equiv \mathrm{Tr}\int dV\,[\frac{\delta\bar{\Gamma}}{\delta\eta}\frac{\delta}{\delta\Phi} + \frac{\delta\bar{\Gamma}}{\delta\Phi}\frac{\delta}{\delta\eta}\\ &+ \int dS\,[\mathrm{Tr}\,\frac{\delta\bar{\Gamma}}{\delta\sigma}\frac{\delta}{\delta c_+} + \mathrm{Tr}\,\frac{\delta\bar{\Gamma}}{\delta c_+}\frac{\delta}{\delta\sigma} + \frac{\delta\bar{\Gamma}}{\delta Y}\frac{\delta}{\delta A} + \frac{\delta\bar{\Gamma}}{\delta A}\frac{\delta}{\delta Y}]\\ &+ \text{ terms }(\bar{\sigma},\bar{c}_+,\bar{A},\bar{Y})\\ &+ \sum_k z_k\partial_{p_k}\end{aligned} \tag{15.25}$$

which fulfills the identities - due to the anticommuting properties of the odd $\Phi\pi$ charge fields -

$$\mathcal{B}_\gamma\,\mathcal{B}(\gamma) = 0 \quad , \quad \forall\gamma \tag{15.26}$$

and

$$\mathcal{B}_\gamma\ \mathcal{B}_\gamma = \int \sum_j \{ - \frac{\delta \mathcal{B}(\gamma)}{\delta F_j} \frac{\delta}{\delta B_j} + \frac{\delta \mathcal{B}(\gamma)}{\delta B_j} \frac{\delta}{\delta F_j} \} \ , \quad \forall \gamma . \tag{15.27}$$

Here B_j stands for the commuting fields Φ, A, σ and F_j for the anti-commuting fields η, Y, c_+; the integration measure is dV, dS or d$\bar{S}$ according to the type of the integral. From (15.27) follows

$$\mathcal{B}_\gamma\ \mathcal{B}_\gamma = 0 \quad , \quad \text{if} \quad \mathcal{B}(\gamma) = 0 \tag{15.28}$$

We shall denote $\mathcal{B}_\gamma$ by $\mathscr{b}$ if γ is the classical action $\bar{\Gamma}^{(0)}$ (15.12) (with conditions (15.15)) without the mass terms. Since $\bar{\Gamma}^{(0)}$ fulfills the Slavnov identity $\mathscr{b}$ is nilpotent due to (15.28):

$$\mathscr{b}^2 = 0 \tag{15.29}$$

The variational operator $\mathscr{b}$ corresponds to the transformations

$$\begin{aligned}
\mathscr{b}\,\Phi &= \frac{\delta\bar{\Gamma}^{(0)}}{\delta\eta} & \mathscr{b}\,\eta &= \frac{\delta\bar{\Gamma}^{(0)}}{\delta\Phi} \\
\mathscr{b}\,c_+ &= \frac{\delta\bar{\Gamma}^{(0)}}{\delta\sigma} & \mathscr{b}\,\sigma &= \frac{\delta\bar{\Gamma}^{(0)}}{\delta c_+} \\
\mathscr{b}\,A &= \frac{\delta\bar{\Gamma}^{(0)}}{\delta Y} & \mathscr{b}\,Y &= \frac{\delta\bar{\Gamma}^{(0)}}{\delta A} \\
\mathscr{b}\,p_k &= z_k
\end{aligned} \tag{15.30}$$

We note that the $\mathscr{b}$-transformations of Φ, c_+ and A are their BRS transformations.

We now apply the action principle (11.5, 11.12) for the Slavnov operator (15.2, 15.24) and obtain

$$\mathcal{B}(\bar{\Gamma}) \sim \Delta\cdot\Gamma = \Delta(\Phi, c_+, A, \eta, \sigma, Y, p_k, z_k) + O(\hbar\Delta) \tag{15.31}$$

where the local insertion Δ has $\Phi\pi$-charge 1 and dimension 4 according to the assignments of Table 5.3.1 and the rules (11.13). Δ is supersymmetric, R- and rigid-invariant, and does not depend explicitly on c_- because of the ghost euqation (15.22).

Acting with $\mathcal{B}_{\bar{\Gamma}}$ on (15.31) we obtain as a consequence of the identity (15.26)

$$0 \sim \mathcal{B}_{\bar{\Gamma}}\Delta + O(\hbar\Delta) \qquad (15.32)$$

Replacing here $\mathcal{B}_{\bar{\Gamma}}$ by $\mathcal{b} \sim \mathcal{B}_{\bar{\Gamma}} + O(\hbar)$ and keeping only the terms of dimension 4 yields the consistency condition

$$\mathcal{b}\Delta = O(\hbar\Delta) \qquad (15.33)$$

We see that for obtaining this consistency condition it was crucial to linearize the action of the Slavnov operator on Γ in the form (15.25) or else, at least to realize that an insertion transforms under $\mathcal{B}_{\bar{\Gamma}}$ as far as BRS-transformations are concerned (cf. the corresponding discussion for BRS-symmetric insertions in Section 12).

15.3 Solution of the consistency condition: the anomaly

The problem of proving the Slavnov identity has thus been reduced to solving the consistency condition (15.33) for local functionals Δ of classical fields, a classical problem.

It is clear that due to the nilpotency of $\mathcal{b}$ any Δ of the form

$$\Delta = \mathcal{b}\hat{\Delta}(\Phi, c_+, A, \eta, \sigma, Y, p_k, z_k) \qquad (15.34)$$

where $\hat{\Delta}$ is is a local insertion of dimension 4 and $\Phi\pi$-charge 0, supersymmetric, rigid- and R-invariant, is a solution of the consistency condition. If we were able to show that any solution had this form then we could prove the Slavnov identity (15.21) at each order. Indeed, redefining Γ_{eff} by

$$\Gamma'_{eff} = \Gamma_{eff} - \hat{\Delta} \qquad (15.35)$$

yields the new vertex functional

$$\Gamma' = \Gamma - \hat{\Delta} + O(\hbar\hat{\Delta}) \qquad (15.36)$$

Applying the Slavnov operator on it and using (15.24, 15.31, 15.25, 15.34) then proves recursively the Slavnov identity:

$$\mathcal{S}(\Gamma') \sim \mathcal{B}(\Gamma') = \mathcal{B}(\Gamma) - \mathcal{B}_{\bar{\Gamma}}\hat{\Delta} + O(\hbar\hat{\Delta})$$

$$\sim \Delta - \mathcal{b}\hat{\Delta} + O(\hbar\Delta) = O(\hbar\Delta) \qquad (15.37)$$

If to the contrary there exists a solution $\Delta = \mathcal{a}$ for the consistency condition such that there is no local $\hat{\Delta}$ with $\mathcal{a} = \mathcal{b}\hat{\Delta}$, $\mathcal{a}$ is a candidate for a possible anomaly. If its numerical coefficient, which has to be calculated explicitly, turns out to be non-zero at some order the anomaly is truely present. It cannot be absorbed as a counterterm like in (15.35-15.37) and the Slavnov identity cannot be established. This means that the theory cannot be defined as a gauge theory, in particular no Green's functions of gauge invariant operators exist - and no unitary S-matrix either (in cases where the theory can be defined on shell). It will turn out that the general solution of the consistency condition allows for exactly one anomaly [IV.10], namely the supersymmetric extension of the well-known chiral anomaly [IV.12, II.12, R.1, R.7].

Let us now go ahead and prove this statement. The consistency condition (15.33) reads, with the $O(\hbar\Delta)$ term ommitted,

$$\mathcal{b}\Delta(\Phi, c_+, A, \eta, \sigma, Y, p_k, z_k) = 0 \qquad (15.38)$$

To solve this is a cohomology problem. Let us call cochains the local functionals of the fields (and of the gauge parameters) of dimension 4, supersymmetric, rigid- and R-invariant. The coboundary operator will then be the nilpotent variational operator $\mathcal{b}$, cocycles the cochains Δ fulfilling (15.38) and coboundaries the cocycles Δ of the form (15.34). Cohomology classes will be defined by the equivalence relation:

$$\Delta_1 \approx \Delta_2 \quad \text{iff} \quad \exists\, \hat{\Delta} \quad \text{such that} \quad \Delta_1 - \Delta_2 = \mathcal{b}\hat{\Delta} \qquad (15.39)$$

A cohomology is called trivial if all cocycles are coboundaries.

The first step in solving (15.38) consists usually [II.12] in expanding Δ in a basis of dimension 4 field monomials. Here the situation is however complicated by the vanishing dimension of the superfield Φ: the basis will have infinitely many elements. A recursive procedure is thus required. We expand Δ as

$$\Delta = \sum_n \Delta_n \tag{15.40}$$

where Δ_n is a homogeneous polynomial of degree n in the fields $\Phi, c_+, A, \eta, \sigma, Y$ and in the gauge parameters p_k, z_k. We also expand the variational operator $\mathscr{b}$ (see (5.31)):

$$\mathscr{b} = \sum_{n \geq 0} \mathscr{b}_n \tag{15.41}$$

where $\mathscr{b}_n$ increases the order in the fields and gauge parameters by an amount of n. For n=0 (use (15.30) keeping only the bilinear terms of $\bar{\Gamma}^{(0)}$ (15.12)), we have explicitly

$$\begin{aligned}
&\mathscr{b}_0 \Phi = c_+ - \bar{c}_+ && \mathscr{b}_0 \eta = \frac{1}{64g^2} \overline{DDDD}\, \Phi \\
&\mathscr{b}_0 c_+ = 0 && \mathscr{b}_0 \sigma = - \overline{DD}\, \eta \qquad \mathscr{b}_0 \bar{\sigma} = DD\, \eta \\
&\mathscr{b}_0 A = 0 && \mathscr{b}_0 Y = \frac{1}{16} \overline{DD}\, \bar{A} \qquad \mathscr{b}_0 \bar{Y} = \frac{1}{15} DDA \\
&\mathscr{b}_0 p_k = z_k && \mathscr{b}_0 z_k = 0
\end{aligned} \tag{15.42}$$

$\mathscr{b}_0$ is nilpotent.

The following lemma will help us to reduce the $\mathscr{b}$-cohomology problem (15.38) to the much simpler one of the $\mathscr{b}_0$-cohomology. We shall use the sign $\overset{o}{\approx}$ for the equivalence relation (15.39) within the $\mathscr{b}_0$-cohomology:

$$\Delta^1 \overset{o}{\approx} \Delta^2 \quad \text{iff} \quad \exists\, \hat{\Delta} \quad \text{such that} \quad \Delta^2 - \Delta^2 = \mathscr{b}_0 \hat{\Delta} \tag{15.43}$$

<u>Lemma 15.1</u> If the general solution of the equation

$$\mathscr{b}_0 \Delta = 0 \tag{15.44}$$

is a $\mathscr{b}_0$-coboundary, i.e.

$$\Delta \overset{o}{\approx} 0 \tag{15.45}$$

then the general solution of the δ-invariance equation

$$\delta\Delta = 0 \tag{15.46}$$

is a δ-coboundary, i.e.

$$\Delta \approx 0 \tag{15.47}$$

N.B. This condition is sufficient, but not necessary as we shall see in explicit examples.

Proof Let us suppose that Δ, expanded according to (15.40), is a coboundary (15.47) up to order n-1 in the fields and gauge parameters:

$$\Delta \approx 0_n = \Delta_n + 0_{n+1} \tag{15.48}$$

where 0_n represents terms of order n and higher. Eq. (15.46) implies

$$0 = \delta\Delta = \delta\Delta_n + 0_{n+1} = \delta_o\Delta_n + Q_{n+1} \tag{15.49}$$

since $\delta - \delta_o$ increases the order by at least 1. Hence

$$\delta_o\Delta_n = 0 \tag{15.50}$$

Then from the hypotheses (15.44, 15.45) there exists a $\hat{\Delta}_n$ such that

$$\Delta_n = \delta_o\hat{\Delta}_n \tag{15.51}$$

Thus

$$\Delta \approx \delta_o\hat{\Delta}_n + 0_{n+1} = \delta\hat{\Delta}_n + 0_{n+1} \approx 0_{n+1} \tag{15.52}$$

which proves the statement at the next order n.

It is shown in Appendix C that the general solution of the $\mathcal{b}_0$-cohomology equation (15.44) is given by

$$\Delta(\Phi, c_+, A, \eta, \sigma, Y, p_k, z_k) \overset{0}{\approx}$$

$$y \{\int dS\ Y\ \tilde{c}_+^2\ A - \int d\bar{S}\ \bar{A}\ \tilde{\tilde{c}}_+^2\ \bar{Y} - \frac{1}{16} \int dV\ \bar{A}(\tilde{\Phi}\tilde{c}_+ - \tilde{\tilde{c}}_+\tilde{\Phi})A\}$$

$$+ l_{abci} \int dS\ A_a A_b A_c c_{+i} + l'_{abci} \int d\bar{S}\ \bar{A}_a \bar{A}_b \bar{A}_c \bar{c}_{+i}$$

$$+ r\ X_3 + r' \bar{X}_3 \qquad (15.53)$$

with

$$X_3 = \mathrm{Tr} \int dS\ c_+\ \overline{DD}D\Phi\ \overline{DD}D\Phi \qquad (15.54)$$

x,y,r and r' are arbitrary coefficients, whereas the coefficients l_{abci} are subject to the constraints

$$i l_{abci}\ f_{ijk} - l_{dbcj}\ T^k_{da} - l_{adcj}\ T^k_{db} - l_{abdj} T^k_{dc} = 0 \qquad (15.55)$$

Similar constraints hold for l'_{abci}. We see that the $\mathcal{b}_0$-cohomology is non-trivial at orders 2 and 3 in the fields. The coefficients x,y,r,r', l and l' are independent of the gauge parameters.

We solve now the $\mathcal{b}$-cohomology problem for Δ expanded according to (15.40), beginning at its lowest order which is 2. $\mathcal{b}$-invariance implies

$$0 = \mathcal{b}\Delta_2 + O_3 = \mathcal{b}_0\Delta_2 + O_3 \qquad (15.56)$$

Hence $\mathcal{b}_0\Delta_2 = 0$ and, from (15.53), $\Delta_2 \overset{0}{\approx} 0$. Thus

$$\Delta = \Delta_2 + O_3 = \mathcal{b}_0\hat{\Delta}_2 + O_3 = \mathcal{b}\hat{\Delta}_2 + O_3$$

$$= \mathcal{b}\hat{\Delta}_2 + \Delta_3 + O_4 \approx \Delta_3 + O_4 \qquad (15.57)$$

Now $\mathcal{b}$-invariance gives, at order 3, $\mathcal{b}_0\Delta_3 = 0$. Thus, from (15.53)

$$\Delta \approx \mathscr{b}_0\hat{\Delta}_3 + rX_3 + r'\bar{X}_3 + 0_4$$

$$\approx rX_3 + r'\bar{X}_3 + \Delta_4 + 0_5 \tag{15.58}$$

and, from $\mathscr{b}$-invariance at order 4,

$$r\mathscr{b}_1X_3 + r'\mathscr{b}_1\bar{X}_3 + \mathscr{b}_0\Delta_4 = 0 \tag{15.59}$$

where the operator $\mathscr{b}_1$ is defined by (15.41) and can be calculated using (15.30) and the trilinear terms of the action (15.12). Eq. (15.59) explicitly reads (at $a_{\omega k} = 0$ for simplicity)

$$\mathscr{b}_0\Delta_4 = - r \, \mathrm{Tr}\int dS \, c^2(\overline{D}\overline{D}D\Phi)^2 - r'\mathrm{Tr}\int d\bar{S} \, \bar{c}^2(DD\overline{D}\Phi)^2 \tag{15.60}$$

One can check that a necessary condition for the r.h.s. to be a $\mathscr{b}_0$-variation is $r' = -r$. Then (15.58) becomes

$$\Delta \approx r(X_3 - \bar{X}_3) + 0_4 = r\mathcal{a} + 0_4$$

$$= r\mathcal{a} + \Delta_4 + 0_5 \tag{15.61}$$

where $\mathcal{a}$ is defined - uniquely up to a $\mathscr{b}$-variation - by

$$\mathscr{b}\mathcal{a} = 0$$

$$\mathcal{a} = \mathrm{Tr}\int dS \, c_+(\overline{D}\overline{D}D\Phi)^2 - \mathrm{Tr}\int d\bar{S} \, \bar{c}_+(DD\overline{D}\Phi)^2 + 0_4 \tag{15.62}$$

and can be calculated iteratively as a power series in Φ, linear in c_+, $\bar{c}_+$. $\mathcal{a}$ is not a $\mathscr{b}$-variation since its lowest order term $X_3 - \bar{X}_3$ is not a $\mathscr{b}_0$-variation, which means that it is an anomaly: $\mathcal{a}$ is the supersymmetric extension of the usual chiral anomaly.

The $\mathscr{b}$-invariance of Δ (15.61) yields now at order 4, $\mathscr{b}_0\Delta_4 = 0$. Then Δ_4 has the form given by the terms in A, Y, c_+ of (15.53). Replacing there $\tilde{\Phi}$ by

$$e^{\tilde{\mathcal{F}}(\Phi)} - 1 = \tilde{\Phi} + 0_2 \tag{15.63}$$

with $\mathcal{F}(\Phi)$ given by (15.1), we can write

$$\Delta \approx r\mathcal{A} + y \{\int dS\ Y\ \tilde{c}_+^2\ A - \int d\bar{S}\ \bar{A}\ \tilde{\tilde{c}}^2\ \bar{Y}$$

$$\frac{1}{16} \int dV\ \bar{A}\ (e^{\tilde{\mathcal{F}}}\ \tilde{c}_+ - \tilde{\bar{c}}_+\ e^{\tilde{\mathcal{F}}} - c_+ + \tilde{\bar{c}}_+)A\}$$

$$+ l_{abci} \int dS\ A_a A_b A_c c_{+i} + l'_{abci} \int d\bar{S}\ \bar{A}_a \bar{A}_b \bar{A}_c \bar{c}_{+i}$$

$$+ O_5 \tag{15.64}$$

One can check now by an explicit computation using (15.30) and the classical action (15.12) that

$$-\frac{1}{16}\,\mathcal{b}\int dV\ \bar{A}A = \frac{1}{16}\int dV\ \bar{A}(\tilde{c}_+ - \tilde{\bar{c}}_+)A$$

$$-\mathcal{b}\int dS\ Y\ \tilde{c}_+ A = \int dS\ Y\ \tilde{c}_+^2\ A - \frac{1}{16}\int dV\ \bar{A}\ e^{\tilde{\mathcal{F}}}\ \tilde{c}_+ A \tag{15.65}$$

$$- 3\ h_{abc} \int dS(\tilde{c}_+ A)_a A_b A_c$$

(and similarly for the terms in $\bar{Y},\bar{c}_+,\bar{A}$) which shows that the y-term in (15.64) is a $\mathcal{b}$-variation, up to terms trilinear in A and $\bar{A}$ which can be absorbed by a redefinition of the coefficients l_{abci} and l'_{abci}. Thus Δ is equivalent to the r.h.s. of (15.64) with $y = 0$. Now, due to dimension and chirality terms trilinear in A or in $\bar{A}$ are only present at the order 4, with the form given in (15.64), so that they must be separately $\mathcal{b}$-invariant. This yields the constraints

$$\frac{i}{2}\ l_{abci}\ f_{ijk} - l_{dbcj}\ T^k_{da} - l_{adcj}\ T^k_{db} - l_{abdj}\ T^k_{dc} = 0 \tag{15.66}$$

(and similarly for l'_{abci}). They are incompatible with the rigid invariance constraints (15.55) - note the difference by a factor 1/2 in the first term - unless $l_{abci} = l'_{abci} = 0$. Hence

$$\Delta \approx r\mathcal{A} + O_5 \quad . \tag{15.67}$$

The $\mathcal{b}_0$-cohomology being trivial at order 5 and higher, Lemma 15.1 can be applied from now on. The general solution of the consistency condition (15.38) is thus

$$\Delta(\Phi, c_+, A, \eta, \sigma, Y, p_k, z_k) \approx r\mathcal{a}(\Phi, c_+, p_k, z_k) \tag{15.68}$$

where $\mathcal{a}$ is the chiral anomaly defined by (15.62).

We note that $\mathcal{a}$ depends on the gauge parameters due to its definition and to the p_k, z_k-dependence of $\mathcal{b}$ (see (15.30) and recall that the classical action depends on them). But the anomaly coefficient r is gauge independent. This can be readily checked: the $\mathcal{b}$-invariance of Δ implies

$$0 = \mathcal{b}(r\mathcal{a}) = (\mathcal{b}r)\mathcal{a} = (z_k \partial_{p_k} r)\mathcal{a} \tag{15.69}$$

hence

$$\partial_{p_k} r = 0 \tag{15.70}$$

15.4 The anomaly in the Slavnov-identity

In the last subsection we have seen that the algebra of BRS-transformations admits an anomaly. It is then crucial to perform an explicit calculation by which one finds out in which models and in which perturbative order the possible anomaly indeed shows up. The algebraic argument above on the potential existence of an anomaly is by construction general enough to treat a rich class of models.

The Slavnov-identity reads according to the above

$$\mathcal{s}(\Gamma) \sim r\mathcal{a} + \mathcal{b}\Delta + O(\hbar^2) \tag{15.71}$$

where the coefficient r and $\mathcal{b}\Delta$ are at least of order $\hbar$ (one loop) and we know from (15.62) that in the expansion in the order of fields $\mathcal{a}$ begins as

$$\begin{aligned} \mathcal{a} &= \mathcal{a}_3 + \mathcal{a}_4 + \ldots \\ \mathcal{a}_3 &= \mathrm{Tr}(\int dS\, \overline{DDD}\Phi\, \overline{DDD}\,\Phi\, c_+ - \int d\bar{S}\, DD\bar{D}\,\Phi\, DD\bar{D}\,\Phi\, \bar{c}_+) \\ &= d_{ijk}(\int dS\, \overline{DDD}\,\Phi^i\, \overline{DDD}\,\Phi^j\, c_+^k - \int d\bar{S}\, DD\bar{D}\,\Phi^i\, DD\bar{D}\,\Phi^j\, \bar{c}_+^k) \end{aligned} \tag{15.72}$$

In order to calculate r we begin with a case where it is zero due to a discrete symmetry. Suppose the theory admits parity

$$\Phi \to -\Phi \qquad c_+ \leftrightarrow \bar{c}_+ \qquad \Gamma(\Phi^P) = \Gamma(\Phi) \tag{15.73}$$

as a symmetry. Then

$$(\mathcal{S}(\Gamma))^P = \mathcal{S}(\Gamma) \tag{15.74}$$

but $\mathcal{a}_3 \to -\mathcal{a}_3$

and consequently r = 0.

This situation is in particular realized for the pure SYM case (i.e. no matter fields present) and this result will be used below. For the general case the most convenient way to compute r is to perform a test which is specific for a contribution in $\mathcal{a}$ which cannot be a variation and not to absorb any part of Δ as a counterterm in the effective Lagrangian. The term $\mathcal{a}_3$ is certainly not a variation and the component expression

$$\int dx\, \varepsilon^{\mu\nu\rho\sigma}\, \partial_\mu v^i_\nu \partial_\rho v^j_\sigma a^k_+ \tag{15.75}$$

contained in it will serve us as guide for finding a good testing operator X. It is isolated by applying

$$\frac{i}{3.2^{10}}\, \varepsilon^{\nu\rho\alpha\beta} \frac{\partial}{\partial p^\alpha} \frac{\partial}{\partial q^\beta} \frac{\delta^3}{\delta v^i_\nu(2)\, \delta v^j_\rho(3)\, \delta a^k_+(1)} \tag{15.76}$$

Therefore

$$r \sim X\, \mathcal{S}(\Gamma)\Big|_{\Theta=0} \tag{15.77}$$

$$X = \frac{i}{3.2^6} \frac{d^{ijk}}{d^2}\, \varepsilon^{\nu\rho\alpha\beta} \frac{\partial}{p^\alpha} \frac{\partial}{q^\beta} (\partial_\Theta \sigma_\nu \partial_{\bar{\Theta}})_2 (\partial_\Theta \sigma_\rho \partial_{\bar{\Theta}})_3 (\partial_\Theta \partial_\Theta)_1 \frac{\delta^3}{\delta c^i_+(1) \delta\Phi^j(2) \delta\Phi^k(3)}$$

where $d^2 \equiv d_{ijk} d^{ijk}$ and we have normalized

$$\mathrm{Tr}\,\tau^i\tau^j = \delta^{ij}$$

$$\mathrm{Tr}\,[\tau^i,\tau^j]\tau^k = if^{ijk} \qquad \mathrm{Tr}\,\tau^{(i}\tau^j\tau^{k)} = d^{ijk} \tag{15.78}$$

It is perhaps worthwhile to repeat that

$$X\,\mathcal{a}_3 = 1 \qquad X\mathcal{b}\Delta = 0 \tag{15.79}$$

since the test X is specific for the non-variation term $\int dx\varepsilon^{\cdots}\partial v\partial va_+$.

We now have to calculate the effect of X on $\mathcal{s}(\Gamma)$ in order to express r in terms of Feyman diagrams. Since the test does not contain matter field derivatives and only one c_+-derivative from $\mathcal{s}(\Gamma)$ only the part $\frac{\delta\bar{\Gamma}}{\delta\eta}\frac{\delta\bar{\Gamma}}{\delta\Phi}$ can contribute. Those parts having a $\bar{\Gamma}_{\Phi\Phi}$ as factor contain f-type couplings or are independent of matter couplings and therefore will vanish in the complete test or due to the fact that pure SYM has no anomaly. I.e.

$$\left.\frac{\delta^3}{\delta c_1^i\,\delta\Phi_2^j\,\delta\Phi_3^k}\mathcal{s}(\Gamma)\right|_{\substack{\text{contribution}\\ \text{to } r}} = \overline{DD}_1\Gamma^{(1)}_{\Phi_1^i\Phi_2^j\Phi_3^k} \equiv \overline{DD}_1\;\Gamma_{ijk}(1,2,3) \tag{15.80}$$

Thus the complete test yields

$$r \sim \frac{i}{3.2^6}\frac{d^{ijk}}{d^2}\varepsilon^{\nu\rho\alpha\beta}\frac{\partial}{\partial p^\alpha}\frac{\partial}{\partial q^\beta}(\partial_\Theta\sigma_\nu\partial_{\bar\Theta})_2(\partial_\Theta\sigma_\rho\partial_{\bar\Theta})_3(\partial_\Theta\partial_\Theta)_1(\overline{DD})_1\Gamma_{ijk}(1,2,3)\Big|_{\Theta=0} \tag{15.81}$$

In the actual computation of Γ_{ijk} one can also drop all diagrams consisting solely of vector lines. As noted at the beginning of this section in a pure SYM $r = 0$, so those contributions cancel. There remain the matter loops,(Fig. 15.1), and we have to choose now a subtraction scheme for their evaluation. We apply the subtraction scheme of section 9 and have only to give the subtraction degrees for the diagrams in question.

Figures: 15.1

1)

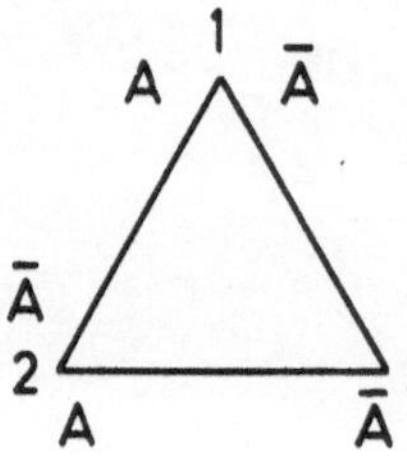

1')

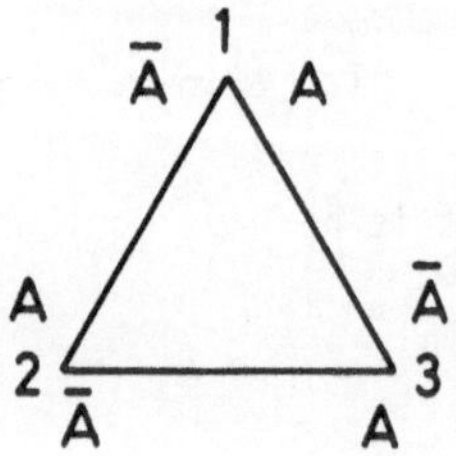

2)

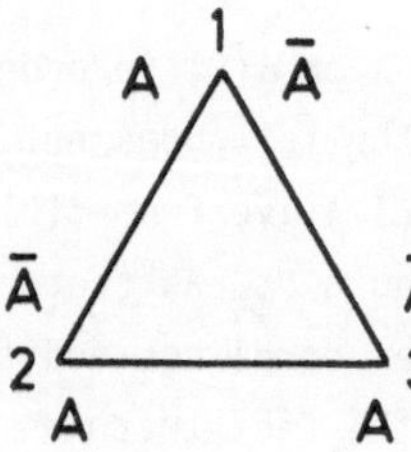

2')

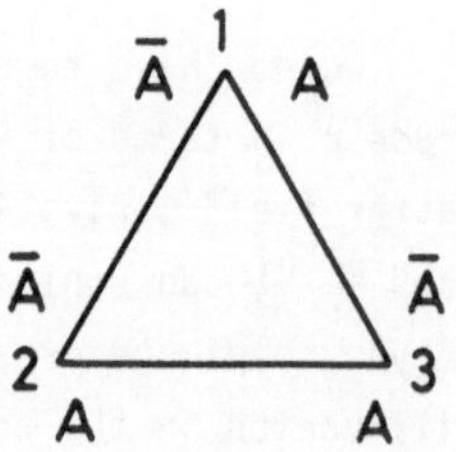

3)

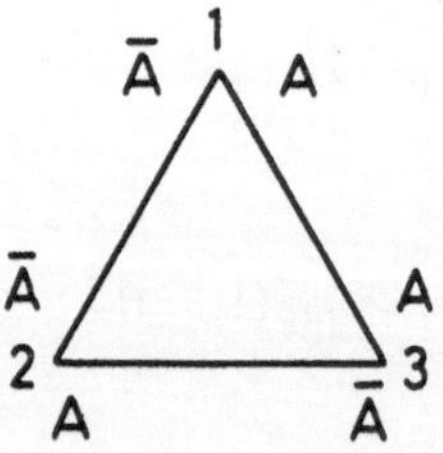

3')

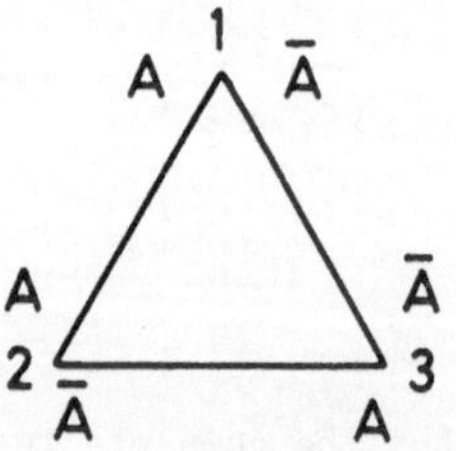

4)

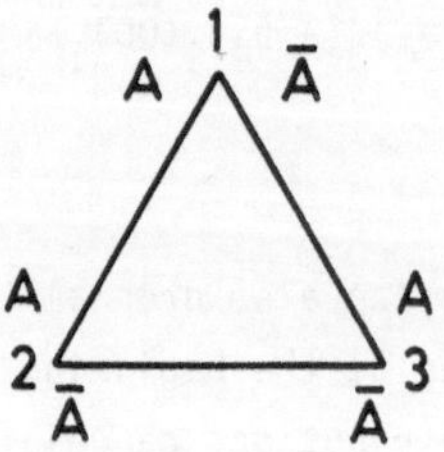

4')

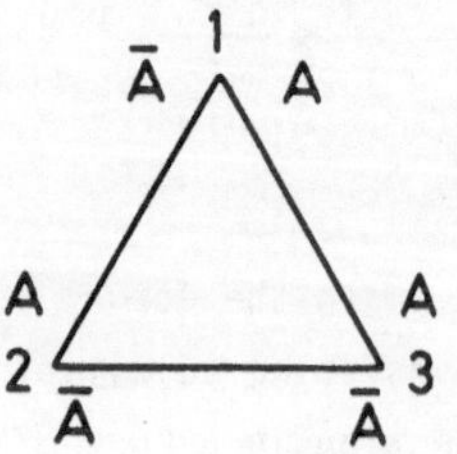

The integrand being expanded in powers of Θ the subtraction degree of the ωth power is

$$\delta_\omega = 4 - 2\cdot 3 + \frac{\omega}{2} = -2 + \frac{\omega}{2} \tag{15.82}$$

and ω is dictated by the test (15.77) to be 8, i.e. $\delta = 2$. An integrand $I^{(\omega)} = I^{(\omega)}(p,q;k)$ has to be subtracted as

$$R = I^{(\omega)}(p,q;k) - I^{(\omega)}(0,0;k)$$
$$-l(\partial_l I)^{(\omega)}(0,0;k) - \frac{1}{2}\, ll(\partial_l \partial_l I)^{(\omega)}(0,0;k) \tag{15.83}$$

$(l = p,q)$

Then

$$\partial p_\alpha \partial q_\beta \nabla\Gamma = \frac{1}{(2\pi)^4}\int dk\left(\partial p_\alpha \partial q_\beta\, I^{(\omega)} - \partial p_\alpha \partial q_\beta\, I^{(\omega)}\Big|_{p=q=0}\right) \tag{15.84}$$

(∇ abbreviates the differentiation with respect to Θ's in (15.81)) and r is obtained by multiplying with the numerical factors given above and taking the limit $p,q \to \infty$. Evaluating at large momenta is enforced by the <u>asymptotic</u> validity of (15.71).

Calculated this way r is the sum of the contributions of the 8 diagrams in Fig. 15.1

$$r = \sum_{\lambda=1}^{4'} r_\lambda \tag{15.85}$$

and we find

$$\begin{aligned} r_2 + r_2' + r_3 + r_3' &= 0 \\ r_4 = r_4' &= 0 \end{aligned} \tag{15.86}$$

i.e. the diagrams containing at least one purely chiral-chiral propagator do not contribute. For the remaining ones we find

$$r_1 = r_1'$$
$$r = r_1 + r_1' = \frac{1}{3\cdot 2^{12}\pi^2}\frac{d^{ijk}}{d^2}\,\mathrm{Tr}\,T^iT^jT^k \tag{15.87}$$

Recall that T^i_{ab} are the generators for a (unitary) representation of the matter fields

$$A \equiv (A_a)$$
$$sA = -\,c^i_+\,T^i\,A \qquad s\bar{A} = \bar{A}\,\bar{c}^i_+\,T^i\ . \tag{15.88}$$

Section 16. Renormalized Supercurrents

In this section we deal with the problem of constructing supercurrents to all orders of perturbation theory. Like in the tree approximation we start with the trace equations

$$-2w_\alpha\Gamma = \Delta_\alpha\cdot\Gamma$$
$$-2\bar{w}_{\dot\alpha}\Gamma = \bar{\Delta}_{\dot\alpha}\cdot\Gamma \tag{16.1}$$

and try then to decompose Δ_α into current plus breaking part

$$\Delta_\alpha = \bar{D}^{\dot\alpha}V_{\alpha\dot\alpha} + 2D_\alpha S\ . \tag{16.2}$$

The new feature is now that Δ_α, V, S are local insertions (cf. sect. 10) with specified subtraction degrees and only the existence of Δ_α is guaranteed directly by the action principle (cf. sect. 11). Decomposing Δ_α according to (16.2) represents just an algebraic problem in those cases where Δ_α is given explicitly (e.g. in the Wess-Zumino model and in SQED) but is much more involved in the general case (i.e. in SYM) where it is virtually impossible to write down Δ_α in n-th order and to go ahead. There a detailed abstract analysis which is based on BRS-invariance has to show that all contributions to Δ_α can indeed be decomposed as desired.

16.1 The Wess-Zumino model

In this model all calculations leading to the renormalized supercurrent can be explicitly performed [I.6]. We need the bilinear renormalized equations of motion as they are given by the action principle.*

$$A \frac{\delta\Gamma}{\delta A} = [\frac{z}{16} \bar{D}\bar{D}(A\bar{A}) + \frac{1}{4} M(s-1)A^2 + \frac{\lambda}{16} A^2]_2^2 \cdot \Gamma \tag{16.1.1}$$

(similarly the conjugate). They follow from

$$\Gamma_{eff} = \int dSL + \int d\bar{S}\bar{L} \tag{16.1.2}$$

where (cf. (11.18), (11.36), (13.34))

$$L = [\frac{z}{32} \bar{D}\bar{D}(A\bar{A}) + \frac{1}{8} M(s-1)A^2 + \frac{1}{48} \lambda A^3]_3^3$$

and the power series

$$z = z(g,M,\mu) = 1 + O(\hbar) \tag{16.1.3}$$

$$\lambda = \lambda(g,M,\mu) = g + O(\hbar)$$

are fixed order by order in $\hbar$ by the normalization conditions

$$< TA(0,\Theta_1)\, \tilde{A}(p,\Theta_2) >^{1PI}\Big|_{\substack{p^2=-\mu^2 \\ \Theta=0 \\ s=1}} = 1/16$$

$$\frac{\partial^2}{\partial\Theta_1^2}\frac{\partial^2}{\partial\Theta_2^2} < TA(0,\Theta_1)\, \tilde{A}(p,\Theta_2)\tilde{A}(q,\Theta_3)>^{1PI}\Big|_{\substack{p^2=q^2=(p+q)^2=-\mu^2 \\ s=1}} = \frac{1}{8}$$

$$\frac{\partial^2}{\partial\Theta_1^2} < TA(0,\Theta_1)\, \tilde{A}(p,\Theta_2) >^{1PI}\Big|_{\substack{p=0 \\ s=1}} = 0 \tag{16.1.4}$$

(The last one is automatic, due to the subtraction scheme, cf. (10.3).)

---

* We use the simplified notation $[\ldots]^\rho_\delta$ instead of $N^\rho_\delta[\ldots]$ for a normal product.

The equations of motion are in fact needed as the contact terms (6.19) with conformal weight n = -2/3 prescribe (employ also the action principle) i.e.

$$-2w_\alpha\Gamma = [-2w_\alpha \Gamma_{eff}]^{7/2}_{7/2}\cdot\Gamma \equiv \Delta_\alpha\cdot\Gamma \qquad (16.1.5.)$$

and explicitly

$$\Delta_\alpha = [-\tfrac{2}{3} D_\alpha(\tfrac{z}{16}\bar{D}\bar{D}(A\bar{A})+\tfrac{1}{4}M(s-1)A^2)+2D_\alpha A(\tfrac{z}{16}\bar{D}\bar{D}(A\bar{A})+\tfrac{1}{4}M(s-1)A^2)]^{7/2}_{7/2} \qquad (16.1.6.)$$

Δ_α can be decomposed as in the tree approximation

$$\Delta_\alpha = \bar{D}^{\dot\alpha}V'_{\alpha\dot\alpha} + 2D_\alpha S \qquad (16.1.7)$$

where

$$V'_{\alpha\dot\alpha} = -\tfrac{z}{6}[D_\alpha A\bar{D}_{\dot\alpha}\bar{A} - AD_\alpha\bar{D}_{\dot\alpha}\bar{A} + \bar{A}\bar{D}_{\dot\alpha}D_\alpha A]^3_3 \ .$$

$$S = -\tfrac{1}{12}[M(s-1)A^2]^3_3 \ .$$

We note therefore that the breaking is already of the standard form S (cf. sect. 6.5), i.e. the supercurrent V' has an interpretation. In order to come closest to the classical approximation and to be able to study the massless limit $s \to 1$ we now use the Zimmermann identity (10.30)

$$[M(s-1)A^2]^3_3\cdot\Gamma = M(s-1)\mathfrak{m}\cdot\Gamma + [\tilde{r}\,\bar{D}\bar{D}(A\bar{A}) + \tilde{u}\bar{D}\bar{D}\bar{A}^2]^3_3\cdot\Gamma$$
$$+ \tilde{x}\,M(s-1)\bar{D}\bar{D}\bar{A} + \tilde{v}\,\Box A \qquad (16.1.8)$$

$$\mathfrak{m}\cdot\Gamma \equiv \frac{1}{1-q}[A^2]^2_2\cdot\Gamma + \tilde{y}\,M(s-1)A$$

$\tilde{a} = \frac{a}{1-q}$ for $a = r, u, v, x, y$.

(We have used also that t = 0, (10.30), as a consequence of R-invariance; cf. sect. 13.5.1.). It yields in addition to the soft term $\mathfrak{m}$ also hard terms which can be differently grouped depending on whether one considers the massless model (s=1) or the massive one ($s\neq1$).

16.1.1 The massless case

The value s = 1 defines the massless model and the breaking term S reads

$$S = -\frac{1}{12}\,[\tilde{r}\bar{D}\bar{D}(A\bar{A}) + \tilde{u}\bar{D}\bar{D}\bar{A}^2]^3_3 - \frac{1}{12}\,\tilde{v}\,\Box A\ . \tag{16.1.9}$$

The $\tilde{u}$, $\tilde{v}$ terms can be absorbed into the current, the $\tilde{r}$ term can be re-written as a B-type term yielding eventually

$$\bar{D}^{\dot{\alpha}}V^{(s=1)}_{\alpha\dot{\alpha}}\cdot\Gamma = -2w_\alpha\Gamma + \tilde{r}\bar{D}\bar{D}D_\alpha B\cdot\Gamma \tag{16.1.10}$$

with

$$V^{(s=1)}_{\alpha\dot{\alpha}}\cdot\Gamma = V'_{\alpha\dot{\alpha}}\cdot\Gamma + [-\frac{1}{3}\,\tilde{r}[D_\alpha,\bar{D}_{\dot{\alpha}}](A\bar{A}) + \frac{2}{3}\,\tilde{u}i\sigma^\mu_{\alpha\dot{\alpha}}\partial_\mu(A^2-\bar{A}^2]^3_3\cdot\Gamma$$
$$+ \frac{1}{24}\,\tilde{v}i\sigma^\mu_{\alpha\dot{\alpha}}\partial_\mu(DDA-\bar{D}\bar{D}\bar{A})$$

$$B = -\frac{1}{2}\,[\bar{A}A]^2_2 \qquad .$$

This supercurrent is strictly conserved

$$\partial^\mu V^{(s=1)}_\mu\cdot\Gamma = iw\Gamma \tag{16.1.11}$$

with canonical r-weight n = -2/3 for the field A. Hence R-current, supersymmetry currents and the symmetric energy-momentum tensor which it contains are all conserved, in correspondence with the strict conservation of these symmetries. All anomalies of the superconformal group can be expressed by the real superfield B. Let us list them (cf. (6.28))

$$\partial^\mu\hat{R}_\mu = 0 + \text{c.t.},$$
trace identity: $\bar{D}^{\dot{\alpha}}\hat{R}_{\alpha\dot{\alpha}} = \tilde{r}\bar{D}\bar{D}D_\alpha B + \text{c.t.},$

$$\partial^\mu\hat{T}_{\mu\nu} = 0 + \text{c.t.},$$
trace identity: $\hat{T}^\lambda{}_\lambda = -\frac{\tilde{r}}{2}\,D\bar{D}\bar{D}DB + \text{c.t.},$

$$\partial^\mu\hat{Q}_{\mu\alpha} = 0 + \text{c.t.},$$
trace identity: $\hat{Q}_\mu{}^\alpha\sigma^\mu_{\alpha\dot{\alpha}} = 2i\tilde{r}DD\bar{D}_{\dot{\alpha}}B + \text{c.t.}$

$$\partial^\mu \hat{M}_{\mu\nu\rho} = 0 + \text{c.t.} \,, \tag{16.1.12}$$

$$\begin{aligned} \partial^\mu \hat{D}_\mu &= \hat{T}^\lambda{}_\lambda + \text{c.t.} \,, \\ &= -\frac{1}{2} \tilde{r} D\bar{D}\bar{D}DB + \text{c.t.} \,, \end{aligned}$$

$$\begin{aligned} \partial^\mu \hat{S}_{\mu\alpha} &= \sigma^\mu_{\alpha\dot{\alpha}} \hat{\bar{Q}}^{\dot{\alpha}}_\mu + \text{c.t.} \,, \\ &= -2i\tilde{r}\bar{D}\bar{D}D_\alpha B + \text{c.t.} \,, \end{aligned}$$

$$\begin{aligned} \partial^\mu \hat{K}_{\mu\nu} &= 2x_\nu \hat{T}^\lambda{}_\lambda + \text{c.t.} \,, \\ &= -\tilde{r} x_\nu D\bar{D}\bar{D}DB + \text{c.t.} \,, \end{aligned}$$

(c.t. abbreviates "contact terms "; $\tilde{r}$ is defined in (16.1.8); B is defined in 16.1.10).

16.1.2 The massive case

For the massive model, $s \neq 1$, it is also of interest to find the superconformal anomalies and in particular to derive the Callan-Symanzik equation from the trace equations of the supercurrent. It will turn out that this is a standard procedure for supersymmetric theories.

Since we aim eventually at

$$\frac{\partial}{\partial g} \Gamma = \left[\frac{\partial \Gamma_{eff}}{\partial g}\right]^4_4 \cdot \Gamma \,, \tag{16.1.13}$$

we introduce first the insertion

$$\left[\frac{\partial L}{\partial g}\right]^3_3 \cdot \Gamma = \left[\frac{1}{32} \frac{\partial z}{\partial g} \bar{D}\bar{D}(A\bar{A}) + \frac{1}{16} \frac{\partial \lambda}{\partial g} A^3\right]^3_3 \cdot \Gamma \quad . \tag{16.1.14}$$

If we try to express the $\tilde{r}$-term in (16.1.8) by $\partial L_{eff}/\partial g$, we see from (16.1.14) that we have to combine the latter one with the equation of motion (16.1.1) in order to get rid of the term $[A^3]^3_3$ which does not show up in (16.1.8). Absorbing similar to the massless case the $\tilde{u}, \tilde{v}, \tilde{x}$-terms in the current and combining appropriately (16.1.14) with (16.1.1) we arrive at the following trace equation

$$\bar{D}^{\dot{\alpha}} V^{(s\neq 1)}_{\alpha\dot{\alpha}} \cdot \Gamma = -2w^{(\gamma)}_\alpha \Gamma - 2D_\alpha S \cdot \Gamma \quad , \tag{16.1.15}$$

where

$$V^{(s\neq 1)}_{\alpha\dot\alpha}\cdot\Gamma = V'_{\alpha\dot\alpha}\cdot\Gamma + [\frac{2}{3}\,\alpha\tilde{u}i\sigma^{\mu}_{\alpha\dot\alpha}\partial_\mu(A^2-\bar{A}^2)]\cdot\Gamma$$

$$+\frac{1}{24}\,\alpha\tilde{v}i\sigma^{\mu}_{\alpha\dot\alpha}\partial_\mu(DDA-\bar{D}\bar{D}\bar{A}) + \frac{2}{3}\,\alpha\tilde{x}\,M(s-1)i\sigma^{\mu}_{\alpha\dot\alpha}\bar{\partial}_\mu(A-\bar{A}) \quad ,$$

$$S = -\frac{1}{12}\,\alpha M(s-1)\mathfrak{m} + \frac{2}{3}\,\beta\partial_g L \quad ,$$

$$w^{(\gamma)}_\alpha = -\frac{2}{3}\,(1+\gamma)D_\alpha(A\,\frac{\delta}{\delta A}) - 2D_\alpha A\,\frac{\delta}{\delta A} \quad .$$

The breaking is of S-type and the contact terms have now become anomalous: instead of $n = -2/3$ there occurs the anomalous R-weight

$$n^\gamma = -\frac{2}{3}\,d^\gamma = -\frac{2}{3}\,(1+\gamma) \qquad (16.1.16)$$

related to the anomalous dimension γ of A. In fact, the notation anticipates already that the Callan-Symanzik functions α,β,γ appear and how they are expressed by the Zimmermann coefficient $\tilde{r}$:

$$\alpha = 1-2\gamma \quad ,$$
$$\beta = 3\gamma\,\frac{\lambda}{\partial_g\lambda} \quad , \qquad (16.1.17)$$
$$\gamma = \frac{2\tilde{r}}{z+4\tilde{r}-\frac{3}{2}\,\lambda\partial_g z/\partial_g\lambda} \quad .$$

The current conservation equation following from (16.1.15) is given by

$$\partial^\mu V^{(s\neq 1)}_\mu\cdot\Gamma = iw^{(\gamma)}\Gamma + i(DDS-\bar{D}\bar{D}\bar{S})\cdot\Gamma \ . \qquad (16.1.18)$$

It indicates breaking of the R-symmetry and the superconformal symmetries by the chiral multiplet S. According to (6.28') we have the following list of conservation equations and trace identities:

$$\partial^\mu\hat{R}_\mu = i(DDS-\bar{D}\bar{D}\bar{S}) + c.t.(\gamma)$$

trace identity: $\bar{D}^{\alpha}\hat{R}_{\alpha\dot\alpha} = -2D_\alpha S + c.t.(\gamma)$

$$\partial^\mu\hat{T}_{\mu\nu} = 0 + c.t.$$

trace identity: $\hat{T}^\lambda{}_\lambda = -\frac{3}{2}\,(DDS+\bar{D}\bar{D}\bar{S}) + c.t.(\gamma)$ (16.1.19)

$$\partial^{\mu}\hat{Q}_{\mu\alpha} = 0 + \text{c.t.} \tag{16.1.19}$$

cont.'d

trace identity: $\hat{Q}_{\mu}{}^{\alpha}\sigma^{\mu}_{\alpha\dot{\alpha}} = 12i\bar{D}_{\dot{\alpha}}\bar{S} + \text{c.t.}\ (\gamma)$

$$\partial^{\mu}\hat{M}_{\mu\nu\rho} = 0 + \text{c.t.}$$

$$\begin{aligned}\partial^{\mu}\hat{D}_{\mu} &= \hat{T}^{\lambda}{}_{\lambda} + \text{c.t.}\\ &= -\tfrac{3}{2}(DDS+\bar{D}\bar{D}\bar{S}) + \text{c.t.}\ (\gamma)\end{aligned}$$

$$\begin{aligned}\partial^{\mu}\hat{S}_{\mu\alpha} &= \sigma^{\mu}_{\alpha\dot{\alpha}}\hat{\bar{Q}}^{\dot{\alpha}}_{\mu} + \text{c.t.}\\ &= -12iD_{\alpha}S + \text{c.t.}\ (\gamma)\end{aligned}$$

$$\begin{aligned}\partial^{\mu}\hat{K}_{\mu\nu} &= 2x_{\nu}\hat{T}^{\lambda}{}_{\lambda} + \text{c.t.}\\ &= -3x_{\nu}(DDS+\bar{D}\bar{D}\bar{S}) + \text{c.t.}\ (\gamma)\end{aligned}$$

c.t. means contact terms, c.t. (γ) contact terms with anomalous contributions. Their space-time integration yields the $\hat{W}$'s of (2.7) with anomalous dimensions, resp. R-weight for A and $\bar{A}$ hence the respective (broken) Ward-identities. In this way the superconformal structure of the full renormalized theory is most concisely expressed: all anomalies in the conservation equations are described by S, all anomalous contributions to contact terms go with γ.

We still have to clarify the status of the integrated R-Ward-identity since the hard S term hints to a hard breaking and similarly the identification of α,β,γ has to be performed. To this end we integrate (16.1.18) over space-time and use (6.18):

$$\hat{W}^{R}(\gamma)\Gamma = -(\int dSS - \int d\bar{S}\bar{S})\cdot\Gamma \tag{16.1.20}$$

Explicitly:

$$\begin{aligned}\hat{W}^{R}(\gamma)\Gamma &= \tfrac{1}{12}\,\alpha M(s-1)(\int dS\,\mathfrak{m} - \int d\bar{S}\,\bar{\mathfrak{m}})\cdot\Gamma\\ &\quad - \tfrac{2}{3}\,\beta[\int dS\partial_{g}L - \int d\bar{S}\partial_{g}\bar{L}]\cdot\Gamma\ .\end{aligned} \tag{16.1.21}$$

The soft term $\mathfrak{m}$ is to be expected from the tree approximation; the term $\partial z/\partial g$ in $\partial_{g}L$ (s. (16.1.14)) drops out since it is real; but the compensation of the trilinear A-terms is non-trivial: they cancel against the anomalous dimension part of the left-hand-side because (s. (16.1.17))

$$\beta = 3\gamma \frac{\lambda}{\partial_g \lambda} \quad . \tag{16.1.22}$$

Hence we confirm again theorem 13.5, namely that the naive conformal R-symmetry (n = -2/3) is like in the tree approximation only broken by soft terms. (16.1.22) can be considered as genuine expression for the non-renormalization theorem of the chiral three-vertex.

By the same technique we shall now identify the Callan-Symanzik functions. In (6.27') we had defined a generalized energy-momentum tensor

$$\hat{T}_{\mu\nu} = -\frac{1}{16}(V_{\mu\nu} + V_{\nu\mu} - 2g_{\mu\nu})V_\lambda{}^\lambda \tag{16.1.23}$$

$$V_{\mu\nu} \equiv [D^\beta, \bar{D}^{\dot\beta}]\sigma_{\mu\beta\dot\beta}\sigma_\nu^{\alpha\dot\alpha}V_{\alpha\dot\alpha}$$

and the dilatational current as a moment of it

$$\hat{D}_\mu = x^\nu \hat{T}_{\mu\nu} \quad . \tag{16.1.24}$$

Then the local dilatational Ward-identity reads

$$\partial^\mu \hat{D}_\mu = \hat{T}_\mu{}^\mu + i x^\nu w_\nu^P \tag{16.1.25}$$

with (s. (6.54))

$$\begin{aligned} w_\nu^P = \frac{1}{16}\big(&-(D D \bar{D}_{\dot\alpha} w_\alpha + \bar{D}\bar{D} D_\alpha w_{\dot\alpha})\sigma_\nu^{\alpha\dot\alpha} \\ &-\sigma_\nu{}^{\beta\dot\beta}[D_\beta, \bar{D}_{\dot\beta}](D^\alpha w_\alpha - \bar{D}_{\dot\alpha}\bar{w}^{\dot\alpha}) \\ &+ 8i\partial_\nu(D^\alpha w_\alpha + \bar{D}_{\dot\alpha}\bar{w}^{\dot\alpha})\big) \end{aligned}$$

and some additional contact terms w^P_{trace} hidden in the trace. Together they form the local dilatational contact terms

$$w^D = w^P_{trace} + x^\nu w_\nu^P \quad , \tag{16.1.26}$$

$$w^P_{trace} = \frac{3}{2} i(D^\alpha w_\alpha + \bar{D}_{\dot\alpha}\bar{w}^{\dot\alpha}) \quad .$$

Space-time integration yields

$$\hat{W}^D = \int dx w^D \tag{16.1.27}$$

hence

$$\hat{W}^D \Gamma = -\frac{3}{2} i [\int dSS + \int d\bar{S}\bar{S}] \cdot \Gamma . \tag{16.1.28}$$

In our concrete example, with S from (16.1.15) and the anomalous contact terms, we have the relation

$$\hat{W}^D(\gamma)\Gamma = \frac{1}{8} i\alpha M(s-1) (\int dS \boldsymbol{m} + \int d\bar{S} \boldsymbol{m}) \cdot \Gamma \tag{16.1.29}$$

$$- i\beta\partial_g (\int dSL + \int d\bar{S}\bar{L}) \cdot \Gamma$$

i.e.

$$\hat{W}^D(\gamma)\Gamma = i\alpha M(s-1)\Delta_0 \cdot \Gamma - i\beta\partial_g \Gamma \tag{16.1.30}$$

where $\Delta_0 \cdot \Gamma \equiv \frac{1}{8} (\int dS \boldsymbol{m} + \int d\bar{S} \bar{\boldsymbol{m}}) \cdot \Gamma$.

The last relation we need comes from naive dimensional analysis:

$$W^D \Gamma = i(M\partial_M + 2\mu^2 \partial_{\mu^2})\Gamma . \tag{16.1.31}$$

(M is the auxiliary mass parameter, cf. section 9; μ^2 is the normalization point, cf. (16.1.4); note that W^D refers to the canonical dimensions.) Taking the zeroth Θ-component of (16.1.30) and combining it with (16.1.31) we arrive at the Callan-Symanzik equation

$$(2\mu^2 \partial_{\mu^2} + M\partial_M + \beta\partial_g - \gamma \mathcal{N})\Gamma = \alpha M(s-1)\Delta_0 \cdot \Gamma . \tag{16.1.32}$$

$$\mathcal{N} \equiv \int dSA \frac{\delta}{\delta A} + \int d\bar{S}\bar{A} \frac{\delta}{\delta \bar{A}}$$

is the leg counting operator. This equation identifies now the functions α, β, γ which we introduced above (16.1.17) as special combinations of the Zimmermann coefficient $\tilde{r}$ and the normalization factors z, λ.

For completeness we discuss still the transition of the Callan-Symanzik equation to the renormalization group equation, i.e. the limit $s \rightarrow 1$. Clearly the right-hand-side of (16.1.32) goes to zero with $s \to 1$, but what happens with the left-hand-side? We wish to show

$$\partial_M \Gamma \Big|_{s=1} = 0 \tag{16.1.33}$$

(the Lowenstein-Zimmermann equation) for then we have derived the renormalization group equation and also the independence of the vertex functions from the subtraction scheme parameter M.

The action principle (11.5) yields

$$M\partial_M \Gamma = [M\partial_M \Gamma_{eff}]^4_4 \cdot \Gamma \ , \tag{16.1.34}$$

hence we study the insertion

$$[M\partial_M L]^3_3 \cdot \Gamma = [\tfrac{1}{32} M\partial_M z \bar{D}\bar{D}(A\bar{A}) + \tfrac{1}{48} M \frac{\partial\lambda}{\partial M} A^3 + \tfrac{1}{8} M(s-1)A^2]^3_3 \cdot \Gamma \tag{16.1.35}$$

We combine it with $[\partial_g L]\cdot\Gamma$, (16.1.14), the Zimmermann identity (16.1.8) and the equation of motion (16.1.1) in such a way that all hard terms are eliminated in favour of $M\partial_M$, ∂_g and $A\delta/\delta A$. Integration over the sum of chiral and antichiral parts then yields

$$(M\partial_M + \bar{\beta}\partial_g - \bar{\gamma}\mathcal{N})\Gamma = \bar{\alpha}(M(s-1)\Delta_0 \cdot \Gamma \tag{16.1.36}$$

where $\bar{\alpha},\bar{\beta},\bar{\gamma}$ are the (unique) solutions of

$$\begin{aligned} &\bar{\alpha} + 2\bar{\gamma} - 1 = 0 \ , \\ &\partial_g \lambda \bar{\beta} - 3\lambda\bar{\gamma} + M\partial_M \lambda = 0 \ , \\ &4\tilde{r}\bar{\alpha} + \partial_g z\bar{\beta} - 2z\bar{\gamma} + 2M\partial_M z = 0 \ . \end{aligned} \tag{16.1.37}$$

(These equations ensure the absence of the hard terms $[\bar{D}\bar{D}(A\bar{A})]^3_3$, $[A^3]^3_3$, $[M(s-1)A^2]^3_3$ in (16.1.36).) But $\bar{\beta},\bar{\gamma}$ can also be directly calculated via normalization conditions (16.1.4) applied to (16.1.36). The result is

$$\bar{\beta} = \bar{\gamma} = 0 \ , \tag{16.1.38}$$

hence from (16.1.37)

$$\bar{\alpha} = 1, \ \partial_M \lambda = 0 \ , \quad M\partial_M z = -2\tilde{r} \tag{16.1.39}$$

i.e. only the kinetic counterterm z-1 has non-trivial M dependence and (16.1.36) reduces to

$$M\partial_M \Gamma = M(s-1)\Delta_0 \cdot \Gamma \tag{16.1.40}$$

At s=1 this is the result (16.1.33) which we wanted to prove.
As an interesting by-product we still note that the Callan-Symanzik functions constructed with the normalization condition (16.1.4) are mass-independent:

$$\begin{aligned} \alpha(g, M^2/\mu^2) &= \alpha(g) \ , \\ \beta(g, M^2/\mu^2) &= \beta(g) \ , \\ \gamma(g, M^2/\mu^2) &= \gamma(g) \ . \end{aligned} \tag{16.1.41}$$

For the proof we differentiate the Callan-Symanzik equation with respect to M at s=1 and use (16.1.33):

$$(\partial_M \beta)\partial_g \Gamma\Big|_{s=1} - (\partial_M \gamma)\mathcal{N}\Gamma\Big|_{s=1} = 0 \tag{16.1.42}$$

On the normalization conditions(16.1.4) this leads to

$$\partial_M \beta = \partial_M \gamma = 0 \ , \tag{16.1.43}$$

whereas $\quad \partial_M \alpha = 0 \qquad\qquad (16.1.44)$

follows from (16.1.17).

16.1.3 Summary

Let us collect our results. Both, massless and massive case permit the construction of a supercurrent with interpretation, i.e. with conserved, symmetric, improved energy-momentum tensor $T_{\mu\nu}$ and conserved, improved supersymmetry currents $Q_{\mu\alpha}$, $\bar{Q}_{\mu\dot{\alpha}}$. In the massless case we can have a strictly conserved R-current too; then the resulting superconformal structure is based on naive R-weights and dimensions and its anomalies are described by the real vectorsuperfield B (16.1.10), (16.1.12). In the massive case another superconformal structure emerges, namely with anomalous dimensions and weights, $n^{\gamma} = -2/3\, d^{\gamma}$ and the anomalies now being given by a chiral field S (16.1.15), (16.1.19). The transition massive-massless is smooth in the sense that in the massless limit the S formulation is also possible, the difference to the B picture coming about by the different choice of contact terms i.e. weights and dimensions. Let us note parenthetically that the different superconformal structures make themselves felt on the supergravity level, where the supercurrent is the source for fields (cf. [R.11]).

16.2. Supersymmetric QED

Let us first recall the effective action of the model (cf. (14.1)). It is in form just the classical action (plus counterterms), but now each term is understood as N_4-insertion.

$$\Gamma_{eff} = \int dSL + \int d\bar{S}\bar{L} \qquad (16.2.1)$$

$$L = z_1 L_1 + \frac{1}{\alpha} L_2 - 8(M^2+b)L_5 + (m+a)L_3 + z_4 L_4$$

$$L_1 = \frac{-1}{256} N_3 [\bar{D}\bar{D}D\Phi\bar{D}\bar{D}D\Phi],$$

$$L_2 = \frac{-1}{256} N_3 [\bar{D}\bar{D}DD\Phi\bar{D}\bar{D}\Phi] = \frac{-1}{256} \bar{D}\bar{D}\, N_2 [DD\Phi\bar{D}\bar{D}\Phi] \equiv \bar{D}\bar{D}\mathcal{L}_2$$

$$L_3 = -\frac{1}{4} N_3 [A_+A_-] ,$$

$$L_4 = \frac{1}{2} \bar{D}\bar{D}\, I_5 , \quad I_5 = \frac{1}{16} N_2 [\bar{A}_+ e^{g\Phi} A_+ + A_- e^{-g\Phi}\bar{A}_-],$$

$$L_5 = -\frac{1}{256} N_3 [\bar{D}\bar{D}(\Phi^2)] .$$

z_1, z_4, a, b are formal power series to be fixed by normalization conditions such that m,M are the physical masses of the matter, respectively vector field.
We shall also use the notation

$$\Gamma_{eff} = \sum_i z_i \Delta_i \qquad (16.2.1')$$

where

$$\Delta_i = \int dSL_i + \int d\bar{S}\bar{L}_i .$$

Next we need the contact terms (6.56)

$$w_\alpha \equiv -2DDD_\alpha\Phi \frac{\delta}{\delta\Phi} + 2D_\alpha\Phi\bar{D}\bar{D} \frac{\delta}{\delta\Phi}$$
$$+n\, D_\alpha (A_+ \frac{\delta}{\delta A_+} + A_- \frac{\delta}{\delta A_-}) + 2D_\alpha A_+ \frac{\delta}{\delta A_+} + 2D_\alpha A_- \frac{\delta}{\delta A_-} \qquad (16.2.2)$$

(recall: $n \equiv n_+ = n_-$ are the R-weights of the matter fields). Now we can proceed and deduce from the action principle (11.9)

$$-2w_\alpha\Gamma = \Delta_\alpha \cdot \Gamma \qquad (16.2.3)$$

where

$$\Delta_\alpha = [-2w_\alpha\Gamma_{eff}]_{7/2} \qquad (16.2.4)$$

i.e. is a local insertion of power counting 7/2. $w_\alpha\Gamma_{eff}$ has just the classical form and can therefore be decomposed in the same way as (6.57)

$$\Delta_\alpha = \bar{D}^{\dot{\alpha}}V'_{\alpha\dot{\alpha}} - 4(n+1)(m+a)D_\alpha L_3 + \frac{64}{3}(M^2+b)D_\alpha L_5 \qquad (16.2.5)$$
$$+ 6\,(n + \frac{2}{3})\bar{D}\bar{D}D_\alpha I_5 - \frac{1}{96\alpha} D_\alpha\bar{D}\bar{D}\, N_2[\Phi(\Psi-\bar{\Psi})]$$

$$V' = N_3\,[-\frac{z_4}{4} nV^1 - \frac{1}{2} z_4 V^2 + \frac{z_1}{16} V^3 + (M^2+b)V^M + \frac{1}{\alpha} V^g],$$

$$V^1_{\alpha\dot{\alpha}} = [D_\alpha,\bar{D}_{\dot{\alpha}}](\bar{A}_+e^{g\Phi}A_+ + A_-e^{-g\Phi}\bar{A}_-)\ ,$$

$$V^2_{\alpha\dot{\alpha}} = D_\alpha(A_-e^{-g\Phi})e^{g\Phi}\bar{D}_{\dot{\alpha}}(e^{-g\Phi}\bar{A}_-)$$
$$-\bar{D}_{\dot{\alpha}}(\bar{A}_+e^{g\Phi})e^{-g\Phi}D_\alpha(e^{g\Phi}A_+)\ ,$$

$$V^3_{\alpha\dot{\alpha}} = \bar{D}\bar{D}D_\alpha\Phi\, DD\bar{D}_{\dot{\alpha}}\Phi\ ,$$

$$V^4_{\alpha\dot{\alpha}} = \frac{1}{4}\sigma^\mu_{\alpha\dot{\alpha}}\partial_\mu(A_+A_- - \bar{A}_-\bar{A}_+)\ ,$$

$$V^M_{\alpha\dot{\alpha}} = -D_\alpha\Phi\bar{D}_{\dot{\alpha}}\Phi + \frac{1}{6}[D_\alpha,\bar{D}_{\dot{\alpha}}]\Phi^2\ ,$$

(V^g is given in (6.57); $\Psi \equiv \bar{D}\bar{D}DD\Phi$). The new feature is that all terms are insertions (normal products, cf. section 10) of the same power counting and that therefore the mass terms appear in their oversubtracted form, their difference from the minimally subtracted ones yielding all the non-naive, the quantum aspects of the problem. Since they are different for the cases where the vector field has a mass or not we shall treat them separately.

16.2.1 Massless vector field*

For $M^2 = 0$ we have an existence problem of the insertions. As noted in section 14.1 there is even in SQED and even in the Feynman gauge $\alpha=1$ a non-trivial off-shell infrared problem, hence we have to check explicitly whether all operators we need exist infraredwise. This check has been performed in [IV.6] for all terms appearing in (16.2.5) and, although individual ones are indeed infrared divergent, the sums V', V^g as well as all terms needed below are not (in the gauge $\alpha=1$ which we use from now on). This means that for $n = -2/3$ we have already a supercurrent with S type breaking. But the breaking S is hard and has a gauge varying piece (the term $D_\alpha \bar{D}\bar{D}\, N_2[\Phi(\Psi-\bar{\Psi})]$) which does not naively vanish between physical states. Similarly the current V' is not gauge invariant (not even when sandwiched with physical states) due to the term $N_3[V^g]$ in it. These defects have still to be cured.

The means to do so are provided by the Zimmermann identities

$$(m+a)L_m\cdot\Gamma = ((m+a)L_3 + r_k L_{kin} + r_g L_4 + r_m \bar{D}\bar{D}\bar{L}_m)\cdot\Gamma \ , \tag{16.2.6}$$

$$N_0[\Phi(\Psi-\bar{\Psi})]\cdot\Gamma = N_2[\Phi(\Psi-\bar{\Psi})]\cdot\Gamma + v_m(L_m-\bar{L}_m)\cdot\Gamma \tag{16.2.7}$$

where (s. (14.11), (14.7))

$$\begin{aligned} L_{kin} &= L_1 + L_2 - L_0 \\ L_0 &= \frac{-1}{256} N_0[\bar{D}\bar{D}DD\Phi\bar{D}\bar{D}\Phi] = \frac{-1}{256}\bar{D}\bar{D}\,\mathcal{L}_o \\ L_m &= -\frac{1}{4} N_2\,[A_+A_-]\ . \end{aligned} \tag{16.2.8}$$

The existence of the <u>unsubtracted</u> insertion L_0, $\mathcal{L}_0$ and $N_0[\Phi(\Psi-\bar{\Psi})]$ which is an ultraviolet problem is shown in [IV.16] on the basis of

---

* For simplicity we normalize $z_1=z_4=1$.

the gauge Ward identities. It is also shown there that L_{kin}, L_m, L_3 and L_4 are formally gauge invariant. (16.2.6) relates the oversubtracted matter mass term L_3 to its minimally subtracted version L_m; (16.2.7) relates the minimally subtracted gauge term $N_2[\Phi(\Psi-\bar{\Psi})]$ to an undersubtracted (i.e. non-subtracted) version of it which vanishes between physical states. With the Zimmermann identities we have isolated the soft mass contribution and the gauge varying breaking part which vanishes between physical states. With the help of

$$D_\alpha \bar{D}\bar{D}L_m = 4i\bar{D}^{\dot{\alpha}} N_3[V^4_{\alpha\dot{\alpha}}] \qquad (16.2.9)$$

we can absorb L_m correction terms in the current. Now we observe that V^g, the gauge varying part of the supercurrent, exists without subtraction. This makes it easy to repair this part of the current: we add and subtract $N_0[V_g]$. All of these manipulations together lead to

$$\bar{D}^\alpha V_{\alpha\alpha}\cdot\Gamma = -2w_\alpha\Gamma - 2D_\alpha S\cdot\Gamma - 6(n+\tfrac{2}{3})\bar{D}\bar{D}D_\alpha I_5\cdot\Gamma \qquad (16.2.10)$$

with

$$V = V_{inv} + V_0$$

$$V_{inv} = V' - \frac{16}{3} i \left(\frac{v_m}{128} + 3(n+1)r_m\right)V^4 - V_0$$

$$V_0 = N_0[V^g]$$

$$S = -2(n+1)\left((m+a)L_m - r_k L_{kin} - \frac{1}{2} r_g \bar{D}\bar{D}I_5\right) + S_0$$

$$S_0 = \frac{1}{192}\bar{D}\bar{D}N_0[\Phi(\bar{\Psi}-\Psi)]$$

V_{inv} is indeed invariant since the gauge variation of $N_3[V_g]$ is linear in Φ hence compensated by the variation of $N_0[V_g]$ to all orders, compatible with renormalization. The supercurrent V altogether is, also to all orders, gauge invariant between physical states. Indeed its gauge varying part V_0 is an unsubtracted normal product and can therefore be shown to vanish in the physical sector by naive manipulations. The breaking S contains also a gauge varying part, S_0, for which the same comment applies. The remaining terms are the soft mass term and two hard breaking terms appearing first in one loop with the Zimmermann coefficients r_k and r_g. Let us divide the further discussion as in the tree approximation into the cases $n = -2/3$, $n = \infty$.

Case $n = -\frac{2}{3}$

The entire breaking is of type S, we have a supercurrent with interpretation, the current is gauge invariant between physical states. In this form, i.e. with canonical contact terms, it gives rise to a superconformal structure governed by the chiral anomaly multiplet S which contains two independent basis elements, L_{kin} and $\bar{D}\bar{D}\, I_5$. This structure exists locally but care has to be taken when one wants to integrate (cf. sect. 14.1, Δ_m does not exist). It is thus preferable to find a form in which one can integrate with impunity. Interestingly enough this version emerges almost automatically when one looks for a superconformal structure with anomalous weights and dimensions i.e. when one tries to derive the Callan-Symanzik equation from local dilatations via the moment construction (cf. the analogous discussion in the Wess-Zumino model, section 16.1.2). The way to proceed is to replace in S of (16.2.10) the hard terms (going with r_k, s_g respectively) by derivatives $\partial_g\Gamma$ and additional contact terms. We therefore write down the trivial identity (use (16.2.1) with $z_1 = z_4 = 1$)

$$(g\partial_g - N_\Phi)L\cdot\Gamma = (g\partial_g aL_3 + L_1 + L_2))\cdot\Gamma \ . \tag{16.2.11}$$

$$N_\Phi \equiv \int dV\Phi\delta_\Phi \ ,$$

the matter field equation

$$A_+ \frac{\delta\Gamma}{\delta A_+} + A_- \frac{\delta\Gamma}{\delta A_-} = (2(m+a)L_3 + 2L_4)\cdot\Gamma \tag{16.2.12}$$

and combine them with the Zimmermann identity (16.2.6). The result is a trace equation with anomalous contact terms:

$$\bar{D}^{\dot\alpha}\tilde{V}_{\alpha\dot\alpha}\cdot\Gamma = -2w_\alpha^{(\gamma)}\Gamma - 2D_\alpha\tilde{S}\cdot\Gamma \ . \tag{16.2.13}$$

Here, the supercurrent $\tilde{V}$ consists again of a gauge invariant part and one having no physical contribution

$$\tilde{V} = \tilde{V}_{inv} + V_o ,$$

$$\tilde{V}_{inv} = N_3 \left[\frac{1}{6} V^1 - \frac{1}{2} V^2 + \frac{1}{16} V^3 + cV^4 + V^g\right] - V_o ,$$

$$c = - \frac{16}{3} i \quad \left(\frac{v_m}{128} + (1+\sigma) r_m\right) . \tag{16.2.14}$$

The breaking is of S type and reads

$$\tilde{S} = \frac{2}{3} \beta \partial_g L - \frac{2}{3} (1+\sigma)(m+a) L_m^* + S_o \tag{16.2.15}$$

L_m^* is the modified local matter mass term

$$L_m^* = L_m + \frac{r_k}{m+a} L_o , \tag{16.2.16}$$

$$L_o = N_o \left[\bar{D}\bar{D}DD\Phi \; \bar{D}\bar{D}\Phi\right] ,$$

which we announced above: it exists also when integrated. The terms $(\partial_g - N_V)L$ and L_m^* are gauge invariant operators in the sense of section 12. The anomalous contact terms are given by

$$w_\alpha^{(\gamma)}\Gamma = - \frac{2}{3} (1+\gamma) D_\alpha \left(A_+ \frac{\delta\Gamma}{\delta A_+} + A_- \frac{\delta\Gamma}{\delta A_-}\right) + 2\left(D_\alpha A_+ \frac{\delta\Gamma}{\delta A_+} + D_\alpha A_- \frac{\delta\Gamma}{\delta A_-}\right)$$

$$- 2 \left(\bar{D}\bar{D}D_\alpha \Phi \frac{\delta\Gamma}{\delta\Phi} - D_\alpha\Phi \; \bar{D}\bar{D} \frac{\delta\Gamma}{\delta\Phi}\right) - \frac{2}{3} \beta (N_\Phi D_\alpha L)\cdot\Gamma. \tag{16.2.17}$$

The coefficients β,γ,σ are the following functions of the Zimmermann coefficients and counterterms

$$\beta = - \frac{1}{2} (1+\sigma) r_k ,$$

$$\gamma = - \frac{1}{2} (1+\sigma) r_g , \tag{16.2.18}$$

$$1+\sigma = \frac{1}{1-r_g + \frac{1}{2} r_k g \partial_g \ln(m+a)} .$$

They will be identified below. The matter part is clearly one which gives correct anomalous R-weights and dimensions to the matter fields, $n^\gamma = -2/3 \, d^\gamma$ (cf. with section 16.1.2): the vector field part consists of the tree contact term part plus the term going with β. This form has to be checked to give still canonical R-weight zero to Φ (a real vector superfield must have this weight if R is unbroken), but nevertheless a non-vanishing anomalous dimension to Φ. Hence we study first its contribution to $\partial^\mu\tilde{V}_\mu$:

$$\text{in} \quad \partial^\mu\tilde{V}_\mu : -\frac{2}{3} i\beta(DDN_\Phi L - \bar{D}\bar{D}N_\Phi \bar{L})\cdot\Gamma \tag{16.2.19}$$

Integration over space-time yields

$$-\frac{2}{3} i\beta(\int dSN_\Phi L - \int d\bar{S}N_\Phi \bar{L})\cdot\Gamma$$

$$= -\frac{2}{3} i\beta\, N_\Phi(\int dSL - \int d\bar{S}\bar{L})\cdot\Gamma \;=\; 0 \quad . \tag{16.2.20}$$

I.e. the R-weight of Φ is the canonical one: $n(\Phi) = 0$. (16.2.21)

Next we study the local dilatations in order to derive the Callan-Symanzik equation. As usual, cf. (16.1.25), the moment construction yields

$$\begin{aligned}\partial^\mu\hat{D}_\mu\cdot\Gamma &= i\hat{W}^D\Gamma + \beta(DDN_\Phi L + \bar{D}\bar{D}N_\Phi\bar{L})\cdot\Gamma \\ &+ \gamma DD(A_\pm \frac{\delta\Gamma}{\delta A_\pm}) + \gamma\bar{D}\bar{D}(\bar{A}_\pm \frac{\delta\Gamma}{\delta\bar{A}_\pm}) - \frac{3}{2}(DD\tilde{S} + \bar{D}\bar{D}\tilde{\bar{S}}) \quad .\end{aligned} \tag{16.2.22}$$

Integrating over space-time and taking the zeroth Θ-component of (16.2.22) we obtain

$$\begin{aligned}\int dx\partial^\mu D_\mu\cdot\Gamma = 0 = iW^D\Gamma &+ \gamma(\int dSA_\pm \frac{\delta\Gamma}{\delta A_\pm} + \int d\bar{S}\bar{A}_\pm \frac{\delta\Gamma}{\delta\bar{A}_\pm}) \\ &- \beta(g\partial_g - N_V)\Gamma \\ &+ (1+\sigma)(m+a)(\int dSL^*_m + \int d\bar{S}\bar{L}^*_m)\cdot\Gamma \;.\end{aligned} \tag{16.2.23}$$

With the relation

$$im\partial_m\Gamma = W^D\Gamma \tag{16.2.24}$$

coming from naive dimensional analysis (16.2.22) becomes the Callan-Symanzik equation

$$(m\partial_m + \beta(g\partial_g - N_\Phi) - \gamma N_A)\Gamma = (1+\sigma)(m+a)\Delta^*_m\cdot\Gamma. \tag{16.2.25}$$

Here

$$\Delta_m^* \equiv \int dSL_m^* + \int d\bar{S}\bar{L}_m^*$$

$$N_A \equiv \int dSA_\pm \frac{\delta}{\delta A_\pm} + \int d\bar{S}\bar{A}_\pm \frac{\delta}{\delta \bar{A}_\pm}$$

and we have identified the Callan-Symanzik functions $\beta, \gamma, 1+\sigma$.

This concludes the treatment of the case with conformal weight $n = -2/3 \rightarrow n^\gamma = -2/3(1+\gamma)$. We have a superconformal structure with anomalies described by a chiral multiplet $\tilde{S}$ (16.2.15) and anomalous weights and dimensions given by γ.

Case $n = \infty$.

The supercurrent (16.2.10) takes the form

$$\begin{aligned} V_{\alpha\dot{\alpha}}^{(\infty)} &:= \lim_{n\to\infty} \frac{1}{n} V_{\alpha\dot{\alpha}} = -\frac{1}{4} V_{\alpha\dot{\alpha}}^1 - 16ir_m V_{\alpha\dot{\alpha}}^4 \\ &= -4\left[D_\alpha, \bar{D}_{\dot{\alpha}}\right] I_5 + 16ir_m \sigma^\mu_{\alpha\dot{\alpha}} \partial_\mu (L_m - \bar{L}_m) \end{aligned} \tag{16.2.26}$$

and the trace equation (16.2.10) reduces (after reshuffling of terms) to

$$\bar{D}\bar{D}\, I_5' \cdot \Gamma = w_5 \Gamma - 2(m+a) L_m \cdot \Gamma + 2r_k L_{kin} \cdot \Gamma \tag{16.2.27}$$

where

$$I_5' = (1 - r_g) I_5 - 2r_m (L_m + \bar{L}_m),$$

$$w_5 = A_+ \frac{\delta}{\delta A_+} + A_- \frac{\delta}{\delta A_-},$$

$$L_{kin} = L_1 + L_2 - L_0 .$$

As noted in section 6 the commutator of these w_5-transformations with supersymmetry is trivial and similarly the supercurrent $V^{(\infty)}$ is in a sense trivial: its first component (correspondingly the $\Theta\bar{\Theta}$-component of I_5') is the axial current belonging to the transformations

$$\delta_R^{(\infty)} \Phi = 0 , \quad \delta_R^{(\infty)} A_\pm = iA_\pm , \quad \delta_R^{(\infty)} \bar{A}_\pm = -i\bar{A}_\pm \tag{16.2.28}$$

but its higher components do not contain any other non-trivial current.

The remarkable property of the current I_5' is that its anomaly $2r_k L_{kin}$ is not renormalized i.e.

$$r_k = \frac{g^2}{16(16\pi)^2} \tag{16.2.29}$$

is of one loop order only (s. (16.2.132) and (16.2.133). The proof of this property will be given below (in section 16.2.4).

16.2.2 Massive vector field

The supercurrent constructed in the preceding subsection for a massless vector field has been seen to exist only for the Feynman gauge $\alpha = 1$. We shall show below that its gauge independence can indeed be proved by starting from an appropriately chosen operator for a massive vector field and going to the limit of vanishing mass. Apart from this motivation the supercurrent for a massive vector field is interesting in its own right since it permits a complete treatment: in a massive theory one can really go on the mass shell, physical states can be truely defined and thus gauge invariance of operators really be checked in more than a formal sense. The task is, exactly as in the massless case, to replace the oversubtracted mass terms by the minimally subtracted ones and to bring the gauge varying terms into a more transparent form. Hence we need the Zimmermann identities

$$(m+a)L_m \cdot \Gamma = ((m+a)L_3 + r_k L_{kin} + r_g L_4 + r_m \bar{D}\bar{D}\bar{L}_m)\cdot\Gamma \ , \tag{16.2.30}$$

Here

$$L_{kin} = L_1 + \frac{1}{\alpha}(L_2 - L_o)$$

and we still work with L_{kin} instead of L_1 alone since we wish to be able to go to the limit $M \to 0$, where L_1 diverges infraredwise.

$$(M^2+b)L_m \cdot \Gamma = ((M^2+b)L_5 + t_4 L_4 + t_m \bar{D}\bar{D}\bar{L}_m)\cdot\Gamma \ . \tag{16.2.31}$$

$$L_M = -\frac{1}{256} N_1[\bar{D}\bar{D}(\Phi^2)]$$

is the minimally subtracted vector mass term.

$$N_o[\Phi(\Psi-\bar{\Psi})]\cdot\Gamma = (N_2[\Phi(\Psi-\bar{\Psi})] + v_m(L_m - \bar{L}_m))\cdot\Gamma \tag{16.2.32}$$

We shall also need the identity

$$(g\partial_g - N_\Phi - 2\alpha\partial_\alpha)L\cdot\Gamma - \frac{2}{\alpha}(L_2 - L_0)\cdot\Gamma$$
$$= (-2L_{kin} - 16(M^2 + b + \frac{1}{2}g\partial_g b)L_5 + g\partial_g aL_3 + (g\partial_g - 2\alpha\partial_\alpha)z_4L_4)\Gamma \ . \qquad (16.2.32)$$

(We have anticipated from sect. 16.2.4 that one can choose: $z_1 = 1$, $\alpha\partial_\alpha b = \alpha\partial_\alpha a = 0$.) The last ingredient is the local counting operator for the matter fields

$$A_+ \frac{\delta\Gamma}{\delta A_+} + A_- \frac{\delta\Gamma}{\delta A_-} = 2(m+a)L_3 + z_4L_4 \ . \qquad (16.2.34)$$

The breaking terms in (16.2.5) can now be rewritten in terms of soft mass terms, $g\partial_g + \ldots$, and contact terms (note: $n = -2/3$)

$$-\frac{4}{3}(m+a)D_\alpha L_3 + \frac{64}{3}(M^2+b)D_\alpha L_5$$
$$= D_\alpha(x_1L_m + x_2L_M + x_3(g\partial_g\ldots) + x_4A_\pm\frac{\delta}{\delta A_\pm}) + y\,\bar{D}\bar{D}\Gamma_m) \ . \qquad (16.2.35)$$

The four coefficients x_i are determined by demanding the absence of the hard terms L_{kin}, L_3, L_4, L_5. The term going with y will eventually be absorbed into the current by using (16.2.9). We shall not write down the actual values for x_i since in the subsection 16.2.4 we shall construct another basis for all formally gauge invariant chiral insertions with dimension three: one which is gauge invariant in the sense of section 12, namely symmetric and α-independent. The terms of the above basis do not have this property. Anticipating this result we may write the breaking as

$$-\frac{3}{2}S = m(1+\sigma)\hat{L}_m + 2\beta\hat{L}_1 - \gamma E + \frac{1}{128\alpha}\bar{D}\bar{D}\,N_o[\Phi(\Psi-\bar{\Psi})] \qquad (16.2.36)$$
$$+ 2M^2L_M$$

here

$$E \equiv A_+\frac{\delta}{\delta A_+} + A_-\frac{\delta}{\delta A_-} \ ,$$

and $\hat{L}_1$, $\hat{L}_m$ are symmetric and α-independent equivalents of L_{kin} and L_m (cf. below (16.2.73), (16.2.74)).

We still have to segregate those parts of L_M, V^M which contribute gauge invariantly. They can be found by introducing the physical vector field ϕ^T

$$\phi^T = \Phi - \phi^L \quad , \tag{16.2.37}$$

$$\phi^L \equiv \frac{\{DD,\bar{D}\bar{D}\}}{16\alpha M^2}\,\Phi \quad .$$

It satisfies

$$w_\Lambda(2)\phi^T(1) = \frac{1}{\alpha M^2}\,(\Box + \alpha M^2)\delta_S(1,2) \tag{16.2.38}$$

i.e. - as an operator - it commutes with the operator $DD\Phi$ (analogously with $\bar{D}\bar{D}\Phi$). The separation of transverse contributions has to be combined with a proper treatment of the remaining longitudinal ones. Now, for any formal product $\mathcal{D}\,\Phi\bar{D}\bar{D}\Phi$ where $\mathcal{D}$ is any differential operator one can in fact define [II.16] a normal product $N_*[\mathcal{D}\,\Phi\bar{D}\bar{D}\Phi]$ which does not contribute between physical states. With the help of this normal product we can decompose

$$N_1[L_M] = \hat{N}_1[L_M^T] + N_*[L_M^L] \ ,$$
$$N_3[V_M] = \hat{N}_3[V_M^T] + N_*[V_M^L] \tag{16.2.39}$$

where

$$L_M^L = \frac{1}{256}\,\bar{D}\bar{D}\,(-2\Phi\phi^L + (\phi^L)^2) \ , \tag{16.2.40}$$

$$V_M^L = -\,D_\alpha\phi^L\bar{D}_{\dot\alpha}\Phi - D_\alpha\Phi\bar{D}_{\dot\alpha}\phi^L - \frac{1}{6}\,[D_\alpha,\bar{D}_{\dot\alpha}](-2\Phi\phi^L + (\phi^L)^2)$$
$$+ D_\alpha\phi^L\bar{D}_{\dot\alpha}\phi^L \quad . \tag{16.2.41}$$

Now the current and the breaking consist of strictly gauge invariant (i.e. symmetric and α-independent) terms plus those which are not gauge invariant but vanish on the physical mass shell.

$$\bar{D}^{\dot\alpha}V_{\alpha\dot\alpha}\cdot\Gamma = -2w_\alpha\Gamma - 2D_\alpha S\cdot\Gamma \tag{16.2.42}$$
$$V = V' + \text{absorp.terms} + (M^2+b)\hat{N}_3[V_M^T] + N_*[V_M^L] + \frac{1}{\alpha}\,N_*[V^g]$$

V' is given in (16.2.5), absorption terms are a multiple of V^4 (cf. (16.2.5) and (16.2.9)); V_M^T and V_M^L are given above and V^g is written in (6.57). S is presented in (16.2.36) with definitions of L_1, L_m given below (16.2.74), (16.2.73) and L_M replaced by (16.2.39).

16.2.3 Massive vector field, massless matter fields

For massive vector fields the convergence criteria of section 9 permit massless chiral fields. We shall shortly discuss this case since it allows for the B-type breaking (cf. section 6.5) for a supercurrent.

Instead of m+a we have now an auxiliary mass m(s-1) and all relations above remain essentially unchanged. The Zimmermann identities (16.2.30), (16.2.31) are replaced by

$$m(s-1)[L_m]^{2}_{2}\cdot\Gamma = [m(s-1)L_m]^{3}_{3}\cdot\Gamma + [\tilde{r}_k L_1 + \tilde{r}_g L_4 + \tilde{r}_m \bar{D}\bar{D}L_m]^{3}_{3}\cdot\Gamma\,, \qquad (16.2.43)$$

$$(M^2+b)[L_M]^{1}_{1}\cdot\Gamma = (M^2+b)[L_5]^{3}_{3}\cdot\Gamma + [\tilde{t}_4 L_4 + \tilde{t}_m \bar{D}\bar{D}L_m]^{3}_{3}\cdot\Gamma \qquad (16.2.44)$$

where we have used L_1 instead of L_{kin} since for a massive vector field L_1 exists. Using (16.2.43) and (16.2.44) for the breaking terms $D_\alpha L_3$ and $D_\alpha L_5$ in (16.2.5) we obtain

$$D_\alpha(-4(n+1)[m(s-1)L_m]^{3}_{3} + \frac{64}{3}(M^2+b)[L_5]^{3}_{3})\cdot\Gamma \qquad (16.2.45)$$

$$= D_\alpha[4(n+1)\tilde{r}_k L_1 + (4(n+1)\tilde{r}_g - \frac{64}{3}\tilde{t}_4)L_4 + (4(n+1)\tilde{r}_m - \frac{64}{3}\tilde{t}_m)\bar{D}\bar{D}L_m]^{3}_{3}\cdot\Gamma$$

$$+ D_\alpha[\frac{64}{3}(M^2+b)[L_M]^{1}_{1} - 4(n+1)m(s-1)[L_m]^{2}_{2}]\cdot\Gamma\,.$$

The gauge varying part S_o, (16.2.10), can be rewritten as B-type breaking:

$$S_o = \frac{1}{192\alpha}\bar{D}\bar{D}N_o[\Phi(\bar{\Psi}-\Psi)] = \frac{1}{192\alpha}\bar{D}\bar{D}N_o[-2\bar{D}\bar{D}\Phi DD\Phi + \Phi\{DD,\bar{D}\bar{D}\}\Phi]$$

$$\equiv \bar{D}\bar{D}N_o B_o\,, \quad \bar{B}_o = B_o\,, \qquad (16.2.46)$$

since there are no symmetric counterterms annihilated by $\bar{D}\bar{D}$.

Finally we observe that

$$\bar{D}\bar{D}D\Phi\bar{D}\bar{D}D\Phi = \bar{D}\bar{D}(D\Phi\bar{D}\bar{D}D\Phi + \bar{D}\Phi DD\bar{D}\Phi + \Phi D\bar{D}\bar{D}D\Phi)$$

$$\equiv \bar{D}\bar{D}B_1 \quad . \tag{16.2.47}$$

This means that at s=1 all S-type breaking terms are of the form

$$S = \bar{D}\bar{D}B \ , \ B = \bar{B} \tag{16.2.48}$$

hence can be rewritten in the B-type breaking version. We thus arrive at the following trace equation

$$\bar{D}^{\dot{\alpha}}V_{\alpha\dot{\alpha}}\cdot\Gamma = -2w_\alpha\Gamma + 6\bar{D}\bar{D}D_\alpha B\cdot\Gamma \tag{16.2.49}$$

with

$$V_{\alpha\dot{\alpha}} = N_3^3[V'_{\alpha\dot{\alpha}}] + 4[D_\alpha,\bar{D}_{\dot{\alpha}}]B$$

V' given by (16.2.5) ,

$$B = -\frac{1}{24}(M^2+b)[\phi^2]_0^0 - \frac{n+1}{128}\tilde{r}_k[B_1]_2^2 + ((n+1)\tilde{r}_g - \frac{16}{3}\tilde{t}_4)[I_5]_2^2$$
$$+(2(n+1)\tilde{r}_m - \frac{32}{3}\tilde{t}_m - \frac{1}{192\alpha}\tilde{v}_m)[L_m+\bar{L}_m]_2^2 + B_0 \ .$$

The main difference of the supercurrent (16.2.49) to the one in the S-type version lies in the term B_1. It is not gauge invariant hence has to be treated like the V^M-term. We employ the field Φ^T (16.2.37) in order to extract the gauge invariant contribution of B_1, and the N_*-product, cf. (16.2.39), in order to have the annihilation of the remaining gauge varying parts between physical states. This yields the final form of the current and the breaking

$$\hat{V} = V\Big|_{\Phi=\Phi^T} + N_*[V], \tag{16.2.50}$$

$$\hat{B} = B\Big|_{\Phi=\Phi^T} + N_*[B]. \tag{16.2.51}$$

What we have constructed herewith is the precise analogue to the B-version of the Wess-Zumino model, section 6.2.1. For massless matter we have a superconformal structure with naive R-weights ($n(\Phi) = 0$, $n(A_\pm)$ arbitrary) and breaking described by a real vector superfield. Current and breaking are gauge invariant between physical states. For $n = -2/3$ we could have sticked to the S-version and hence the analogy to the Wess-Zumino model is complete: there are indeed two different energy-momentum tensors possible even in the presence of a gauge invariance.

16.2.4 The gauge invariance of the supercurrent

For massless vector field the supercurrent has been constructed as formally gauge invariant operator in Feynman gauge ($\alpha=1$). It is thus very important to check its potential α-dependence and, in fact, to show that it is a gauge invariant operator in the sense of section 12. In order to do so we construct gauge invariant local insertions in the massive model and study then the limit of vanishing vector field mass [IV.12].

Before entering into the actual derivation of gauge invariant operators, we have to ensure the α-independence of the physical S-matrix. For Γ_{eff} (16.2.1) the action principle (11.5) yields

$$\begin{aligned}\partial_\alpha Z &= [\partial_\alpha \Gamma_{eff}]\cdot Z \qquad (16.2.52)\\ &= [\partial_\alpha z_1\Delta_1 - \tfrac{1}{\alpha^2}\Delta_2 + \partial_\alpha a\Delta_3 + \partial_\alpha z_4\Delta_4 - 8\partial_\alpha b\Delta_5]\cdot Z \quad .\end{aligned}$$

With the Zimmermann identity

$$\mathcal{L}_0 = \mathcal{L}_2 + \frac{r}{2} I_5 + r_m(L_m + \bar{L}_m) \qquad (16.2.53)$$

where $\mathcal{L}_2$ is defined in (16.2.1), $\mathcal{L}_0$ is its unsubtracted version, or, integrated

$$\Delta_0 = \Delta_2 + r\Delta_4 \, , \qquad (16.2.54)$$

we find

$$\partial_\alpha Z = [\partial_\alpha z_1\Delta_1 - \frac{1}{\alpha^2}\Delta_0 + \partial_\alpha a\Delta_3 + (\partial_\alpha z_4 + \frac{r}{\alpha^2})\Delta_4 - 8\partial_\alpha b\Delta_5]\cdot Z. \qquad (16.2.55)$$

Hence for the choice of counterterms such that

$$z_1 = 1, \ \partial_\alpha a = \partial_\alpha b = 0, \ z_4 = -\int_1^\alpha d\beta \frac{r}{\beta^2} + 1 \qquad (16.2.56)$$

the desired relation is obtained:

$$\partial_\alpha Z = - \frac{1}{\alpha^2} \Delta_0 \cdot Z \ . \tag{16.2.57}$$

It shows that the physical S-matrix is α-independent since the α-variation is an unphysical insertion, Δ_0, vanishing between physical states.

A generalization needed below is the corresponding expression for the matrix elements of an operator Q. They are α-independent if one can show that

$$\partial_\alpha (Q\cdot Z) = - \frac{1}{\alpha^2} \Delta_0 \cdot Q \cdot Z \tag{16.2.58}$$

(the right-hand-side of (16.2.58) represents a double insertion.)

The appropriate tool for finding gauge invariant insertions are gauge invariant <u>differential</u> operators i.e. differential operators ∂_λ which commute with the Ward-identities (14.5), (14.6) and are α-independent. Candidates are to be found amongst $m\partial_m$, $M\partial_M$, $g\partial_g$, $\alpha\partial_\alpha$ and the counting operators N_Φ, N_A. Before actually combining them it is useful to observe that Δ_M - which is, of course, not symmetric - is in fact α-independent. For the proof we start from the local Zimmermann identity

$$\mathcal{L}_0 N_0[\Phi^2]\cdot Z = \{\mathcal{K}_2 + \frac{r}{2} I_5 + r_m(L_m + \mathcal{L}_m)\} N_0[\Phi^2]\cdot Z \ , \tag{16.2.59}$$

hence have

$$\Delta_0 \cdot N_0[\Phi^2]\cdot Z = \{\Delta_2 + r\Delta_4\}\ N_0[\Phi^2]\cdot Z \ , \tag{16.2.60}$$

and therefore

$$- \frac{1}{\alpha^2} \Delta_0 N_0[\Phi^2] Z = \partial_\alpha (N_0[\Phi^2]\cdot Z). \tag{16.2.61}$$

After integration this is the desired result

$$- \frac{1}{\alpha^2} \Delta_0 \Delta_M \cdot Z = \partial_\alpha (\Delta_M \cdot Z) \ . \tag{16.2.62}$$

If we wish to build up gauge invariant operators from the above differential operators we may therefore cancel unsymmetric pieces with the help of Δ_M and the Zimmermann identity (s. (16.2.31))

$$(M^2 + b)\Delta_M = (M^2 + b)\Delta_5 + t_4\Delta_4 \,. \tag{16.2.63}$$

As long as Δ_M is to be multiplied by α-independent coefficients, the α-independence is maintained. A short calculation shows now that

$$(m\partial_m + M\partial_M + \varkappa_3 M^2\Delta_M)Z = [(m+a)\Delta_3 + 16t_4\Delta_4]\cdot Z \tag{16.2.64}$$

i.e. is symmetric. It is also α-independent since

$$\varkappa_3 = 16\ (1 + \frac{b}{M^2}) \tag{16.2.65}$$

is α-independent and $m\partial_m$, $M\partial_M$ commute with ∂_α hence have the same property. Analogously it is seen that

$$\begin{aligned} -\frac{1}{2}GZ \equiv -\frac{1}{2}(g\partial_g - N_\phi - 2\alpha\partial_\alpha + \varkappa_1 M^2\Delta_M)Z &= [\Delta_1 + \frac{1}{\alpha}(\Delta_2 - \Delta_0) \\ -\frac{1}{2}g\partial_g a\Delta_3 - \frac{1}{2}(g\partial_g z_4 &+ \frac{M^2}{M^2+b}\varkappa_1 t_4)\Delta_4]\cdot Z \end{aligned} \tag{16.2.66}$$

$$\varkappa_1 = -16 + \frac{8}{M^2}(g\partial_g b - 2b) \tag{16.2.67}$$

is symmetric. To see its α-independence requires also a little calculation:

$$\begin{aligned} \partial_\alpha(G'Z) = (G'-2)\partial_\alpha Z &= -G'\frac{1}{\alpha^2}\Delta_0 Z + \frac{2}{\alpha^2}\Delta_0 Z \\ &= -\frac{1}{\alpha^2}\Delta_0 G'Z \end{aligned} \tag{16.2.68}$$

($G' = G - \varkappa_1 M^2\Delta_M$), i.e. G is α-independent since G', x_1 and Δ_M are so.

Hence we define as gauge invariant insertions finally

$$m\,\hat{\Delta}_3 Z := (m\partial_m + M\partial_M + \varkappa_3 M^2\Delta_M)Z \tag{16.2.69}$$

$$\hat{\Delta}_1 Z := -\frac{1}{2}GZ\,. \tag{16.2.70}$$

In the context of the supercurrent we need local gauge invariant insertions. Let us first construct $\hat{L}_m$, the gauge invariant representative for the minimally subtracted local matter mass term. We begin with a Zimmermann identity on the level of vectorial densities:

$$\mathcal{L}_0 \cdot L_m \cdot Z = [\mathcal{L}_2 + r\mathcal{L}_4 + r_m(L_m + \bar{L}_m)] L_m \cdot Z \tag{16.2.71}$$

which implies

$$\partial_\alpha (L_m \cdot Z) = -\frac{1}{\alpha^2} \Delta_0 \cdot L_m \cdot Z \tag{16.2.72}$$

i.e. the α-independence of L_m. We normalize the relative coefficient between $\hat{L}_m$ and L_m such, that

$$m\hat{L}_m = (m+a)L_m \tag{16.2.73}$$

(which is possible since a is α-independent). The integrands $\hat{L}_1$, $\hat{L}_3$ of $\hat{\Delta}_1, \hat{\Delta}_3$ can differ from gauge invariant local insertions only by $\bar{D}\bar{D}\bar{L}_m$, hence by suitable normalization of u_1, u_3 they are gauge invariant and we can define

$$\hat{L}_1 := L_1 + \frac{1}{\alpha}(L_2 - L_0) - \frac{1}{2} g\partial_g a L_3 - \frac{1}{2}\left(g\partial_g z_4 + \frac{M^2}{M^2+b} \varkappa_1 t_4\right) L_4 + u_1 \bar{D}\bar{D}\bar{L}_m \tag{16.2.74}$$

$$m\hat{L}_3 := (m+a)L_3 + 16 t_4 L_4 + u_3 \bar{D}\bar{D}\bar{L}_m \quad . \tag{16.2.75}$$

Since $\bar{D}\bar{D}\bar{L}_m$ is already gauge invariant, u_1 and u_3 can even be chosen to vanish at $\alpha=1$. The last gauge invariant operator is provided by

$$EZ \equiv -\left(J_+ \frac{\delta}{\delta J_+} + J_- \frac{\delta}{\delta J_-}\right) Z \ , \tag{16.2.76}$$

the local version of the matterlegs counting operator. $\hat{L}_1$, $\hat{L}_3$, $\bar{D}\bar{D}\bar{L}_m$, E constitute a basis for all gauge invariant chiral, local operators of dimension three. Thus there is a Zimmermann identity

$$m(1+\sigma)\hat{L}_m = m\hat{L}_3 - 2\beta\hat{L}_1 - \gamma E + u\bar{D}\bar{D}\bar{L}_m \tag{16.2.77}$$

with α-independent coefficients.
As the notation anticipates this relation is nothing but a local Callan-Symanzik equation. The proof and the identification of the coefficients consists in the integration of (16.2.77) and insertion of (16.2.69), (16.2.70):

$$m(1+\sigma)\hat{\Delta}_m = m\hat{\Delta}_3 - 2\beta\hat{\Delta}_1 - \gamma N_A, \tag{16.2.78}$$

$$(m\partial_m + M\partial_M + \beta(g\partial_g - N_\Phi - 2\alpha\partial_\alpha) - \gamma N_A)Z$$
$$= (m(1+\sigma)\hat{\Delta}_m - M^2(\varkappa_3 + \beta\varkappa_1)\Delta_M)Z \ . \tag{16.2.79}$$

After all of these preparations the α-dependence of the supercurrent is easily discussed. Using (16.2.77) in combination with (16.2.30) and (16.2.74), (16.2.75) we have the form of $\beta,\gamma,1+\sigma$ as function of Zimmermann coefficients and counterterms:

$$\beta = -\frac{1}{2}(1+\sigma)r_k ,$$

$$\gamma = -\frac{1}{2z_4}(1+\sigma)\, r_g + \frac{1}{2z_4}(16 + \beta\varkappa_1 \frac{M^2}{M^2+b})t_4 + \frac{1}{2z_4}\beta g\partial_g\, z_4 \ , \tag{16.2.80}$$

$$2\gamma+\sigma = \beta g\partial_g \ln(m+a) .$$

Case $n = -\frac{2}{3}$

We recall the trace equation (16.2.42)

$$\bar{D}^{\dot{\alpha}} V_{\alpha\dot{\alpha}}\Gamma = -2w_\alpha\Gamma - 2D_\alpha S\cdot\Gamma \tag{16.2.81}$$

with the current

$$V = V' + \text{absorp.terms} + (M^2+b)\hat{N}_3[V_M^T] + N_*[V_M^L] + \frac{1}{\alpha}N_*\,[V^g] \tag{16.2.82}$$

and the breaking

$$-\frac{3}{2}S = m(1+\sigma)\hat{L}_m + 2\beta\hat{L}_1 + \gamma E$$
$$+ 2M^2(\hat{N}_1[L_M^T] + N_*[L_M^L]) + \frac{1}{128\alpha}\bar{D}\bar{D}N_0[\Phi(\Psi-\bar{\Psi})] \tag{16.2.83}$$
$$\equiv -\frac{3}{2}(\hat{S} + S_M + S_0 - \frac{2}{3}\gamma E) .$$

Equation (16.2.61) tells us that $N_1[L_M]$ is α-independent, hence between physical states so is $\hat{N}_1[L_M^T]$ (s. (16.2.39)). Since $\hat{N}_1[L_M^T]$ is also symmetric, S is symmetric and α-independent, i.e. gauge invariant between physical states. The r.h.s. of equation (16.2.81) is gauge invariant between physical states and there are no terms $v_{\alpha\dot{\alpha}}$ with the characteristics of the current satisfying $\bar{D}^{\dot{\alpha}} v_{\alpha\dot{\alpha}} = 0$. Hence $V_{\alpha\dot{\alpha}}$ is gauge invariant between physical states. Since the N_*-terms do not contribute there we may call the first three terms of (16.2.82) the gauge invariant supercurrent of the massive theory

$$V_{inv} \equiv V' + \text{absorp.terms} + (M^2+b)\hat{N}_3[V_M^T]. \tag{16.2.84}$$

Let us now discuss the limit $M \to 0$ with which we claimed to control the α-independence of the supercurrent with massless vector field. The breaking (16.2.83)

$$S = \hat{S} + S_M + S_0 + \frac{2}{3}\gamma E \tag{16.2.85}$$

is composed of the gauge invariant part $\hat{S}$, the term S_M which vanishes in the limit $M \to 0$ (for the proof s. App. of [IV.12]), the term S_0 which coincides with the one of the massless theory and the contact terms. What we still have to show is that for $M \to 0$ $\hat{S}$ goes into $\tilde{S}-S_0$ of (16.2.15) and that the function γ or more generally, (16.2.80) also goes into (16.2.18). Since $\hat{S}$ and the coefficients β,γ,σ are α-independent we can control the limit at any convenient point, e.g., at $\alpha = 1$.

$$\begin{aligned} -\frac{3}{2}\hat{S} &= m(1+\sigma)\hat{L}_m + 2\beta\hat{L}_1 \\ &= (m+a)(1+\sigma)L_m + 2\beta(L_1+L_2-L_0) - \beta g\partial_g a L_3 \\ &\quad -\beta(g\partial_g z_4 + \varkappa_1 t_4)L_4 + \text{absorption terms.} \end{aligned} \tag{16.2.86}$$

Since (s. (16.2.56))

$$z_4\Big|_{\alpha=1} = 1 \tag{16.2.87}$$

and $\lim\limits_{M\to 0} t_4 = 0$, (16.2.88)

(for the proof of (16.2.88) s. [IV.12] app.)
we find

$$\lim_{M\to 0} (-\frac{3}{2}\hat{S}) = (m+a)(1+\sigma)L_m^* - \beta(g\partial_g - N_V)L \qquad (6.2.89)$$

where

$$L_m^* = L_m - \frac{2\beta}{(1+\sigma)(m+a)} L_o = L_m + \frac{r_k}{m+a} L_o \qquad (16.2.90)$$

coincides with (16.2.16), and we arrived at the value of S in the massless vector case (16.2.15). β,γ,σ (16.2.80) go over into (16.2.18) hence breaking and contact terms go over into those of the massless vector theory. Thus the current goes also α-indpendently into its massless vector counterpart and one can check explicitly that e.g.

$$\lim_{M\to 0} M^2 V^M = 0 \qquad (16.2.91)$$

(s. app. of [IV.12]). This finishes the proof of α-independence (between physical states) of the supercurrent (16.2.13).

Case n = ∞

The equation to be studied (cf. 16.2.27)) reads

$$z_4 \bar{D}\bar{D} I_5 \cdot Z = \quad EZ - 2(m+a)L_3 \cdot Z \,. \qquad (16.2.92)$$

First we replace L_3 by its gauge invariant counterpart $\hat{L}_3$, (16.2.75), then we use the gauge invariant Zimmermann identity (16.2.77) and obtain

$$\bar{D}\bar{D} I_5' \cdot Z = (E - 2m\hat{L}_m - 2r_k\hat{L}_1 + \frac{1}{2} r_k g\partial_g \ln(m+a)E)Z \quad , \qquad (16.2.93)$$

where in I_5' we have absorbed all terms $\bar{D}\bar{D}\bar{L}_m$, L_4 having appeared under way and where we also divided by $(1+\sigma)$. Anticipating from section 16.2.4 that r_k is α-independent and of one-loop order we conclude that the r.h.s. of (16.2.93) is α-independent i.e. gauge invariant. I_5' is thus the gauge invariant renormalized axial current belonging to the transformations (6.69). With the help of

$$EZ = (-2\, z_4 L_4 - 2(m+a)L_3)Z \qquad (16.2.94)$$

and the definition (16.2.74) we may rewrite (16.2.93)

$$\bar{D}\bar{D}I_5'\cdot Z = (E - 2m\hat{L}_m - 2r_k(L_{kin} - \frac{1}{2}(g\partial_g z_4 + g\partial_g \ln(m+a))L_4 + u_1\bar{D}\bar{D}\Gamma_m))Z .$$

The term $2r_k(L_{kin}\cdots)$ is thus still gauge invariant and may be defined as a new basis element

$$\hat{L}_1' = L_{kin} - \frac{1}{2}(g\partial_g z_4 + \varkappa_1 t_4 + \frac{1}{2} z_4 g\partial_g \ln(m+a))L_4 + u_1\bar{D}\bar{D}\Gamma_m \qquad (16.2.95)$$

replacing $\hat{L}_1$. With it the anomalous "trace" equation for I'_5 takes finally the form

$$\bar{D}\bar{D}I_5'\cdot Z = (E - 2m\hat{L}_m + 2r_k\hat{L}_1')Z . \qquad (16.2.93')$$

In this formulation one can easily check that the limit $M \to 0$ exists and coincides with the result (16.2.27) for massless vector field.

16.2.5 The non-renormalization of the axial anomaly

In a consistent gauge theory with rigid Abelian γ_5-symmetry the conservation equation for the corresponding current has the general form

$$\partial^\mu j_\mu^{\ 5}\cdot\Gamma = w_5\Gamma + r\,\mathfrak{A}\cdot\Gamma \qquad (16.2.96)$$

where $j_\mu^{\ 5}$ is the axial current and $\mathfrak{A}$ a hard term, the anomaly, which is also a total divergence

$$\mathfrak{A} = \partial^\mu B_\mu . \qquad (16.2.97)$$

Usually w_5 commutes with the respective gauge Ward-identity, hence the difference $\partial^\mu j_\mu^{\ 5} - \mathfrak{A}$ is gauge invariant. The non-renormalization theorem for the anomaly then states that in <u>any</u> renormalization scheme and <u>any</u> gauge a definition of $j_\mu^{\ 5}$ and $\mathfrak{A}$ as individually gauge invariant operators can be found such that r is of one loop only.

$$r = \hat{r}\hbar . \qquad (16.2.98)$$

The relevance of this theorem in perturbation theory originates from the fact that non-perturbatively one is in many cases able to give an a priori geometric and topological meaning to the relation between $j_\mu{}^5$ and $\mathfrak{A}$ (by the so-called index theorems). There r must be of one loop. Hence, if perturbation theory is suppósed to be smoothly related to the non-perturbative theory one better finds in perturbation theory the non-renormalization theorem. Since these two ends of the theory are very well separated it is no question that separate proofs for them are indeed required; i.e. one cannot derive the non-renormalization theorem for the perturbation theory from the index theorem without additional hypothesis.

The proof of (16.2.96) and (16.2.97) in perturbation theory proceeds in two steps [IV.13]. First, the separaration of Q in

$$w_5\Gamma = Q\cdot\Gamma \tag{16.2.99}$$

into $\partial^\mu j_\mu{}^5, \mathfrak{A}$ (and possibly soft terms in cases where the rigid γ_5-invariance is softly broken) has to be performed. This is an essentially algebraic problem whose solution relies on the different gauge variation properties of $j_\mu{}^5$ and B_μ: $j_\mu{}^5$ is gauge invariant, whereas B_μ is not and can thus be characterized eventually by its gauge variation. Second, it has to be proved that r is of one loop order only. This property originates from the fact that at some stage in the definition of B_μ one arrives at an operator which is naively definable i.e. has finiteness properties. Its normalization can thus be fixed once and forever and requires no readjustment in higher orders of perturbation theory.

In the present context, SQED, we first have to find out which current is a candidate for the non-renormalization theorem. All possible currents are contained in (16.2.5), labelled by n, the R-weight of the matter fields. The Zimmermann identities (16.2.30), (16.2.31) show that as "basic" elements in the breaking occur only L_{kin}, L_4, $\bar{D}\bar{D}\mathcal{L}_m$.

Out of these at most L_{kin} shows the phenomenon of becoming naively gauge invariant after application of derivatives: recall (16.2.46)

$$\bar{D}\bar{D}B_1 \equiv \bar{D}\bar{D}(D\Phi\bar{D}\bar{D}D\Phi + \bar{D}\Phi DD\bar{D}\Phi + \Phi D\bar{D}\bar{D}D\Phi) = \bar{D}\bar{D}D\Phi\bar{D}\bar{D}D\Phi, \tag{16.2.100}$$

whereas L_4, $\bar{D}\bar{D}\mathcal{L}_m$ are naively gauge invariant since I_5, $\mathcal{L}_m$ have already this property. Hence, in a (partial) current conservation equation L_{kin} should be the breaking term, all others should already be absorbed in the current. From (16.2.10) one can see that this is impossible for any n without destroying the interpretation of V as a supercurrent: absorbing I_5 from S creates a B term. Hence abandoning the interpretation of V as supercurrent we turn to the easiest case, namely $n = \infty$, and study the "trace" equation (16.2.93') for the current I_5' which has L_{kin} as its only hard breaking term. We are thus led to try the proof of a non-renormalization theorem for equation (16.2.93')

$$\bar{D}\bar{D}I_5'\cdot\Gamma = w_5\cdot\Gamma - 2(m+a)L_m\cdot\Gamma - 2r_k\hat{L}_1'\cdot\Gamma \qquad (16.2.101)$$

($w_5\Gamma$ are the Legendre transformed contact terms EZ).

According to the above line of argument the first task is to charakterize the breaking algebraically. The action principle yields

$$w_5\Gamma = Q\cdot\Gamma \qquad (16.2.102)$$

with Q being a chiral insertion of dimension three. The softly broken conservation law

$$W_5\Gamma = \int dx(DDw_5 - \bar{D}\bar{D}\bar{w}_5)\Gamma = \text{soft term} \qquad (16.2.103)$$

then permits in Q only hard terms of the form

$$Q = \bar{D}\bar{D}J_5 \; . \qquad (16.2.104)$$

Hence we study

$$w_5\Gamma = \bar{D}\bar{D}J_5\cdot\Gamma + \text{soft terms} \; , \qquad (16.2.105)$$

where J_5 is real, has dimension two, R-weight zero and we know that

$$\begin{aligned} &\bar{D}\bar{D}\, w_{\wedge}\, J_5 = 0, \\ &DD\, w_{\bar{\wedge}}\, J_5 = 0, \end{aligned} \qquad (16.2.106)$$

since w_5 commutes with the gauge Ward-identities (14.6). The list of all terms J_5 (which are not invariant) is the following

$$J_5 = a_1\, \Phi D\bar{D}\bar{D}D\Phi + a_2(D\Phi\bar{D}\bar{D}D\Phi + \bar{D}\Phi DD\bar{D}\Phi)$$

$$+ a_3(D\bar{D}\Phi D\bar{D}\Phi + \bar{D}D\Phi\bar{D}D\Phi) + a_4\, D\bar{D}\Phi\bar{D}D\Phi$$

$$+ a_5(D\bar{D}\bar{D}\Phi D\Phi + \bar{D}DD\Phi\bar{D}\Phi) + a_6\, DD\Phi\bar{D}\bar{D}\Phi$$

$$+ a_7\Phi\{DD,\bar{D}\bar{D}\}\Phi \; . \qquad (16.2.107)$$

Imposing (16.2.106) leads to

$$a_2 = a_1 \quad , \quad a_i = 0 \quad i \neq 1,2 \; . \qquad (16.2.108)$$

We may therefore write the general solution as

$$J_5 = rB_1 + j_5 \qquad (16.2.109)$$

where

$$j_5 = aI_5 + b(L_m + \bar{L}_m)$$

consists of gauge symmetric terms.

The gauge variation B_1^{Λ} of B_1 is linear in the quantized field hence can be defined naively and defines the operator B_1 up to symmetric terms. Furthermore, J_5 is unique since there do not exist improvement terms for it. Suppose now that

$$J_5 = r'B_1' + a'I_5 + b'(L_m + \bar{L}_m) \qquad (16.2.110)$$

is constructed with a B_1'

$$B_1' = B_1 + y_1 I_5 + y_2(L_m + \bar{L}_m) \qquad (16.2.111)$$

which has the same variation B_1^{Λ} as B_1 but represents another quantum extension of B_1 (classic) than B_1. In this case, inserting (16.2.111) into (16.2.110) we find

$$J_5 = r'B + (a' + y_1 r')I_5 + (b' + y_2 r')(L_m + \bar{L}_m) \qquad (16.2.112)$$

and comparing with (16.2.109) we see that

$$r = r' \ , \tag{16.2.113}$$

i.e. r is uniquely determined by B_1^{Λ} and thus independent of the representative B_1 chosen. This fact suggests also that r is α-independent which we now wish to prove. We know already (cf. (16.2.73)) that L_m is α-independent, just because there is no other symmetric chiral insertion of dimension two. Similarly, since I_5 can mix with $L_m + \bar{L}_m$, there is an α-independent

$$\hat{I}_5 = (1+u_5)I_5 + u_m(L_m + \bar{L}_m) \ . \tag{16.2.114}$$

For the α-dependence of B_1 we note first

$$\partial_\alpha(w_\Lambda B\cdot\Gamma) = \partial_\alpha(w_{\bar{\Lambda}} B\cdot\Gamma) = 0 \ , \tag{16.2.115}$$

hence $\partial_\alpha(B\cdot\Gamma)$ can differ from the term vanishing on shell only by <u>symmetric</u> insertions

$$\partial_\alpha(B\cdot\Gamma) = -\frac{1}{\alpha^2}\Delta_0\cdot B\cdot\Gamma + (x_5\hat{I}_5 + x_m(\hat{L}_m + \hat{\bar{L}}_m))\cdot\Gamma \ . \tag{16.2.116}$$

Thus, there is an α-independent $\hat{B}$

$$\hat{B} = B + b_5\hat{I}_5 + b_m(\hat{L}_m + \hat{\bar{L}}_m) \ . \tag{16.2.17}$$

(Note that the relative coefficient between B and $\hat{B}$ is <u>one</u>.) We may therefore rewrite (16.2.109) in a gauge independent basis

$$J_5 = r\hat{B} + \hat{a}\hat{I}_5 + \hat{b}(\hat{L}_m + \hat{\bar{L}}_m) \ , \tag{16.2.118}$$

with the coefficient r unchanged. The fact that $\bar{D}\bar{D}J_5$ is α-independent shows that all coefficients $r, \hat{a}, \hat{b}$ have the same property.

Having characterized gauge independently the anomaly and its coefficient we proceed now to the second part of the problem: the proof that r is of one-loop only [IV.13, IV.15]. For this we recall the Callan-Symanzik equation

$$C\Gamma \equiv (\mu\partial_\mu + \beta(g\partial_g - N_\Phi - 2\alpha\partial_\alpha) - \gamma N_A)\Gamma = \text{soft}, \tag{16.2.119}$$

where $\mu\partial_\mu \equiv m\partial_m + M\partial_M$, and note the important relations

$$[C,w_5] = 0 \quad , \tag{16.2.120}$$

$$[C,W_\Lambda] = \beta W_\Lambda \tag{16.2.121}$$

(W_Λ is the gauge Ward-identity operator (14.6)) expressing the respective symmetry of C. Acting with (16.2.120) on Γ and recalling (16.2.105) we find

$$\bar{D}D(CJ_5 \cdot \Gamma) = \text{soft} \tag{16.2.122}$$

as a constraint on CJ_5. Since there are no terms satisfying (16.2.122) trivially, it implies

$$CJ_5 \cdot \Gamma = \text{soft}. \tag{16.2.123}$$

We now apply (16.2.121) to (16.2.123) which yields

$$C(W_\Lambda J_5 \cdot \Gamma) - \beta W_\Lambda (J_5 \cdot \Gamma) = \text{soft} . \tag{16.2.124}$$

Using (16.2.118) we see

$$W_\Lambda J_5 \cdot \Gamma = rB_1^\Lambda , \tag{16.2.125}$$

hence (16.2.124) implies

$$C(rB_1^\Lambda) - \beta r B_1^\Lambda = 0 , \tag{16.2.126}$$

or

$$(\mu\partial_\mu + \beta g \partial_g - 2\alpha\partial_\alpha) r - 2r\beta = 0 . \tag{16.2.127}$$

Since r is dimensionless and α-independent this means

$$\beta(g\partial_g - 2) r = 0 , \tag{16.2.128}$$

and since β starts with order $\hbar$, we have finally

$$(g\partial_g - 2) r = 0 , \tag{16.2.129}$$

i.e. r is of order g^2.

For the explicit calculation of r we identify it first with a Zimmermann coefficient. We note that $\bar{D}\bar{D}\hat{B}$, $\bar{D}\bar{D}\hat{I}_5$, $\hat{L}_m$ used in (16.2.118) is just another basis than $\hat{L}_1$, $\hat{L}_m$, E which had been used in section 16.2.3. Hence we may write instead of

$$r\bar{D}\bar{D}\hat{B} = rL_1 + \ldots$$

$$r\bar{D}\bar{D}\hat{B} = r(L_1 + \frac{1}{\alpha}(L_2 - L_0)) + \ldots \qquad (16.2.130)$$

by changing appropriately coefficients of $\bar{D}\bar{D}I_5$ and $\bar{D}\bar{D}\hat{L}_m$. Comparing with (16.2.93') we can identify

$$r = 2r_k \qquad (16.2.131)$$

where r_k is the coefficient in the Zimmermann identity (16.2.6) or (16.2.30). Hence its value is given by

$$r_k = -32\ (m+a) < T\int dS_3 N_2[A_+A_-(3)]\tilde{\Phi}(p,1)\Phi(o,2)>\Big|^{1PI}_{p=0} \quad . \qquad (16.2.132)$$

At order g^2 the diagrams contributing to this vertex function are simply

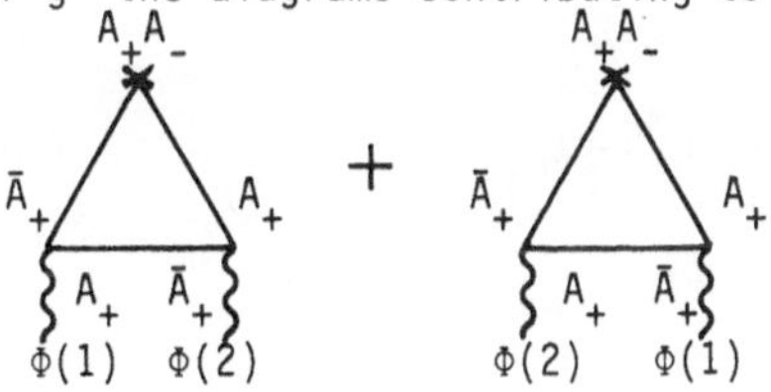

and they yield as numerical value

$$r = 2r_k = \frac{g^2}{8(16\pi)^2} \quad . \qquad (16.2.133)$$

16.3 Supersymmetric Yang-Mills theory

16.3.1 General preparation

In the Wess-Zumino model and in SQED the renormalized supercurrent was obtained by using explicitly Zimmermann identities for oversubtracted mass terms and a subsequent reorganization of the result. All manipulations could be performed on a few terms in a direct fashion. The case of SYM is much more involved. Since BRS-invariance is a non-linear symmetry the construction of BRS-invariant operators is non-trivial. We have seen already in the tree approximation, sect. 6.4, how many terms depending on external fields and ghosts were needed in order to describe completely the off-shell structure In higher orders a somewhat abstract procedure is therefore necessary [IV.14].

We think of a general SYM-theory maintaining parity (hence being consistent, cf. sect. 15.4) with matter fields in the adjoint, resp. a unitary representation and we shall assume all fields to be massive but neglect all soft terms i.e. treat the theory only far off-shell, in the Euclidean region (indicated by ~ in the following). Considering mass zero limits would necessitate the study of infrared anomalies (s. chapt. V) which has not yet been carried out in detail for local operators.

At this stage we suppose to have established the following symmetries (asymptotically in momentum space):

rigid gauge invariance: $W_\omega \Gamma \sim 0$ (16.3.1)

conformal R-invariance: $W_R \Gamma \sim 0$ (16.3.2)

supersymmetry: $W_\alpha \Gamma \sim 0, \ \bar{W}_{\dot\alpha} \Gamma \sim 0$ (16.3.3)

BRS-invariance: $\mathcal{S}(\Gamma) \sim 0$ (16.3.4)

(c.f. (15.18) - (15.21) and the corresponding discussion).

Our aim is to construct a BRS-invariant supercurrent V fulfilling a trace equation

$$\bar{D}^{\dot\alpha}[V_{\alpha\dot\alpha} \cdot \Gamma] \sim -2w_\alpha \Gamma - 2D_\alpha [S \cdot \Gamma] \quad , \tag{16.3.5}$$

where the anomaly S is a BRS-invariant, chiral insertion of dimension three and the contact terms w_α were given in (6.79) and commute with rigid, R, supersymmetry and BRS transformations. This construction will be performed in section 16.3.2.

As in the previous cases, a suitable choice of a basis for S will show its connection with the Callan-Symanzik equation (section 16.3.3), although this connection is not straightforward since the usual moment construction of the dilatation current does not yield at once the true dilatation Ward identity, as we saw in section 6.4 (s. (6.78)). Restrictions on the anomaly S, hence on the coefficients of the Callan-Symanzik equation, will also be seen to follow from R invariance (16.3.2), which can be expressed via the contact terms w_α by (6.72):

$$W_R\Gamma = \int dx w\Gamma\Big|_{\Theta=0} \sim 0 \ , \tag{16.3.6}$$

$$w \ \equiv \ D^\alpha w_\alpha - \bar{D}_{\dot{\alpha}}\bar{w}^{\dot{\alpha}} \ . \tag{16.3.7}$$

Let us finally mention another important point,which will be fully discussed in section (16.3.4). Starting directly from R-invariance (16.3.6) one infers the existence of a conserved supercurrent V'_μ:

$$iw\Gamma \sim \partial^\mu[V'_\mu\cdot\Gamma] \tag{16.3.8}$$

its $\Theta=0$ component being the conserved axial current associated with R-invariance. V'_μ fulfills a trace identity of the form

$$\begin{aligned} &\bar{D}^{\dot{\alpha}}[V'_{\alpha\dot{\alpha}}\cdot\Gamma] \sim -2w_\alpha\Gamma + \bar{D}\bar{D}D_\alpha[B\cdot\Gamma] \\ &B = \bar{B} \ , \end{aligned} \tag{16.3.9}$$

which is related to the former one (16.3.5) by the redefinition

$$\begin{aligned} &V_{\alpha\dot{\alpha}} = V'_{\alpha\dot{\alpha}} - \frac{2}{3}[D_\alpha,\bar{D}_{\dot{\alpha}}]B \\ &S \ \ = \frac{1}{6}\bar{D}\bar{D}B \ . \end{aligned} \tag{16.3.10}$$

The essential difference between the two formulations lies in the BRS properties: V and S are BRS-invariant, V' and B are not.

16.3.2 The BRS-invariance of current and breaking

In order to establish the trace identity (16.3.5) let us first formulate the problem in terms of the functional Z $(J, J_A, \xi_\pm, \rho, Y, \sigma)$ for Green's functions (see Appendix B). The trace identity thus reads

$$\bar{D}^{\dot{\alpha}}[V_{\alpha\dot{\alpha}}\cdot Z] \sim -2\frac{\hbar}{i}w_\alpha Z - 2D_\alpha[S\cdot Z] \tag{16.3.11}$$

where the contact terms w_α are obtained from (6.79) by Legendre transformation:

$$w_\alpha Z = 2\{-\bar{D}\bar{D}D_\alpha J\frac{\delta}{\delta J} + D_\alpha J\bar{D}\bar{D}\frac{\delta}{\delta J} - \bar{D}\bar{D}D_\alpha\rho\frac{\delta}{\delta\rho} + D_\alpha\rho\bar{D}\bar{D}\frac{\delta}{\delta\rho}$$
$$-\xi_- D_\alpha\frac{\delta}{\delta\xi_-} - \sigma D_\alpha\frac{\delta}{\delta\sigma} + D_\alpha\xi_+\frac{\delta}{\delta\xi_+}\}Z$$
$$+ w_\alpha^{\text{matter}} Z \; . \tag{16.3.12}$$

The properties of w_α we shall use are:
(the first follows from R-invariance (16.3.6))

$$D^\alpha w_\alpha - \bar{D}_{\dot{\alpha}}\bar{w}^{\dot{\alpha}} = w \quad , \quad \int dxwZ \sim 0 \; , \tag{16.3.13}$$

$$\bar{D}\bar{D}w_\alpha = 0 \quad , \tag{16.3.14}$$

$$[\mathcal{S}, w_\alpha] = 0 \; , \tag{16.3.15}$$

$$[\mathcal{G}(1), w_\alpha(2)] = -2\frac{i}{\hbar}\delta^S(1,2)D_\alpha\xi_+ \; ,$$
$$[\bar{\mathcal{G}}(1), w_\alpha(2)] = -\frac{1}{64\alpha}D\,\bar{D}\bar{D}\bar{\delta}^S(2,1)\bar{D}\bar{D}\frac{\delta}{\delta\rho(2)} \; . \tag{16.3.16}$$

Here $\mathcal{S}$ is the Slavnov identity operator (c.f. (B.23), with obvious change for the α-gauge)

$$\mathcal{S}Z \equiv -\text{Tr}\int dVJ\frac{\delta Z}{\delta\rho}$$
$$+\int dS\{\text{Tr}\xi_+\bar{D}\bar{D}\frac{\delta}{\delta J} + \text{Tr}\xi_-\frac{\delta}{\delta\sigma} - J_A\frac{\delta}{\delta Y}\}$$
$$-\text{c.c.} \quad \sim 0 \quad , \tag{16.3.17}$$

and $\mathcal{G}$, $\bar{\mathcal{G}}$ are the operators appearing in the ghost equation (5.97)

$$\mathcal{G}Z \equiv [-\frac{1}{128\alpha}\bar{D}\bar{D}DD\frac{\delta}{\delta\rho} + \frac{i}{\hbar}\xi_+]\,Z \sim 0 \; ,$$
$$\bar{\mathcal{G}}Z \equiv [\quad \frac{}{128\alpha}DD\bar{D}\bar{D}\frac{\delta}{\delta\rho} + \frac{i}{\hbar}\bar{\xi}_+]\,Z \sim 0 \; . \tag{16.3.18}$$

w_α also commutes with supersymmetry. These commutators with W_α, $\bar{W}_{\dot\alpha}$ and W_R define w_α as a superfield of R-weight n = -1.

The action principle

$$-2w_\alpha Z \sim \frac{i}{\hbar}\Delta_{\dot\alpha} Z \tag{16.3.19}$$

defines a dimension 7/2 insertion Δ_α which is a superfield with R-weight n = -1 and, due to (16.3.13)-(16.3.16), fulfills

$$\int dx[D^\alpha\Delta_\alpha - \bar{D}_{\dot\alpha}\bar{\Delta}^{\dot\alpha}]\cdot Z \sim 0 \tag{16.3.20}$$

$$\bar{D}\bar{D}[\Delta_\alpha Z] \sim 0 \tag{16.3.21}$$

$$\mathscr{s}[\Delta_{\dot\alpha} Z] \sim 0 \tag{16.3.22}$$

$$\begin{aligned} &\mathcal{G}(1)[\Delta_\alpha(2)\cdot Z] \sim 4\delta^S(1,2)D_\alpha\xi_+ Z \\ &\bar{\mathcal{G}}(1)[\Delta_\alpha(2)\cdot Z] \sim -\frac{\hbar}{i}\frac{1}{32\alpha}D_\alpha\bar{D}\bar{D}\bar{\delta}^S(2,1)\bar{D}\bar{D}\frac{\delta Z}{\delta\rho(2)} \ . \end{aligned} \tag{16.3.23}$$

In particular Δ_α is BRS-invariant (16.3.22) but satisfies the inhomogeneous ghost equations (16.3.23). A special solution of the latter which also fulfills (16.3.20)-(16.3.22) is provided by the point splitted insertion

$$\begin{aligned} \Delta^\varepsilon_\alpha(z)\cdot Z = \left(\frac{\hbar}{i}\right)^2 \frac{1}{68\alpha}\Bigg\{ &\frac{\delta}{\delta\xi_+(z_+)} D_\alpha\bar{D}\bar{D}DD\frac{\delta}{\delta\rho(z_-)} \\ &- D_\alpha\bar{D}\bar{D}\frac{\delta}{\delta\bar{\xi}_+(z_+)}\bar{D}\bar{D}\frac{\delta}{\delta\rho(z_-)} - \bar{D}\bar{D}\frac{\delta}{\delta J(z_+)}D\bar{D}\bar{D}DD\frac{\delta}{\delta J(z_-)} \\ &+ (z_+ \leftrightarrow z_-)\Bigg\} Z \end{aligned} \tag{16.3.24}$$

where $z_\pm = (x \pm \varepsilon, \Theta, \bar\Theta)$, ε space-like. For any point-splitted insertion of the form

$$\Delta^\varepsilon(z)\cdot Z = \left(\frac{\hbar}{i}\right)^2\frac{1}{2}\left\{\frac{\delta}{\delta\zeta_1(z_+)}\frac{\delta}{\delta\zeta_2(z_-)} + \frac{\delta}{\delta\zeta_1(z_-)}\frac{\delta}{\delta\zeta_2(z_+)}\right\} Z \tag{16.3.25}$$

(ζ_i = source of the elementary or composite field φ_i)

we can define the normal product

$$:\Delta(z): \equiv :\varphi_1(z)\varphi_2(z): = \lim_{\varepsilon\to 0} \mathrm{FP}\ \Delta^{\varepsilon}(z) \tag{16.3.26}$$

by subtracting from Δ^{ε} the singular part of its Wilson-Zimmermann short distance expansion [IV.16,17]. Let us note that if Δ^{ε} fulfills some identity like (16.3.20)-(16.3.23), then $:\Delta:$ shares this property. In this way we can obtain from (16.3.24) the insertion

$$:\Delta^4_\alpha: = \frac{1}{32\alpha} :(c_-\, D_\alpha\bar{D}\bar{D}DDQ + D_\alpha\bar{D}\bar{D}\bar{c}_-\bar{D}\bar{D}Q - \bar{D}\bar{D}\Phi D_\alpha\bar{D}\bar{D}DD\Phi): \tag{16.3.27}$$

as a special solution of (16.3.23) with (16.3.20)-(16.3.22) fulfilled. It coincides in the treee approximation with Δ^4_α (6.85). One easily checks that it can be decomposed

$$:\Delta^4_\alpha: = \frac{1}{16\alpha}\left[\bar{D}\bar{D}D_\alpha :B^2: - \bar{D}^{\dot{\alpha}}(:V^2_{\alpha\dot{\alpha}}: + :V^3_{\alpha\dot{\alpha}}:)\right] \tag{16.3.28}$$

where $:B^2:$, $:V^2:$ and $:V^3:$ are the normal product versions - according to the definition (16.3.26) - of the classical objects given in (6.88, 6.89). They are BRS invariant, $:B^2:$ is real and $:V^{2,3}:$ are axial vector fields. We can rewrite (16.3.28) as

$$:\Delta^4_\alpha: = \bar{D}^{\dot{\alpha}}\tilde{V}_{\alpha\dot{\alpha}} + 2\,D_\alpha\,\tilde{S} \tag{16.3.29}$$

with

$$\tilde{V}_{\alpha\dot{\alpha}} = -\frac{1}{16\alpha}\left(:V^2_{\alpha\dot{\alpha}}: + :V^3_{\alpha\dot{\alpha}}: - \frac{2}{3}[D_\alpha,\bar{D}_{\dot{\alpha}}]:B^2:\right) ,$$

$$\tilde{S} = -\frac{1}{96\alpha}\bar{D}\bar{D}:B^2: , \tag{16.3.30}$$

$$:B^2: = :DD\Phi\bar{D}\bar{D}\Phi + DDQc_- + \bar{D}\bar{D}Q\bar{c}_-: .$$

Then (16.3.19) becomes

$$-2w_\alpha Z \sim \frac{i}{\hbar}\,\bar{D}^{\dot{\alpha}}\tilde{V}_{\alpha\dot{\alpha}}Z + 2D_\alpha\tilde{S}\cdot Z + \Delta'_\alpha\cdot Z \quad , \tag{16.3.31}$$

where Δ'_α fulfills the conditions (16.3.20)-(16.3.22) and the homogeneous ghost equations

$$\mathcal{G}(1)[\Delta'_\alpha(2)\cdot Z] \sim 0 , \quad \bar{\mathcal{G}}(1)[\Delta'_\alpha(2)\cdot Z] \sim 0 . \tag{16.3.32}$$

Let us now go back to the formulation in terms of the vertex functional $\Gamma(\Phi,A,c_\pm,\rho,Y,\sigma)$. Eq. (16.3.31) reads

$$-2w_\alpha\Gamma \sim :\Delta_\alpha^4:\cdot\Gamma(\Phi,A,c_\pm,\rho,Y,\sigma) + \Delta'_\alpha\cdot\Gamma\ (\Phi,A,c_+,\eta,Y,\sigma)\ , \tag{16.3.33}$$

where we made explicit that $\Delta'_\alpha\cdot\Gamma$ depends on c_-,ρ only in the combination (5.109)

$$\eta = \rho + \frac{1}{128\alpha}\,(DDc_- + \bar{D}\bar{D}\bar{c}_-). \tag{16.3.34}$$

This is due to the homogeneous ghost equation (16.3.32) and holds also for Γ but not for $w_\alpha\Gamma$. $\Delta'_\alpha\cdot\Gamma$ fulfills the same condition, (16.3.20), whereas (16.3.22) becomes

$$\mathcal{B}_{\bar{\Gamma}}[\Delta'_\alpha\cdot\Gamma] \sim 0 \tag{16.3.35}$$

with $\bar{\Gamma}$ and $\mathcal{B}_{\bar{\Gamma}}$ defined in (5.98, 5.113).

We shall now show by induction that $\Delta'_\alpha\cdot\Gamma$ can be expressed as

$$\Delta'_\alpha\cdot\Gamma = \bar{D}^{\dot\alpha}[V'_{\alpha\dot\alpha}\cdot\Gamma] + 2D_\alpha[S'\cdot\Gamma] \tag{16.3.36}$$

where $V'_{\alpha\dot\alpha}$ and S' are axial vector with R-weight 0 resp. chiral with R-weight -2 insertions of dimension 3, obey the homogeneous ghost equation (hence depend on η) and are BRS invariant

$$\mathcal{B}_{\bar{\Gamma}}[V'_{\alpha\dot\alpha}\cdot\Gamma] \sim \mathcal{B}_{\bar{\Gamma}}[S'\cdot\Gamma] \sim 0\ . \tag{16.3.37}$$

This is of course true at order zero as we have seen in section 6, with the result (c.f. (6.100, 6.88))

$$S'\cdot\Gamma = -\frac{2}{3}B^1 + O(\hbar)\ . \tag{16.3.38}$$

Let us assume (16.3.36) to hold up to and including the order n-1:

$$\Delta'_\alpha\cdot\Gamma \sim \bar{D}^{\dot\alpha}[V'_{\alpha\dot\alpha}\cdot\Gamma] + 2D_\alpha[S'\cdot\Gamma] + \hbar^n\,\Delta_\alpha^{(n)} + O(\hbar^{n+1}) \qquad (16.3.39)$$

where $V'_{\alpha\dot\alpha}$ and S' obey the BRS-symmetry (16.3.37) to all orders. The local functional $\Delta_\alpha^{(n)}(\Phi,A,c_+,\eta,Y,\sigma)$ represents the - still unknown - contributions of order n.

The constraints (16.3.35), (16.3.20), (16.3.21) yield for $\Delta_\alpha^{(n)}$ ($\mathcal{b} \equiv \mathcal{B}_{\Gamma_{class}}$, see (15.30))

$$\mathcal{b}\,\Delta_\alpha^{(n)} = 0 \ , \qquad (16.3.40)$$

$$\bar{D}\bar{D}\Delta_\alpha^{(n)} = 0 \ , \qquad (16.3.41)$$

$$\int dx(D^\alpha\Delta_\alpha^{(n)} - \bar{D}_{\dot\alpha}\bar{\Delta}^{(n)\dot\alpha}) = 0 \ . \qquad (16.3.42)$$

The general solution of (16.3.41), (16.3.42) is found in Appendix E to be

$$\Delta_\alpha^{(n)} = \bar{D}^{\dot\alpha}V_{\alpha\dot\alpha}^{(n)} + 2D^\beta S_{\alpha\beta}^{(n)} \ , \qquad (16.3.43)$$

where $V_{\alpha\dot\alpha}^{(n)}$ is an axial vector superfield and $S_{\alpha\beta}^{(n)}$ a chiral superfield, both being local functionals of Φ,A,c_+,η,Y,σ. The constraint (16.3.40) applied to (16.3.43) yields the equation

$$\bar{D}^{\dot\alpha}\mathcal{b}V_{\alpha\dot\alpha}^{(n)} + 2D^\beta\mathcal{b}S_{\alpha\beta}^{(n)} = 0 \qquad (16.3.44)$$

which we have to solve. As constraints on all solutions we have the respective dimensions (dimV = dimS = 3), the R-weights (n(V) = 0, n(S) = -2) and the fact that S is a chiral field. One lists all possible S-terms according to the number of derivatives which they contain and finds after quite some calculation that only the following four types of solutions exist:

(A) $\mathcal{b}\,S_{\alpha\beta}^{(n)} = -\frac{1}{2}\bar{D}\bar{D}\bar{T}_{\alpha\beta}$, $T_{\dot\alpha\dot\beta}$ chiral ,

$$\mathcal{b}\,V_{\alpha\dot\alpha}^{(n)} = 2(D^\beta\bar{D}_{\dot\alpha}\bar{T}_{\alpha\beta} + \bar{D}^{\dot\beta}D_\alpha T_{\dot\alpha\dot\beta}) \ , \qquad (16.3.45)$$

(B) $$\mathfrak{s}\, S^{(n)}_{\alpha\beta} = -\frac{1}{2}\bar{D}\bar{D}D_\beta U_\alpha ,$$

$$\mathfrak{s}\, V^{(n)}_{\alpha\dot{\alpha}} = -DD\bar{D}_{\dot{\alpha}}U_\alpha + \bar{D}\bar{D}D_\alpha \bar{U}_{\dot{\alpha}} ,$$

(C) $$\mathfrak{s}\, S^{(n)}_{\alpha\beta} = -\frac{1}{2}\bar{D}\bar{D}D^\gamma U_{(\alpha\beta\gamma)} , \qquad (16.3.45\ \text{cont.'d})$$

$$\mathfrak{s}\, V^{(n)}_{\alpha\dot{\alpha}} = 2\, D^\beta \bar{D}_{\dot{\alpha}} D^\gamma U_{(\alpha\beta\gamma)} - 2\bar{D}^{\dot{\beta}} D_\alpha \bar{D}^{\dot{\gamma}} \bar{U}_{(\dot{\alpha}\dot{\beta}\dot{\gamma})} ,$$

(D) $$\mathfrak{s}\, S^{(n)}_{\alpha\beta} = -\bar{D}\bar{D}D_\alpha \bar{D}_{\dot{\alpha}} W_\beta{}^{\dot{\alpha}} , \quad \bar{W}^{\dot{\alpha}}{}_\beta = -W_\beta{}^{\dot{\alpha}} ,$$

$$\begin{aligned} \mathfrak{s}\, V^{(n)}_{\alpha\dot{\alpha}} = {} & (DD\bar{D}\bar{D} + \bar{D}\bar{D}DD - 6\, D\bar{D}\bar{D}D) W_{\alpha\dot{\alpha}} \\ & + 4\, \bar{D}_{\dot{\alpha}} D_\alpha \bar{D}_{\dot{\beta}} D^\beta W_\beta{}^{\dot{\beta}} + 4\, D_\alpha \bar{D}_{\dot{\alpha}} D^\beta \bar{D}_{\dot{\beta}} W_\beta{}^{\dot{\beta}} \\ & - 4\, \bar{D}_{\dot{\alpha}} DD\bar{D}_{\dot{\beta}} W_{\dot{\alpha}}{}^{\dot{\beta}} - 4\, D_\alpha \bar{D}\bar{D} D^\beta W_{\beta\dot{\alpha}} , \end{aligned}$$

which may be summarized by

$$\mathfrak{s}\, S^{(n)}_{\alpha\beta} = (\mathcal{D}_S U)_{\alpha\beta} ,$$
$$\mathfrak{s}\, V^{(n)}_{\alpha\dot{\alpha}} = (\mathcal{D}_V U)_{\alpha\dot{\alpha}} , \qquad (16.3.46)$$

where $\mathcal{D}_s$ and $\mathcal{D}_V$ have the property

$$\bar{D}^{\dot{\alpha}} (\mathcal{D}_V A)_{\alpha\dot{\alpha}} + 2D^\beta (\mathcal{D}_s A)_{\alpha\beta} = 0 \qquad (16.3.47)$$

for any superfield A.

To prove that $\mathfrak{s}V^{(n)}$, $\mathfrak{s}S^{(n)}$ as given above are indeed variations is a cohomology problem similar to the one solved in section 15 for the BRS-invariance for Γ. The result is that the expressions on the r.h.s. of (A)...(D) are indeed variations. Hence in (16.3.46)

$$U = \mathfrak{s}\hat{U} \qquad (16.3.48)$$

and

$$S^{(n)}_{\alpha\beta} = (\mathcal{D}_s \hat{U})_{\alpha\beta} + \mathfrak{s}\text{-inv.} ,$$
$$V^{(n)}_{\alpha\dot{\alpha}} = (\mathcal{D}_V \hat{U})_{\alpha\dot{\alpha}} + \mathfrak{s}\text{-inv.} . \qquad (16.3.49)$$

When we insert this result in (16.3.43) the $\hat{U}$-terms cancel due to (16.3.47) which means that (16.3.43) holds with $V^{(n)}$ and $S^{(n)}$ $\mathscr{b}$-invariant. $V^{(n)}$ and $S^{(n)}$ can be expanded in their respective basis of $\mathscr{b}$-invariant local functionals with the appropriate dimension (d=3) and quantum numbers:

$$\{V^i_{\alpha\dot{\alpha}}\} \quad \text{and} \quad \{S^i_{\alpha\beta}\} = \{\varepsilon_{\alpha\beta}S^i\} \quad . \tag{16.3.50}$$

Since the explicit basis of V is not needed later on, we shall not elaborate on it. For the basis of S we first observe that there is no symmetric $S_{(\alpha\beta)}$ fulfilling all requirements - a fact anticipated in writing (16.3.50) - so that

$$S^{(n)}_{\alpha\beta} = \varepsilon_{\alpha\beta}S^{(n)} \tag{16.3.51}$$

can be expanded in the basis $\{S^i\}$. The latter consists in two classes of terms:

a) $S^i = \frac{1}{6}\bar{D}\bar{D}B^i$ with

$$B^1 = \mathscr{b}(\eta\Phi) \quad , \quad B^{1,k} = \mathscr{b}(\eta\Phi^{2k+1}),\ k = 1,2,\ldots \tag{16.3.52}$$

$$B^3 = \bar{A}_+ e^{\tilde{\Phi}} A_+ + A_- \bar{e}^{\tilde{\Phi}} \bar{A}_- \quad , \quad B^4 = \text{Tr}\, e^{-\Phi}\bar{A}e^{\Phi}A$$

b) $S^5 = \text{Tr}\, F^\alpha F_\alpha$, $S^6 = \text{Tr}\, A^3$, $S^7 = \mathscr{b}\text{Tr}(\sigma c_+)$.

(Note that the BRS-invariant B^2 (6.88) does not appear in the list since it does not fulfill the ghost equation.) The classical basis $\{S^i\}$ can be extended to a quantum basis

$$\{\hat{S}^i\cdot\Gamma(\Phi,A,c_+,\eta,Y,\sigma) : \mathscr{B}_{\bar{\Gamma}}[\hat{S}\cdot\Gamma] \sim 0\} \tag{16.3.53}$$

identical or equivalent to the basis (16.3.52) at zeroth order. A particular choice for it will be given in the next subsection. The basis for V is similarly extended to a basis of $B_{\bar{\Gamma}}$-invariant insertions $\hat{V}^i_{\alpha\dot{\alpha}}\cdot\Gamma$.

Since $V^{(n)}$ and $S^{(n)}$ contribute in (16.3.39) at order n, one can replace in their expansion the classical basis elements (16.3.52) by their quantum extensions, the difference being of order n+1. This defines $\mathcal{B}_{\bar{\Gamma}}$-invariant insertions $\hat{V}^{(n)}\cdot\Gamma$ and $\hat{S}^{(n)}\cdot\Gamma$ which, when substituted in (16.3.43) and then in (16.3.39), gives rise to

$$\Delta'_{\alpha}\cdot\Gamma \sim \bar{D}^{\dot{\alpha}}[V'_{\alpha\dot{\alpha}}\cdot\Gamma] + 2D_{\alpha}[S'\cdot\Gamma] + O(\hbar^{n+1}) \tag{16.3.54}$$

($\hat{V}^n$ and $\hat{S}^n$ being absorbed in V' and S').

This shows that (16.3.39) holds at next order in n together with the invariance property (16.3.37) and therefore finishes the inductive proof of (16.3.36).

We can finally write the supercurrent trace identity (16.3.31) or (16.3.33) as

$$\begin{aligned} &\bar{D}^{\dot{\alpha}}[V_{\alpha\dot{\alpha}}\cdot\Gamma] \sim -2w_{\alpha}\Gamma - 2D_{\alpha}[S\cdot\Gamma] \\ &V_{\alpha\dot{\alpha}} = \tilde{V}_{\alpha\dot{\alpha}} + V'_{\alpha\dot{\alpha}}\ , \quad S = \tilde{S} + S' \end{aligned} \tag{16.3.55}$$

with $\tilde{V}$, $\tilde{S}$ defined by (16.3.30), and V', S' being $\mathcal{B}_{\bar{\Gamma}}$-invariant and fulfilling the homogeneous ghost equations.

16.3.3 Renormalized supercurrent and Callan-Symanzik equation

Our aim is now to construct a basis for S' in (16.3.55) similar to the one found in the previous cases (section 16.2, 16.3) which allowed us to make the connection between the supercurrent anomalies and the Callan-Symanzik equation. But now we have to take care of the fact that the moment construction (see (16.1.24)-(16.1.28)) leads to a dilatation Ward identity operator W'^{D} (6.77) corresponding to dimensions (2,0,3) for (Φ,ρ,c_-) instead of their canonical dimensions (0,2,1). To the latter corresponds the true dilatation Ward identity operator W^D (6.78)

$$\begin{aligned} &W^D\Gamma = W'^{D}\Gamma + 2i\mathcal{N}_-\Gamma\ , \\ &\mathcal{N}_- \equiv N_{\Phi} - N_{\rho} + N_{c_-} + N_{\bar{c}_-}\ , \\ &N \equiv \int\varphi\delta_{\varphi} \qquad \text{(counting operator)}\ . \end{aligned} \tag{16.3.56}$$

The analog of eq.(16.1.29) reads

$$W'_D\cdot\Gamma \sim -\frac{3}{2}\, i[\int dS S + \int d\bar{S}\bar{S}]\cdot\Gamma\ . \tag{16.3.57}$$

Let us first rewrite the breaking S in the trace identity (16.3.55) in the form

$$S = S_- + S'' \tag{16.3.58}$$

with

$$S_- = \frac{1}{6}\bar{D}\bar{D}B_- \quad , \quad B_- = -\frac{1}{16\alpha} :B^2: + 4 :B^1: \quad . \tag{16.3.59}$$

$:B^2:$ was defined in the last subsection (see (16.3.30) by the point-splitting procedure (16.3.25), (16.3.26).
$:B^1:$ is defined here in the same way. Both are quantum expansions of the classical B^1, B^2 in (6.88). The precise definition of $:B^1:$ is

$$:B^1:\cdot\Gamma = \Phi\frac{\delta\Gamma}{\delta\Phi} - \rho\,\frac{\delta\Gamma}{\delta\rho} + \frac{1}{128\alpha} :\Phi\{DD,\bar{D}\bar{D}\}\Phi + Q(DDc_- + \bar{D}\bar{D}\bar{c}_-):\cdot\Gamma \, . \tag{16.3.60}$$

From the reality of B_- follows

$$[\int dSS_- - \int d\bar{S}\bar{S}_-]\cdot\Gamma = 0 \, . \tag{16.3.61}$$

One can check moreover that

$$[\int dSS_- + \int d\bar{S}\bar{S}_-]\cdot\Gamma \sim \frac{4}{3}\,\mathcal{N}_-\Gamma \quad ,$$

$$\mathcal{N}_- \equiv N_\Phi - N_\rho - N_{c_-} - N_{\bar{c}_-} \quad . \tag{16.3.62}$$

The last result requires the validity of the ghost equations (16.3.18) inside the symbols : :, which we check on an example. Let (using (16.3.25), (16.3.26))

$$\Delta \;\; = :c_-\bar{D}\bar{D}DDQ: (z) \quad , \tag{16.3.63}$$

$$\begin{aligned}\Delta\cdot Z &= \frac{1}{2}\lim_{\varepsilon\to 0} \mathrm{FP}\,\left(\frac{\hbar}{i}\right)^2 \left[\frac{\delta}{\delta\xi_+(z_+)}\,\bar{D}\bar{D}DD\,\frac{\delta}{\delta\rho(z_-)} + (z_+\leftrightarrow z_-)\right] Z \\ &\sim \frac{1}{2}\lim_{\varepsilon\to 0} \mathrm{FP}\,\frac{\hbar}{i}\,128\alpha\,\left[\frac{\delta}{\delta\xi_+(z_+)}\,\xi_+(z_-) + (z_+\leftrightarrow z_-)\right]Z \\ &= -\,128\alpha\,\frac{\hbar}{i}\,\xi_+\,\frac{\delta Z}{\delta\xi_+} \quad .\end{aligned} \tag{16.3.64}$$

Or:

$$\Delta\cdot\Gamma = 128\alpha\, c_- \frac{\delta\Gamma}{\delta c_-} \quad . \tag{16.3.65}$$

For the last equality in (16.3.64) we have used the vanishing of the anticommutator term

$$\{\frac{\delta}{\delta\xi_+(z_+)} \, , \, \xi_+(z_-)\} = \delta^S(z_+,z_-) = 0$$

following from $\Theta_+ = \Theta_- = \Theta$.

From its definition S_- reduces at zeroth order to the full breaking S we have computed in the tree approximation (see (6.100, 6.103, 6.105)). Hence the remainder $S''\cdot\Gamma$ in (16.3.58) is of order $\hbar$. Let us expand it in a basis of $\mathcal{B}_{\bar{\Gamma}}$-invariant chiral insertion S^i defined by

$$[\int dS\hat{S}^i + \int d\bar{S}\hat{\bar{S}}^i]\cdot\Gamma = \nabla^i\Gamma \tag{16.3.66}$$

where ∇^i stands symbolically for a basis of symmetric operators as the one constructed in Appendix D (with obvious changes when going to the α-gauge):

$$\begin{aligned}
\{\nabla^i\} &= \{\mathcal{N}_\Phi, \mathcal{N}_{A_+}, \mathcal{N}_A, \mathcal{N}_+, \partial_g, \partial_h, \partial_{a_k}\} \\
\mathcal{N}_\Phi &= N_\Phi - N_\rho - N_{c_-} - N_{\bar{c}_-} + 2\alpha\partial_\alpha \\
\mathcal{N}_{A_+} &= N_{A_+} + N_{A_-} - N_{Y_+} - N_{Y_-} + \text{conj.} \\
\mathcal{N}_A &= N_A - N_Y + \text{conj.} \\
\mathcal{N}_+ &= N_{c_+} - N_\sigma + \text{conj.}
\end{aligned} \tag{16.3.67}$$

(all ∇^i commute with the ghost equations and with parity). One can convince oneself that the definition (16.3.66) determines the $\hat{S}^i$ uniquely [IV.14] since the ∇^i form a basis of symmetric operators. The expansion of S" then reads

$$S''\cdot\Gamma = \frac{2}{3} \sum_i \alpha_i \hat{S}^i\cdot\Gamma \, , \quad \alpha_i = O(\hbar) \quad . \tag{16.3.68}$$

Inserting it in (16.3.58) and then in (16.3.56), (16.3.57) gives the anomalous dilatation Ward identity

$$W^D\Gamma \sim -i \sum_i \alpha_i \nabla^i \Gamma \quad , \tag{16.3.69}$$

where we have used (16.3.66) and (16.3.62). We observe that the correction term $2\mathcal{N}_-\Gamma$ in (16.3.56) is exactly compensated by the S_- contribution to S (16.3.58), due to the property (16.3.62). This was indeed our motivation for the construction of S_- via the point-splitting technique.

As usual, naive dimensional analysis tells us

$$im\partial_m \Gamma = W^D\Gamma \tag{16.3.70}$$

(here m comprises <u>all</u> mass parameters of the theory). Hence combining (16.3.69) and (16.3.70) yields, with a slight change of notation for the coefficients α_i, the Callan-Symanzik equation

$$\begin{aligned}[m\partial_m &+ \beta_g\partial_g + \beta_h\partial_h + \sum_k \beta_k\partial_{a_k} \\ &- \gamma_\Phi\mathcal{N}_\Phi - \gamma_{A_+}\mathcal{N}_{A_+} - \gamma_A\mathcal{N}_A - \gamma_+\mathcal{N}_+]\Gamma \sim 0 \ .\end{aligned} \tag{16.3.71}$$

Restrictions on the coefficients $\alpha_i = \{\beta,\gamma\}$ result from asymptotic R-invariance (16.3.2):

$$0 \sim W^R\Gamma \sim - [\int dSS - \int d\bar{S}\bar{S}]\cdot\Gamma \tag{16.3.72}$$

(c.f. (16.1.20) and following equations for the case of the chiral model). Due to (16.3.61) the S_- contribution to (16.3.72) disappears, which leaves the homogeneous equations

$$\sum_i \alpha_i [\int dS\hat{S}^i - \int d\bar{S}\hat{\bar{S}}^i]\cdot\Gamma \sim 0 \quad . \tag{16.3.73}$$

The α_i being of order $\hbar$ at least the restrictions at their lowest order are quickly established by replacing the $\hat{S}^i$ by their classical limit S^i. The latter are obtained by applying definition (16.3.66) to the classical action (see (6.80) and section 5):

$$S^g = \frac{1}{128g^3}\,\mathrm{Tr}F^{\alpha}F_{\alpha}\ ,\quad S^h = \mathrm{Tr}A^3\ ,$$

$$S^{a_k} = \mathscr{b}\,\bar{D}\bar{D}\mathrm{Tr}\eta(\Phi^{2k+1} + O(\Phi^{2k+3}))\ ,$$

$$S^{\Phi} = \mathscr{b}\,\bar{D}\bar{D}\mathrm{Tr}\eta\Phi\ , \tag{16.3.74}$$

$$S^{+} = -\mathrm{Tr}(\sigma c_+ c_+) + YP + \mathrm{Tr}\bar{D}\bar{D}(\eta Q),$$

$$S^{A_+} = 2z_1\bar{D}\bar{D}(\bar{A}_{\pm}e^{\pm\tilde{\Phi}}\,A_{\pm})\quad ,$$

$$S^{A} = 2z_2\bar{D}\bar{D}\mathrm{Tr}(e^{-\Phi}\bar{A}e^{\Phi}A) + 3h\mathrm{Tr}A^3\quad .$$

Hence (16.3.73) yields at the one loop order - S^g, S^{a_k}, S^{Φ}, S^A-$3hS^h$ giving identically vanishing contributions -

$$\beta_h = 3h\gamma_A + O(\hbar^2)\quad ,\quad \gamma_+ = O(\hbar^2)\quad . \tag{16.3.75}$$

The restriction implied by (16.3.73) at higher orders will depend in a complicated way on the normalization conditions. Like in the Wess-Zumino model (16.1.22), such restrictions are the genuine expression for the non-renormalization theorem for hard chiral vertices.

We may summarize these results by saying that the supercurrent still contains all the information on the superconformal symmetries, but due to its non-improved character (occurrence of the term S_-) it has to be extracted with some additional work: one always has to treat separately the S_- contribution in the breaking. Since there exists a local quantum extension of it, (16.3.59), this is only a technical complication but no question of principle. The supercurrent still yields all relations amongst the anomalies of the superconformal group.

16.3.4 The "conserved" supercurrent

The BRS-invariant supercurrent V constructed in the last subsection is not conserved. Indeed its trace identity (16.3.5) yields, using the definition (16.3.7),

$$\partial^{\mu}V_{\mu}\cdot\Gamma \sim iw\Gamma + i(DDS - \bar{D}\bar{D}\bar{S})\cdot\Gamma\quad . \tag{16.3.76}$$

(Recall that S has operator dimension three, i.e. is hard.)

In particular its $\Theta = 0$ component is a non-conserved axial current. On the other hand a conserved axial current j_μ^R associated with R-invariance exists, fulfilling

$$\partial^\mu j_\mu^R \cdot \Gamma \sim i w^R \Gamma \equiv i w \Gamma \Big|_{\Theta=0} \tag{16.3.77}$$

as a consequence of the R-Ward identity (16.3.2). (I.e. its non-conservation is soft.) We define the conserved supercurrent V'_μ as the - unique - supersymmetric extension of $j_\mu R$:

$$V'_\mu \cdot \Gamma \Big|_{\Theta=0} = j_\mu^R \cdot \Gamma \quad . \tag{16.3.78}$$

The supersymmetric extension of (16.3.77) gives then

$$\partial^\mu V'_\mu \cdot \Gamma \sim i w \Gamma \quad . \tag{16.3.79}$$

The currents j_μ^R, V'_μ are uniquely defined by (16.3.77), (16.3.79), since there is no improvement term, i.e. no axial vector insertion a_μ with

$$\partial^\mu a_\mu \cdot \Gamma = 0 \quad . \tag{16.3.80}$$

In order to look for a relation of the current V' with the BRS-invariant supercurrent V let us try first to derive a trace identity. Defining the insertion B_α by

$$\bar{D}^{\dot{\alpha}} V'_{\alpha\dot{\alpha}} \cdot \Gamma \sim -2 w \alpha \Gamma + B_\alpha \cdot \Gamma \tag{16.3.81}$$

we see from the conservation law (16.3.79) that it must obey the identity

$$D^\alpha B_\alpha \cdot \Gamma - \bar{D}_{\dot{\alpha}} \bar{B}^{\dot{\alpha}} \cdot \Gamma = 0 \tag{16.3.82}$$

together with (see (16.3.14))

$$\bar{D}\bar{D} B_\alpha \cdot \Gamma = 0 \quad . \tag{16.3.83}$$

B_α is given in the tree approximation by (6.100):

$$B_\alpha = \bar{D}\bar{D}D_\alpha B$$
$$B = 4\,(B^1 - \frac{1}{64\alpha} B^2)\ ,\ B = \bar{B} \qquad (16.3.84)$$

B^1, B^2 given by (6.88).

We shall now show by induction that the general solution of (16.3.82) and (16.3.83) is given at all orders by

$$B_\alpha \cdot \Gamma = \bar{D}\bar{D}D_\alpha B \cdot \Gamma \qquad (16.3.85)$$

with $B = \bar{B}$, $\dim(B) = 2$.

The proof is similar to the one used in subsection 16.3.2 to establish the trace identity. Let us assume (16.3.85) to hold up to and including order n-1:

$$B_\alpha \cdot \Gamma = \bar{D}\bar{D}D_\alpha B \cdot \Gamma + \hbar^n B_\alpha^{(n)} \cdot \Gamma + O(\hbar^{n+1}) \quad . \qquad (16.3.86)$$

Conditions (16.3.82) and (16.3.83) yield at order n the constraints

$$D^\alpha B_\alpha^{(n)} - \bar{D}_{\dot\alpha} \bar{B}^{(n)\dot\alpha} = 0 \ ,$$
$$\bar{D}\bar{D}B_\alpha^{(n)} = 0 \quad . \qquad (16.3.87)$$

Their general solution is given in Appendix E:

$$B_\alpha^{(n)} = \bar{D}\bar{D}D_\alpha B^{(n)} + D^\beta S_{(\alpha\beta)}^{(n)} \qquad (16.3.88)$$

with $\bar{B}^{(n)} = B^{(n)}$.

Here $S_{(\alpha\beta)}^{(n)}$ is chiral, symmetric in its indices, of dimension 3 and has R-weight -1. By checking all possibilities one concludes that there is no such object:

$$S_{(\alpha\beta)}^{(n)} = 0 \ . \qquad (16.3.89)$$

Substituting (16.3.88) in (16.3.87) and redefining $B \to B + \hbar^n B^n$ establishes (16.3.87) at its next order, which ends the proof of (16.3.85).

This result yields the looked for trace identity (16.3.8)

$$\bar{D}^{\dot{\alpha}} V'_{\alpha\dot{\alpha}} \cdot \Gamma \sim -2w_{\alpha}\Gamma + \bar{D}\bar{D}D_{\alpha} B \cdot \Gamma \qquad (16.3.90)$$

with $B = \bar{B}$, $\dim B = 2$.

The relation to the trace identity (16.3.5) is then immediately given by the redefinition (16.3.10).

Considering now the BRS-invariance of B and V' we recall subsection 16.2 for the abelian case (cf. (16.2.48)) where we have seen that

$$F^{\alpha}F_{\alpha} = \bar{D}\bar{D}(D\Phi\bar{D}\bar{D}D\Phi + \bar{D}\Phi DD\bar{D}\Phi + \Phi D\bar{D}\bar{D}D\Phi) \equiv \bar{D}\bar{D}B \qquad (16.3.91)$$

is of the desired form. But B, hence V', is not BRS-invariant. This holds a fortiori in the present non-Abelian case.

CHAPTER V.

RENORMALIZATION : SOFT ANOMALIES

The analysis of the renormalization properties performed in the preceding chapter led for all completely massive theories in the class of our examples to a satisfactory result. Of the examples containing massless particles, the Wess-Zumino model, SQED and S'QED still permitted an immediate treatment in the loop expansion, for the former two also all the desired vertex functions of operators were constructed without too severe an obstruction from the infrared region. (By this we mean that they exist for non-exceptional momenta.) For the O'Raifeartaigh model, however, we could prove the supersymmetry Ward-identity only up to an IR-anomaly and for SYM we did, because of the $1/(k^2)^2$-problem, not even attempt to go beyond the analysis of the deep Euclidean region. In this chapter we shall attack these problems. For the O'Raifeartaigh model we shall find that the IR-anomaly can be absorbed by going over from the $\hbar$-expansion to an expansion in $\sqrt{\hbar}\ \ln \hbar$. The validity of a strict Ward-identity implies then a well-determined mass generated by radiative corrections: a mass "sum-rule" [V.2]. For pure SYM an IR-regulator will be introduced which breaks supersymmetry, but maintains BRS-invariance. One constructs then Green's functions of BRS-symmetric operators and shows that they are independent of the regulator and thus have also supersymmetry [V.5,6].

SECTION 17. MASS GENERATION - THE O'RAIFEARTAIGH MODEL

In section 13.3.2 we have seen that within the $\hbar$-expansion supersymmetry cannot be maintained in higher orders (in fact starting with one loop). The supersymmetry Ward-identity contains a non-vanishing contribution on the right-hand-side of dimensions $\rho = \delta = 5/2$:

$$W_\alpha \Gamma = N^{5/2}_{5/2}\,[U_\alpha]\cdot\Gamma \quad . \tag{17.1}$$

Since U_α starts with

$$U_\alpha = uW_\alpha \int dx \bar{A}_0 A_0 + O(\hbar^2) \quad , \tag{17.2}$$

one is tempted nevertheless to absorb the breaking as a counterterm in Γ_{eff} and one is then lead to consider along with the divergent diagram of fig. 13.3.2

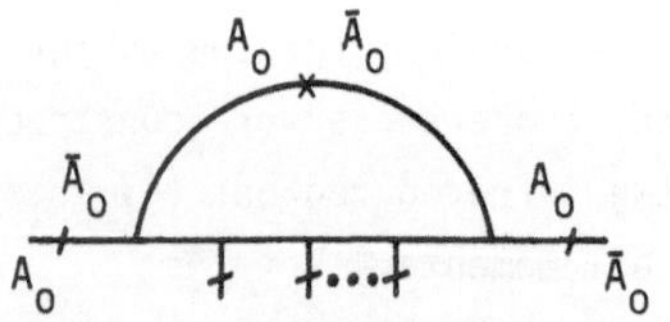

FIGURE 13.3.2

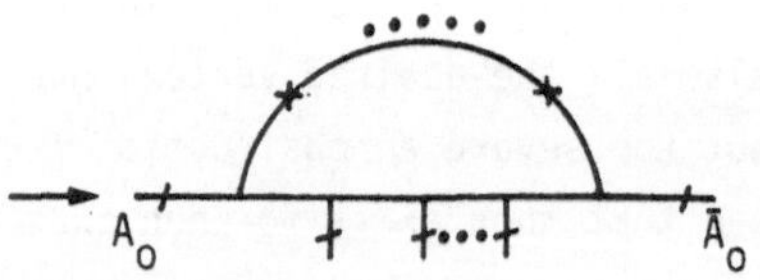

FIGURE 17.1

all other divergent diagrams of comparable magnitude: $\hbar$ <u>not</u> taken as a formal expansion parameter all the diagrams of fig. 17.1 are non-negligible higher order contributions, which due to the IR-divergences become as important as the one of fig. 13.3.2 and hence have to be taken into account at the same approximation order. This suggests [V.1] partially summing the perturbation series, and thus a shifting of the pole in the $\langle A_0 \bar{A}_0 \rangle$ propagator to a non-zero value: a mass is generated for the A_0 field. Its value is determined by the Ward-identity (17.1). Of course with this summation one has left the perturbative framework which before served to construct the Ward-identity. It is preferable to work within one scheme within which one establishes the Ward-identity <u>and</u> decribes the mass generation. This has been proposed in [IV.8] and then applied to the above model in [V.2] in a modified form. We are going to present now the main ideas.

The supersymmetry Ward-identity is supposed to determine the mass of the A_0-field. We formalize this by introducing already in the tree approximation a mass μ^2 for A_0 and choosing normalization conditions such that the A_0-propagator has a pole at μ^2: blindly putting a mass for A_0 breaks supersymmetry, but the hope is that theorem 13.3 controls the breaking. Let us be explicit; we first add to the tree approximation (13.37) a breaking term:

$$\Gamma^{\eta}_{tree} = \Gamma^{(0)} + \Delta^{\eta} ,$$

$$\Delta^{\eta} \equiv \frac{1}{64} \int dV \hat{\eta} \bar{A}_o A_o \quad , \tag{17.3}$$

$$\hat{\eta} = \eta - \frac{1}{4} u\Theta^2\bar{\Theta}^2 \quad , \quad u = \mu^2 + O(\hbar) \quad ,$$

$$\eta = C + \Theta\chi + \bar{\Theta}\bar{\chi} + \frac{1}{2}\,\Theta^2 M + \frac{1}{2}\,\bar{\Theta}^2\bar{M} + \Theta\sigma^{\mu}\bar{\Theta}v_{\mu} + \frac{1}{2}\,\bar{\Theta}^2\Theta\lambda + \frac{1}{2}\Theta^2\bar{\Theta}\bar{\lambda} + \frac{1}{4}\,\Theta^2\bar{\Theta}^2 D .$$

The reason for this peculiar choice will become clear by considering the Ward-identities which hold for (17.3):

$$W^{\eta}_{\alpha}\Gamma^{\eta} = 0 , \qquad \bar{W}^{\eta}_{\dot{\alpha}}\Gamma^{\eta} = 0 , \qquad W^{\eta}_{R}\Gamma^{\eta} = 0 , \tag{17.4}$$

i.e. η being a shifted superfield with R-weight zero makes Γ^{η} invariant under shifted Ward-identity operators:

$$W^{\eta}_{x} = W_x - i\int dV \delta_x \hat{\eta} \frac{\delta}{\delta\hat{\eta}} \quad , \qquad (x = \alpha,\dot{\alpha},R)$$

$$W_x = -i\int dS \delta_x A_k \frac{\delta}{\delta A_k} - i\int d\bar{S} \delta_x \bar{A}_k \frac{\delta}{\delta\bar{A}_k} \quad . \tag{17.5}$$

Furthermore at $\eta = 0$ we have

$$W_{\alpha}\Gamma = -\frac{iu}{2} \int dV \Theta_{\alpha}\bar{\Theta}^2 \left.\frac{\delta\Gamma^{\eta}}{\delta\eta}\right|_{\eta=0} , \tag{17.6}$$

$$W_R\Gamma = 0 ,$$

$$\Gamma \equiv \Gamma^{\eta}\Big|_{\eta=0} \tag{17.7}$$

and in the tree approximation

$$W_{\alpha}\Gamma_{tree} = \mu^2\int dx \psi_{o\alpha}\bar{A}_o(x) . \tag{17.8}$$

The external field η controls the covariance of the breaking in the course of renormalization, this is the guiding principle for its introduction (cf. section 13.2). Indeed, the Ward-identities (17.4) permit a generalization to all orders. We introduce as Γ_{eff}:

$$\Gamma^{\eta}_{eff} = \Gamma^{\eta}_{tree} + \frac{M(s-1)}{8} (\int dSA_o^2 + \int d\bar{S}\bar{A}_o^2)$$
$$+ \sum_{i=0}^{4} a_i\Delta_i + a_\eta\Delta^\eta + a_F\Phi_o$$
$$+ \sum_n \sum_i a_{ni}(\Gamma_{ni} + \Gamma^P_{ni}) + \Gamma_{kin.} \quad , \qquad (17.9)$$

where

$$\Delta_k = \frac{1}{16} [\int dV\bar{A}_k A_k]^4_4 \quad , \qquad k = 0,1,2$$
$$\Delta_3 = \frac{1}{4} \; [\int dSA_1A_2 + \int d\bar{S}\bar{A}_1\bar{A}_2]^4_4 \quad ,$$
$$\Delta_4 = \frac{1}{32} [\int dSA_oA_1^2 + \int d\bar{S}\bar{A}_o\bar{A}_1^2]^4_4 \quad , \qquad (17.10)$$
$$\Phi_o = \frac{1}{4} \; (\int dSA_o + \int d\bar{S}\bar{A}_o)$$

and Γ_{ni} are all possible monomials in the components of* A_o, A_1 and η which are independent from Δ_i, Δ^η and are I-invariant with UV-dimension $d \leq 4$, IR-dimension ≥ 4 (cf. table). They are classified according to their R-weight n:

$$W_R\Gamma_{ni} = n\,\Gamma_{ni} \; . \qquad (17.11)$$

Γ^P_{ni} denotes the parity conjugate of Γ_{ni}, it has R-weight -n. Γ_{lin} are linear terms in the quantized fields A_k; they may have IR-dimension < 4.

Table 17.1. UV-, IR-dimensions d,r

	x	Θ	∂_Θ	s-1	s	A_o	ψ_o	F_o	A_1	A_2	η
d	-1	-1/2	+1/2	1	1	1	3/2	2	1	1	0
r	-1	-1/2	+1/2	1	0	2	3/2	2	2	1	0

Theorems 13.3 and 13.5 guarantee now supersymmetry and R-invariance (17.4), indeed strict invariance since all soft terms can be absorbed (no IR-anomaly for susy, no clash with normalization conditions for R-invariance). The parameters a_i ($i = 0,\dots, 4$), a_η, a_F of Γ_{eff} and

*Radiative corrections do not depend on A_2, since the interaction Δ_4 involves only A_o and A_1.

f, u occurring in the Ward-identity operators (17.5) (cf. (3.41)) are fixed by the normalization conditions (3.45) - (3.49) taken at an Euclidean point $\varkappa^2$, $\varkappa^2 < 0$, together with

$$\Gamma_{A_0\bar{A}_0}(p^2 = \mu^2) = 0 \tag{17.12}$$

which fixes the pole of the A_0-propagator to be at μ^2.

The η-dependent Ward-identities (17.4) imply to all orders (17.6) and (17.7) and thus describe a theory with soft breaking of supersymmetry:

$$\begin{aligned} W_\alpha\Gamma &= -\frac{iu}{2}\int dV\Theta_\alpha\bar{\Theta}^2 \left.\frac{\delta\Gamma^\eta}{\delta\eta}\right|_{\Theta=0} , \\ &= uU_\alpha\Gamma \end{aligned} \tag{17.13}$$

(UV-, IR-dimension of U_α : $d = r = 5/2$). u is a function of the parameters of the explicitly broken theory and is a formal power series in $\hbar$:

$$u(\mu,\varkappa,m,\xi,g) = \mu^2 + \hbar u_1(m,\xi,g) + \sum_{n\geq 2}\hbar^n u_n(\mu^2,\varkappa,m,\xi,g) . \tag{17.14}$$

Its one-loop value u_1 does not depend on μ^2 [IV.5]

$$u_1 = \ldots\ g^2\xi^2/m^2 \tag{17.15}$$

(the dots stand for a numerical coefficient).

In order to make out of (17.13) a strict Ward-identity, i.e. one with vanishing right-hand-side, one puts

$$u(\mu,\varkappa,m,\xi,g) = 0 \tag{17.16}$$

and solves for μ. The solution $\mu^*(\varkappa,m,\xi,g)$ of (17.16) is inserted in Γ thus obtaining a generating functional Γ^* fulfilling the Ward-indentity for spontaneously broken supersymmetry

$$W^*_\alpha\Gamma^* = 0 \quad , \quad \bar{W}^*_{\dot{\alpha}}\Gamma^* = 0 \tag{17.17}$$

(the superscript * denotes the replacement $\mu \to \mu^*$). In this way one remains in a strictly perturbative framework, although the expansion has drastically changed in performing the replacement $\mu \to \mu^*$. In fact it is governed by small mass behaviour [V.3,4] of Feynman diagrams (or Green's functions) since the mass of the A_o-field has become $\hbar$-dependent and has to be taken into account when the $\hbar$-expansion is performed. Since for $\mu = 0$ Γ exists - it only leads to an IR-anomaly - there exists [IV.8, V.3,4] a formal expansion of Γ in μ with $\ln \mu$ dependent coefficients:

$$\Gamma(\mu) = \Gamma(o) + \sum_{m \geq 1} \mu^m \Gamma_m(\ln \mu) \ , \tag{17.18}$$

where each term is itself a formal power series in $\hbar$. The function u (17.14) has a similar expansion

$$u(\mu) = \mu^2 + \hbar u_1 + \sum_{n \geq 2} \hbar^n (u_n(o) + \sum_{m \geq 1} \mu^m u_{n,m}(\ln \mu)) . \tag{17.19}$$

Therefore (17.16) can be solved iteratively in a formal power series of the form

$$\mu^*(\varkappa, m, \xi, g) = \hbar^{1/2} \mu_1^* + \sum_{n \geq 3} \hbar^{n/2} \mu_n^*(\ln \hbar) \ , \tag{17.20}$$

with $\mu_1^* = (-u_1^{1/2})$ given by (17.15).

Substituting (17.20) into (17.18) one obtains the functional Γ^*:

$$\Gamma^* = \Gamma_0 + \hbar \Gamma_2^* + \sum_{n \geq 3} \hbar^{n/2} \Gamma_n^*(\ln \hbar) \ , \tag{17.21}$$

which fulfills the Ward-identities (17.17) in the sense of formal power series in $\sqrt{\hbar} \ln \hbar$. The peculiar form of the lowest order term in (17.21) comes from (17.15) (μ^2-independence of u_1) and the fact that

$$\Gamma(\mu) = \Gamma(o) + O(\mu^2) \qquad \text{(in the tree approximation).}$$

One might wonder whether the replacement $\mu \to \mu^*$ does in fact not completely destroy the relation between number of loops in a Feynman diagram and powers of $\hbar$. Let us check that this is not the case by counting $\hbar$'s in a connected diagram. An ordinary propagator contributes $\hbar$, a vertex of the tree interaction $\hbar^{-1}$, for the entire diagram there is an overall factor $\hbar$ since $Z = \exp i/\hbar\, Z_c$ and we start from the Gell-Mann-Low formula. Suppose in the diagram considered there is a $A_0\bar{A}_0$-line with n insertions of A_0-mass counter-terms, then the $\hbar$-dependent A_0-mass will contribute

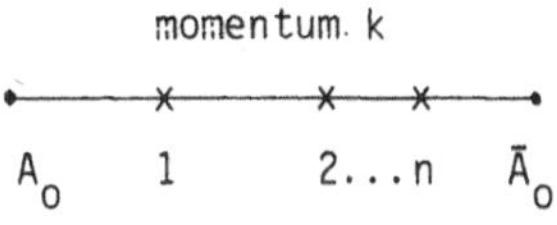

$$: \int dk \frac{1}{(k^2-\mu^{*2}\hbar)^{n+1}} (\ldots.) \sim (\mu^{*2}\hbar)^{1-n} \tag{17.22}$$

$(n \geq 1)$* with $\ln \mu^2\hbar$-dependent coefficient (we are only counting powers). Hence e.g. in the diagram of figure 17.1 with power x in $\hbar$ carried by the insertion we have a total $\hbar$-power: $n(x-1) + 2$.

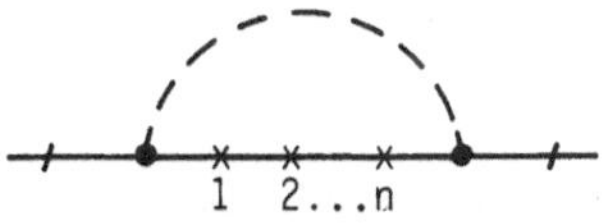

FIGURE 17.1

Starting from the explicitly broken theory where A_0 has mass μ, its mass counter-terms are the value of $\Gamma^{(1)}_{A_0\bar{A}_0}$ at $p^2 = 0$, $s = 1$. Due to the on-shell normalization we have

$$\Gamma_{A_0\bar{A}_0} = p^2 - \mu^2 + \hbar(p^2 - \mu^2)\gamma(p^2,\mu^2,\ldots) + O(\hbar^2) \; , \tag{17.23}$$

i.e. at $p^2 = 0$, $s = 1$

$$\Gamma^{(1)}_{A_0\bar{A}_o} = -\hbar\mu^2\gamma(0,\mu^2,\ldots.) \; . \tag{7.24}$$

Thus the counter-term carries $\hbar$-power $x = 2$ if we replace μ^2 by μ^{*2}: We do not lose the connection between $\hbar$-powers and number of loops in a diagram.

---

*In the case $n = 0$ the $\hbar$ in the denominator does not contribute to the overall $\hbar$-power of the diagram.

Similarly if there occurs a $\int A_0^3$-counter-term in the explicitly broken theory it is at least multiplied by $\hbar^{1/2}$ in the transition to the spontaneously broken version and again the powers in $\hbar$ do not decrease.

Another check whether μ^* is indeed a physical mass is provided by the renormalization group equation. If μ^* satisfies such an equation it is a physical quantity. We shall now present this check [V.2].

Let us recall from section 12 that an insertion $\tilde{\Delta}$ is called symmetric if

$$W_x^\eta \tilde{\Delta}\Gamma^\eta = 0 \quad , \quad x = \alpha, \dot{\alpha}, R \tag{17.25}$$

analogously a (functional) differential operator ∇ is called symmetric if

$$[W_x^\eta, \nabla] = 0 \quad . \tag{17.26}$$

The insertions Δ_i $(i = 0,\dots,4)$, Δ_η and Φ_0 of equation (17.10) are a basis for the symmetric insertions of UV-dimension $d \leq 4$ in the tree approximation. Their IR-dimension is $r \geq 4$, hence there is no convergence problem. They can be extended to a basis of symmetric insertions at every order in $\hbar$:

$$\begin{aligned} \tilde{\Delta}_a &= \Delta_a + O(\hbar) \quad , \quad a = 0,\dots,4,\eta \\ \tilde{\Phi} &= \Phi_o \quad , \end{aligned} \tag{17.27}$$

Φ_o remains trivially symmetric since it is linear in the fields. The standard construction proceeds analogously to the above construction of an invariant Γ^η: one couples Δ_a in the tree approximation to an external field η_a with appropriate dimension and quantum numbers and renormalizes the theory containing these external fields such that (17.25) holds.

For differential operators (17.26) provides the instruction how to obtain them: starting from a differential operator $\lambda\partial_\lambda$(λ = M,$\varkappa$,g,m are the parameters of the model, the variation of which will interest us) we may render it symmetric by adding suitable derivatives with respect to ξ and μ^2 to it:

$$\lambda\tilde{\partial}_\lambda = \lambda\partial_\lambda + b_\lambda \xi\partial_\xi + c_\lambda \mu^2 \partial_{\mu^2} \tag{17.28}$$

with the conditions

$$\begin{aligned} (\lambda\partial_\lambda + b_\lambda \xi\partial_\xi + c_\lambda \mu^2\partial_{\mu^2})\, f &= 0 \quad , \\ (\lambda\partial_\lambda + b_\lambda \xi\partial_\xi + c_\lambda \mu^2\partial_{\mu^2})\, u &= 0 \quad . \end{aligned} \tag{17.29}$$

(Recall from (3.41) and (17.3), (17.5) that W_x depends on f and u for $x = \alpha,\dot{\alpha}$).

Quite analogously we may symmetrize the counting operators

$$\begin{aligned} N_k &= \int dSA_k \frac{\delta}{\delta A_k} + \int d\bar{S}\bar{A}_k \frac{\delta}{\delta \bar{A}_k} \quad , \qquad k = 0,1,2 \\ N_\eta &= \int dV\eta \frac{\delta}{\delta\eta} \end{aligned} \tag{17.30}$$

by adding suitable functions

$$\begin{aligned} \tilde{N}_0 &= N_0 + b_0 \xi\partial_\xi + c_0 \mu^2 \partial_{\mu^2} \quad , \\ \tilde{N}_\eta &= N_\eta + b_\eta \xi\partial_\xi + c_\eta \mu^2 \partial_{\mu^2} \quad , \\ \tilde{N}_{1,2} &= N_{1,2} \end{aligned} \tag{17.31}$$

which are determined from

$$\begin{aligned} (\ 1 - b_0 \xi\partial_\xi - c_0 \mu^2\partial_{\mu^2})\, f &= 0 \quad , \\ (\ 0 + b_0 \xi\partial_\xi + c_0 \mu^2\partial_{\mu^2})\, u &= 0 \quad , \\ (\ 0 + b_\eta \xi\partial_\xi + c_\eta \mu^2\partial_{\mu^2})\, f &= 0 \quad , \\ (-1 + b_\eta \xi\partial_\xi + c_\eta \mu^2\partial_{\mu^2})\, u &= 0 \quad . \end{aligned} \tag{17.32}$$

It is to be noted that the system (17.29), (17.32) is soluble for b,c in the tree approximation where $f = 4\xi/g$ and $u = \mu^2$ hence it is soluble to all orders in $\hbar$. The 7 symmetric differential operators (17.28), (17.30) applied to the generating functional Γ^η yield via the action principle, section 11, symmetric insertions which can be expanded in the basis (17.27). Eliminating the 5 insertions $\tilde{\Delta}_a$ between the 7 equations so obtained, one gets 2 partial differential equations:

$$\begin{aligned} &(M\tilde{\partial}_M + \sigma_g\tilde{\partial}_g + \sigma_m m\tilde{\partial}_m - \sum_0^2 \nu_k\tilde{N}_k - \nu_\eta\tilde{N}_\eta)\Gamma^\eta = \sigma_0\Phi_0 \ , \\ &(\varkappa\tilde{\partial}_\varkappa + \beta_g\tilde{\partial}_g + \beta_m m\tilde{\partial}_m - \sum_0^2 \gamma_k\tilde{N}_k - \gamma_\eta\tilde{N}_\eta)\Gamma^\eta = \alpha\Phi_o \ . \end{aligned} \tag{17.33}$$

Inserting (17.28) and (17.31) into (17.33) the equations become

$$\begin{aligned} &(M\partial_M + \sigma_g\partial_g + \sigma_m m\partial_m + \sigma_\xi\xi\partial_\xi + \sigma_\mu\mu^2\partial_{\mu^2} - \sum_0^2 \nu_k N_k - \nu_\eta N_\eta)\Gamma^\eta = \sigma_o\Phi_o \ , \\ &(\varkappa\partial_\varkappa + \beta_g\partial_g + \beta_m m\partial_m + \beta_\xi\xi\partial_\xi + \beta_\mu\mu^2\partial_{\mu^2} - \sum_0^2 \gamma_k N_k - \gamma_\eta N_\eta)\Gamma^\eta = \alpha\Phi_o \ , \end{aligned} \tag{17.34}$$

with the conditions

$$\begin{aligned} &(M\partial_M + \sigma_g\partial_g + \sigma_m m\partial_m + \sigma_\xi\xi\partial_\xi + \sigma_\mu\mu^2\partial_{\mu^2} + \nu_0)\ f = 0 \ , \\ &(M\partial_M + \sigma_g\partial_g + \sigma_m m\partial_m + \sigma_\xi\xi\partial_\xi + \sigma_\mu\mu^2\partial_{\mu^2} + \nu_0)\ u = 0 \end{aligned} \tag{17.35}$$

and other similar ones where σ,ν are replaced by β γ and M is replaced by $\varkappa$.

The coefficients of the partial differential equations can be determined by use of the normalization conditions (3.45) - (3.49) at $\varkappa^2$ and (17.12) at μ^2 and since those are all M-independent we find

$$\sigma = \nu = 0 \ , \qquad \text{(for all indices)} \tag{17.36}$$

also $\gamma_\mu = \alpha = 0$ (since we normalize μ^2 on shell).
Hence:

$$\partial_M\Gamma^\eta = 0 \ , \quad \partial_M f = 0 \ , \quad \partial_M u = 0 \ . \tag{17.37}$$

These equations state that at s = 1 the Green's functions are independent of the auxiliary mass M, the same being true for f and u (the functions occurring as shifts in the Ward-identity operators). With $\mathcal{D} \equiv \varkappa\partial_\varkappa + \beta_g\partial_g + \beta_m m\partial_m + \beta_\xi \xi\partial_\xi$ we also have

$$(\mathcal{D} - \sum_0^2 \gamma_k N_k - \gamma_\eta N_\eta)\Gamma^n = 0 , \tag{17.38}$$

$$\begin{aligned}(\mathcal{D} + \gamma_0)\ f &= 0 ,\\ (\mathcal{D} + \gamma_\eta)\ u &= 0 .\end{aligned} \tag{17.39}$$

These are the renormalization group equations, which govern the $\varkappa$-dependence of the respective quantities.

Thus far we have considered the explicitly broken theory; in order to go over to the spontaneously broken one we have to put $\eta = 0$ and to perform the replacement $\mu \to \mu^*$ (17.16), (17.20). From (17.39) we obtain as renormalization group equations for u, using (17.16)

$$\begin{aligned} 0 &= [\mathcal{D}u(\mu,\varkappa,m,\xi,g)]_{\mu=\mu^*} ,\\ &= \mathcal{D}^*u(\mu^*(\varkappa,m,\xi,g),\varkappa,m,\xi,g)\\ &\quad - \mathcal{D}^*\mu^*(\varkappa,m,\xi,g)\ [\frac{\partial u}{\partial\mu}]_{\mu=\mu^*} . \end{aligned} \tag{17.40}$$

The first term vanishes due to (17.16) to which one has applied a differential operator, hence

$$\mathcal{D}^*\mu^*(\varkappa,m,\xi,g) = 0, \tag{17.41}$$

since $\partial u/\partial\mu$ is non-vanishing a $\mu = \mu^*$.

Recall

$$\mathcal{D}^* = \varkappa\partial_\varkappa + \beta^*_g\partial_g + \beta^*_m\partial_m + \beta^*_\xi\partial_\xi .$$

(17.41) expresses the fact that the generated mass μ^* does not depend on the scale of the normalization mass $\varkappa$ and confirms μ^* as being a physical parameter.

Similarly one obtains the renormalization group equation for Γ^* from (17.38)

$$[(\mathcal{D} - \sum_0^2 \gamma_k N_k)\Gamma(\mu,\varkappa,m,\xi,g)]_{\mu=\mu^*} = 0 \; ,$$

$$(\mathcal{D}^* - \sum_0^2 \gamma_k^* N_k)\Gamma^* - \mathcal{D}^*\mu^* \left[\frac{\delta\Gamma}{\delta\mu}\right]_{\mu=\mu^*} = 0 \; , \qquad (17.42)$$

$$(\mathcal{D}^* - \sum_0^2 \gamma_k^* N_k)\Gamma^* = 0 \; .$$

One can also show that the perturbative solution (17.20) is compatible with the exact solution of (17.41), [V.2].

This concludes our discussion of the vertex functions and global invariances of the O'Raifeartaigh model.

SECTION 18: THE OFF-SHELL INFRARED PROBLEM IN SYM

18.1. Statement of the problem. Tree approximation

As the example in section 9 (fig. 9.2, p. 141) shows there arises an off-shell infrared problem in supersymmetric gauge theories as soon as one works with the superfield formalism. If existence of each individual diagram is desired - which was the basis for deriving the action principle - it seems to be unavoidable to use an auxiliary mass as an infrared cut-off. We shall therefore abandon part of the symmetries i.e. break them and try to study the breaking independent features. Since the infrared problem is caused by the pure gauge field terms in the vector-superfield propagator the hope is that gauge invariant quantities exist infraredwise. BRS invariance defines which objects are candidates for being physical and thus should not be violated. We therefore propose to break supersymmetry explicitly, but softly by mass terms for the dangerous gauge sector, then to construct the candidates for physical quantities and to show that they indeed exist in particular by being independent of the infrared regulators introduced [V.5,6]. There is of course an immediate difficulty, namely that the regularizing mass terms have to be introduced in a manner which is compatible with the BRS invariance. But here the general solution of the Slavnov identity, cf. section 5.3, will help us. We have seen that the replacement (cf. 5.115) in $\bar{\Gamma}$ of Φ by $\hat{\Phi}$

$$\Phi_i \to \hat{\Phi}_i = c_1\Phi_1 + \sum_{k\geq 2} c_{\omega k} S^{\omega}_{i(i_1\ldots i_k)}\Phi^{i_1}\ldots\Phi^{i_k} \tag{18.1}$$

which is a kind of generalized wave function renormalization, left invariant the Slavnov identity. It maintains this property even if we permit c_1 to develop a Θ-dependent part - which breaks of course supersymmetry:

$$c_1 \to c_1 + \frac{1}{2}\mu^2\Theta^2\bar{\Theta}^2 \; . \tag{18.2}$$

The Θ-independent new c_1 is now an ordinary wave function renormalization in which we are not interested for the moment, hence we put $c_1 = 1$ in the following. Looking into (18.1) it is clear that $\Phi \to (1 + \frac{1}{2}\mu^2\Theta^2\bar{\Theta}^2)\Phi$ means $D \to D + 2\mu^2 C$ (in $\bar{\Gamma}$). Let us check what that means in the tree approximation, in particular for the (C,D) bilinear terms. $\bar{\Gamma}$ is the gauge invariant part (cf. (5.93)) hence

$$\bar{\Gamma}_{bil}(C,D) = \frac{1}{2g^2}\int dx\,[D + 2\mu^2 C + \Box C]^2 \; . \tag{18.3}$$

In the (α,B) gauge formulation we may add the same gauge fixing term (B-dependent term) as previously (see eqs. (15.3, 15.7)). Thus the complete bilinear part of Γ in the fields C,D and B and in the $\Phi\Pi$ -fields reads

$$\begin{aligned}\Gamma_{bil}(C,D,B,c_+,c_-) &= \bar{\Gamma}_{bil}(C,D)\\ &+ \mathrm{Tr}\int dx[-f(D-\Box C) + \frac{\alpha}{2}f^2 - 4if_-(\Box + \mu^2)b_+] \; ,\end{aligned} \tag{18.4}$$

where f, f_-, b_+ are the components of the superfields B, c_-, c_+ which are coupled to C and D:

$$\begin{aligned} f &= \mathrm{Re}[B|_{\Theta^2\text{-comp.}}] \; ,\\ f_- &= \mathrm{Re}[c_-|_{\Theta^2\text{-compon.}}]\\ b_+ &= \mathrm{Im}[c_+|_{\Theta^0\text{-comp.}}] \; .\end{aligned} \tag{18.5}$$

The resulting free propators are now massive, with mass μ^2:

$$
\begin{aligned}
&\langle TCC\rangle = K(\alpha-g^2) \quad &&, \langle TCD\rangle = K[(\alpha+g^2)k^2 - 2\alpha\mu^2)], \\
&\langle TDD\rangle = K[\alpha(k^2-2\mu^2)^2 - g^2k^4] \quad &&, \langle TCf\rangle = 2K(k^2-\mu^2)\ , \\
&\langle TDf\rangle = 2K(k^2-\mu^2)(k^2-2\mu^2) \quad &&, \langle Tff\rangle = 0\ , \qquad (18.6)\\
&\langle Tf_-b_+\rangle = -K(k^2-\mu^2)\ , \\
&\text{with} \quad K \equiv -\frac{i\hbar}{4}\ [k^2-\mu^2 + i\varepsilon]^{-2}\ .
\end{aligned}
$$

Having tamed now the infrared divergences off-shell by breaking supersymmetry softly and not violating BRS we have to control the breaking, in particular its covariance, by a manageable tool. As in the O'Raifeartaigh model we introduce an external vectorsuperfield whose highest component is shifted:

$$u' = u + \frac{1}{2}\mu^2\Theta^2\bar{\Theta}^2\ , \qquad (18.7)$$

u is a gauge singlet superfield of dimension zero and R-weight zero. With the field u' included in the Ward identity operator the supersymmetry Ward identity reads

$$W_\alpha\Gamma \equiv W^h_\alpha\Gamma - 2i\mu^2\int dx\ \frac{\delta\Gamma}{\delta u^\alpha_\lambda} = 0 \quad , \qquad (18.8)$$

where u_λ is the $\bar{\Theta}^2\Theta$-component of u and the homogeneous Ward identity operator is

$$W^h_\alpha = -i\sum_\varphi \int dz\,\delta_\alpha\varphi\,\frac{\delta}{\delta\varphi} \qquad (18.9)$$

(here φ stands for all superfields of the theory).

In order to establish the non-physical character of the IR cut-off μ^2 and of the supersymmetry breaking (18.8) we shall extend the BRS-doublet trick explained in sections 5.5 and 15. Namely we shall show that it is possible to treat μ^2 as a gauge parameter and the superfield u as a gauge object as well. We therefore allow μ^2 and u to vary like gauge parameters under BRS:

$$s\mu^2 = \nu^2 \quad , \quad s\nu^2 = 0 \quad ,$$
$$su = v \quad , \quad sv = 0 \quad , \tag{18.10}$$

where ν^2 is an anticommuting parameter and v an anticommuting superfield of dimension and R-weight zero, to both of which $\Phi\pi$ charge 1 is assigned. The shifted field u' (18.7) transforms as

$$su' = v' = v + \frac{1}{2}\nu^2\theta^2\bar{\theta}^2 \quad . \tag{18.11}$$

We shall thus impose the Slavnov identity - we shall work in the (α,B)-gauge, see (15.2, 15.3) -

$$\begin{aligned} s(\Gamma) \equiv\ & \mathrm{Tr}\int dV \frac{\delta\Gamma}{\delta\rho}\frac{\delta\Gamma}{\delta\Phi} \\ & + \int dS\{\mathrm{Tr}[B\frac{\delta\Gamma}{\delta c_-} + \frac{\delta\Gamma}{\delta\sigma}\frac{\delta\Gamma}{\delta c_+}] + \frac{\delta\Gamma}{\delta Y}\frac{\delta\Gamma}{\delta A}\} \\ & + \int d\bar{S}\{\mathrm{Tr}[\bar{B}\frac{\delta\Gamma}{\delta\bar{c}_-} + \frac{\delta\Gamma}{\delta\bar{\sigma}}\frac{\delta\Gamma}{\delta\bar{c}_+}] + \frac{\delta\Gamma}{\delta\bar{Y}}\frac{\delta\Gamma}{\delta\bar{A}}\} \\ & + \sum_k z_k\partial_{p_k}\Gamma + \int dVv\ \frac{\delta\Gamma}{\delta u} = 0 \ , \end{aligned} \tag{18.12}$$

where p_k represents all gauge parameters, μ^2 included, and z_k their BRS variations.

Because of the shift (18.11) and in order to preserve the compatibility of the supersymmetry Ward identity with the Slavnov identity we also have to modify (18.8) as

$$W_\alpha\Gamma = W_\alpha^h\Gamma - 2i\int dx[\mu^2\frac{\delta}{\delta u_\lambda^\alpha} + \nu^2\frac{\delta}{\delta v_\lambda^\alpha}]\Gamma = 0 \quad . \tag{18.13}$$

The functionals obeying this Ward identity depend on the shifted superfields u',v'.

The Slavnov identity (18.12) assures the gauge independence of the physical quantities. In particular:

$$\partial_{\mu^2} Z_{phys}(q) = 0 \ , \tag{18.14}$$

$$\frac{\delta}{\delta u} Z_{phys}(q) = 0 \tag{18.15}$$

for Z_{phys} (q) the generating functional of the Green's functions of gauge invariant operators Q_r with sources q_r (see 5.164). Moreover (18.15) implies that Z_{phys} (defined at $\nu^2 = 0$) obeys the exact supersymmetry Ward identity

$$W_\alpha^h Z_{phys}(q) = -i \sum_r \int dz \delta_\alpha q_r \frac{\delta}{\delta q_r} Z_{phys}(q) = 0 . \tag{18.16}$$

Our task is to show the existence of a vertex functional Γ satisfying the Slavnov identity (18.12) and a set of normalization conditions fixing Γ in such a way that the free action posesses the required μ^2-mass terms as in (18.3, 18.4).

We do this now in the tree approximation, i.e. we look for the general classical Γ satisfying the Slavnov and Ward identities (18.12, 18.13). In this way we shall obtain the most general parametrization of the theory. We shall see in particular that a supplementary set of gauge parameters shows up.

The discussion of section 5.5 summarized in section 15.1 is practically unchanged. First the (α,B)-gauge condition (15.3)

$$\frac{\delta\Gamma}{\delta B} = \alpha\bar{D}\bar{D}\bar{B} + \frac{1}{8}\bar{D}\bar{D}DD\Phi + \frac{1}{2}\chi\bar{D}\bar{D}\bar{c}_- \tag{18.17}$$

leads again via the Slavnov identity (18.12) to the ghost equation (15.22):

$$\mathcal{G}\Gamma \equiv [\frac{\delta}{\delta c_-} + \frac{1}{8}\bar{D}\bar{D}DD\frac{\delta}{\delta\rho}]\Gamma = -\frac{1}{2}\chi\bar{D}\bar{D}\bar{B} . \tag{18.18}$$

As a consequence, again, (see (15.7))

$$\begin{aligned}&\Gamma(\Phi,c_+,A,\rho,\sigma,Y,c_-,B,p_1,z_1,u,v) = \\ &= \bar{\Gamma}(\Phi,c_+,A,\eta,\sigma,Y,p_1,z_1,u,v) \\ &+ \mathrm{Tr}\int dV[BDD\Phi + \bar{B}\bar{D}\bar{D}\Phi + \alpha B\bar{B} + \frac{1}{2}\chi(c_-\bar{B} + \bar{c}_-B)]\end{aligned} \tag{18.19}$$

with $\qquad \eta = \rho - \frac{1}{8}(DDc_- + \bar{D}\bar{D}\bar{c}_-)$.

The Slavnov identity takes then the form (c.f. (15.24))

$$\mathcal{S}(\Gamma) = \mathcal{B}(\bar{\Gamma}) \equiv \mathrm{Tr}\int dV \frac{\delta\bar{\Gamma}}{\delta\eta}\frac{\delta\bar{\Gamma}}{\delta\Phi}$$

$$+ \int dS[\mathrm{Tr}\frac{\delta\bar{\Gamma}}{\delta\sigma}\frac{\delta\bar{\Gamma}}{\delta c_+} + \frac{\delta\bar{\Gamma}}{\delta Y}\frac{\delta\bar{\Gamma}}{\delta A} \quad + \text{terms } (\bar{\sigma},\bar{c}_+,\bar{Y},\bar{A})$$

$$+ \sum_k z_k \partial_{p_k} \bar{\Gamma} + \int dVv \frac{\delta\bar{\Gamma}}{\delta u} = 0 \qquad (18.20)$$

and the linearized Slavnov operator (c.f. (15.25)) reads

$$\mathcal{B}_{\bar{\Gamma}} = \mathrm{Tr}\int dV \left[\frac{\delta\bar{\Gamma}}{\delta\eta}\frac{\delta}{\delta\Phi} + \frac{\delta\bar{\Gamma}}{\delta\Phi}\frac{\delta}{\delta\eta}\right]$$

$$+ \int dS[\mathrm{Tr}\frac{\delta\bar{\Gamma}}{\delta\sigma}\frac{\delta}{\delta c_+} + \mathrm{Tr}\frac{\delta\bar{\Gamma}}{\delta c_+}\frac{\delta}{\delta\sigma} + \frac{\delta\bar{\Gamma}}{\delta Y}\frac{\delta}{\delta A} + \frac{\delta\bar{\Gamma}}{\delta A}\frac{\delta}{\delta Y}]$$

$$+ \text{terms } (\bar{\sigma},\bar{c}_+,\bar{Y},\bar{A})$$

$$+ \sum_k z_k \partial_{p_k} + \int dVv \frac{\delta}{\delta u} \quad . \qquad (18.21)$$

The identities (15.26 - 15.28) remain valid.

The (p_k, z_k, u, v)-independent action $\bar{\Gamma}_S^0$ (15.8) is still a solution of the Slavnov identity (18.20). From this special solution we deduce the general solution (see (15.12))

$$\bar{\Gamma}^{(0)}(\Phi,c_+,A,\eta,\sigma,Y,p_k,z_k,u,v) =$$

$$= \bar{\Gamma}_S^0(\hat{\Phi},\hat{c}_+,\hat{A},\hat{\eta},\hat{\sigma},\hat{Y}) + \sum_k z_k \{\mathrm{Tr}\int dV\eta G_k(\Phi,u',p)$$

$$+ \int dS[h_k^c(p)\mathrm{Tr}(\sigma c_+) + h_k^A(p)YA] + \text{c.c.}\}$$

$$+ \mathrm{Tr}\int dVv'\eta F(\Phi,u',p) \qquad (18.22)$$

by performing in $\bar{\Gamma}_S^0$ the substitutions (see (15.1, 15.14))

$$\hat{\Phi} = (\mathcal{F},u',p) =$$

$$= c_1(u',p) + \sum_{k\geq 2}\sum_{\omega=0}^{\omega(k)} c_{\omega k}(u',p) s^{\omega}_{i(i_1\ldots i_k)}\Phi_{i_1}\ldots\Phi_{i_1} \quad ,$$

$$(18.23)$$

$$\hat{\eta} = \frac{\delta}{\delta\hat{\Phi}} \operatorname{Tr}\int dV \eta \mathcal{F}^{-1}(\hat{\Phi},u',p)\Big|_{\hat{\Phi}=\mathcal{F}(\Phi,u',p)} ,$$

$$\hat{c}_{+} = t_c(p)c_{+} \quad , \quad \hat{\sigma} = \frac{1}{t_c(p)}\,\sigma,$$

$$\hat{A} = t_A(p)A \quad , \quad \hat{Y} = \frac{1}{t_A(p)}\,Y \qquad (18.23 \text{ cont.'d})$$

and by adding (z_k,v')-dependent terms. The functions appearing in the latter are determined by the Slavnov identity, with the result (see (15.14))

$$G_k(\Phi,u',p) = -\partial_{p_k} \mathcal{F}^{-1}(\hat{\Phi},u'p)\Big|_{\hat{\Phi}=\mathcal{F}(\Phi,u',p)} ,$$

$$h_k^c(p) = -\partial_{p_k} \ln t_k(p) \quad , \quad h_k^A(p) = \partial_{p_k} \ln t_A(p) \, , \qquad (18.24)$$

$$F(\Phi,u',p) = -\partial_u \mathcal{F}^{-1}(\hat{\Phi},u',p)\Big|_{\hat{\Phi}=\mathcal{F}(\Phi,u',p)} .$$

We have used rigid invariance (15.4) - hence the tensorial structure of the function $\mathcal{F}$ in (18.23) - and supersymmetry (18.13) - which implies that Γ depends on the shifted superfields u' and v'. We notice that the only sensible change as compared with section 15 is the u-dependence of the function $\mathcal{F}$: its coefficients $c_{\omega k}$ are arbitrary functions of u' and p_k. The arbitrary functions t_c and t_A defined in (18.23) do not depend on u due to the chirality of the superfields c_+ and A.

In order to fix the action we must still impose normalization conditions - which are a generalization of the conditions (5.180 - 5.183):

1) $\Gamma_{D_i D_j} = \frac{1}{4g^2}\,\delta_{ij}$

2) $\Gamma_{\sigma_i^F c_{+j}^A c_{+k}^A} = -4\, f_{ijk}$

3) $\Gamma_{\eta_i^D c_{+j}^A} = -4\, \delta_{ij}$ (8.25)

4) $\Gamma_{u^c \eta_i^D c_{+j}^A} = 4\,\delta_{ij}$ (18.25 cont.'d)

5) $S_k^{\omega}\Gamma_{\eta^D c_+^A (C)^{k-1}} = -4(k-1)!a_{\omega k}, \ k \geq 2$

6) $S_k^{\omega}\Gamma_{u^c \eta^D c_+^A (C)^{k-1}} = -4(k-1)!e_{\omega k}, \ k \geq 2$

7) $S_k^{\omega}\Gamma_{(u^c)^m \eta^D c_+^A (C)^{k-1}} = 0 \ , \quad m \geq 2, \ k \geq 1$.

The conditions on the matter field vertices will be given in the next subsection for the specific model considered there.

The conditions 4), 6) and 7) which involve the external field u are new. Condition 4) fixes the amplitude of u, condition 6) the couplings linear in u' and condition 7) the coupling non-linear in u' - arbitrarily set to zero since they are irrelevant in what concerns the control of the soft breaking of supersymmetry (18.13). To the contrary the linear terms in u', since they survive at u = o, must be taken into account. In particular the infinite set of parameters $e_{\omega k}$ defined by condition 6) belongs to the parametrization of the theory and is on the same footing as the $a_{\omega k}$'s defined by condition 5). They will be included in the set of gauge parameters together with their BRS variations $y_{\omega k} = se_{\omega k}$. Thus the complete set of gauge parameters p_k and their BRS variations z_k is

$$(p_k) = (\alpha, a_{\omega k}, e_{\omega k}, \mu^2), \ (z_k) = (\chi, x_{\omega k}, y_{\omega k}, \nu^2) \ . \qquad (18.26)$$

We must still check that the normalization conditions (18.25) indeed fix the action with the desired mass terms. We first observe that they determine the functions c_1, $c_{\omega k}$, t_c and t_A:

$$t_c(p) = t_A(p) = 1 \ , \qquad (18.27)$$

$$c_1(u',p) = (1-u')^{-1} \qquad (18.28)$$

whereas $c_{\omega k}$ (u',p) is the solution of the system of equations

$$-k[c_1(u',p)]^{-1} + f_{\omega k}(c_{\omega' k'}; \ k'<k) = a_{\omega k} + e_{\omega k} u' \ . \qquad (18.29)$$

These results are found as in section 5 (see (5.179, 5.138 - 5.140)) with the u-field dependence taken into account.

Eq. (18.28) - which follows from the 3rd and 4th normalization condition - implies

$$c_1 = 1 + \frac{1}{2}\mu^2\theta^2\bar{\theta}^2 \qquad \text{at } u = 0 \,. \tag{18.30}$$

Hence

$$\tilde{\Phi} = (1 + \frac{1}{2}\mu^2\theta^2\bar{\theta}^2)\Phi + O(u) + O(\Phi^2) \tag{18.31}$$

yields just the bilinear action (18.3, 18.4) leading to the propagators (18.6) with mass μ^2.

We finally notice that no normalization conditions are required for defining α and μ^2: α is defined by the gauge condition (18.17) and μ^2 by the softly broken supersymmetry Ward identity (18.13).

18.2 Higher orders: Solution of the cohomology.

As for all models already discussed in this book we define the present theory by a set of Ward (or Slavnov) identities expressing the imposed symmetries and by a set of normalization conditions fixing the parameters left free once the Ward identities are fulfilled. We saw in the last section that a unique theory resulted in the tree approximation, and we have now to extend this result to the higher orders of the perturbation series. We successively achieved this in section 15 for the asymptotic theory (mass effects neglected) provided the chiral anomaly was absent. Our aim now is to prove exact Ward and Slavnov identities instead of the approximate identities of section 15. We shall restrict ourselves to models of gauge superfields Φ_i interacting with massless chiral matter superfields A_a [V.5,6]. They are given in the classical approximation by the action $\bar{\Gamma}^0$ (18.22), with

$$\begin{aligned}\bar{\Gamma}^0_s &= \Gamma_{YM}(\Phi) + \mathrm{Tr}\int dV\eta Q_s(\Phi,c_+) - \mathrm{Tr}\int dS\sigma c_+^2 - \mathrm{Tr}\int d\bar{S}\bar{\sigma}\bar{c}_+^2 \\ &+ \int dV\bar{A}e^{\tilde{\Phi}}A + h_{abc}\int dS A_a A_b A_c + \bar{h}_{abc}\int d\bar{S}\bar{A}_a\bar{A}_b\bar{A}_c \\ &- \int dSY\tilde{c}_+A - \int d\bar{S}\bar{A}\tilde{\bar{c}}_+\bar{Y}\end{aligned} \tag{18.32}$$

($\tilde{\Phi} = \tilde{\Phi}_i T^i$, $\tilde{c}_+ = c_{+i} T^i$, T^i_{ab} = generators of the gauge group in the matter field representation; see section 15.1 for the notation).

$\bar{\Gamma}^{(0)}$ was fixed by the normalization conditions (18.25) to which we add now those for the matter field vertices:

$$
\begin{aligned}
&8) \quad \Gamma_{F_a A_b A_c} = -4\, h_{abc} \quad , \quad \Gamma_{\bar{F}_a \bar{A}_b \bar{A}_c} = -4\, \bar{h}_{abc} \; , \\
&9) \quad \Gamma_{F_a \bar{F}_b} = 16\, \delta_{ab} \; , \qquad\qquad\qquad\qquad\qquad\qquad (18.33) \\
&10) \quad \Gamma_{Y^F_a c_{+i} A_b} = 4\, T^i_{ab} \quad , \quad \Gamma_{\bar{A}_c \bar{c}_{+i} \bar{Y}^F_b} = 4\, T^i_{ab}
\end{aligned}
$$

where A and F are the Θ^0- and Θ^2-components of the chiral superfield A.

Before proceeding further let us remark that the classical potential, a function of the $\Theta = 0$ component A of the superfield A, is bounded from below by zero, a minimum zero defining a supersymmetric vacuum. Depending on the group and the representation T there can be a degenerate set of such vacua. The vacua corresponding to non-zero values of A lead to a spontaneous break-down of gauge invariance, some gauge and/or matter supermultiplets becoming massive. We shall concentrate in the following on the phase defined by the A = o vacuum, and let aside the problem of its stability with respect to the other phases, a question which would push us beyond the scope of perturbation theory.

Let us give now the list of the requirements being imposed on the vertex functional Γ and defining the theory:

(1) Gauge condition (18.17) and ghost equation (18.18).

(2) Rigid invariance (15.4).

(3) Conformal R-invariance (15.17) (exact), which is obeyed by the classical action (18.32) due to the absence of matter mass terms.

(4) Softly broken supersymmetry (18.13).

(5) Slavnov identity (18.12).

(6) Normalization conditions (18.25, 18.33).

The normalization conditions are now given at some Euclidean normalization point $\varkappa^2$ of momentum space. Moreover, for the conditions 1) in (18.25) and 8) in (18.33) the gauge parameters must be fixed at some normalization value (μ_0^2, α_0, $a_{0\omega k}$, $e_{0\omega k}$). The reason is that the corresponding invariant counter terms Δ_1 and Δ_8 (see (D.8) in Appendix D) do not allow for gauge parameter dependent coefficients. We shall come back to this point at the end of the subsection 18.3.

Since we have massless as well as massive fields due to the IR cut-off mass μ^2 we have to define carefully the subtraction procedure. We shall use the Lowenstein-Zimmermann procedure (momentum space subtractions with auxiliary mass) described in Chapter III and work in component fields, the supersymmetry being explicitly broken by the cut-off μ^2. An inspection of the propagators (all the massive ones are listed in eqs. (18.6)) leads to the assignments of Table 18.1 for the IR- and UV-dimensions r and d of each field (if a dimension is given for a superfield, the dimensions of its components are computed by assigning dimension -1/2 (IR and UV) to the Θ-variables). The number m appearing in the table is arbitrary and does not contribute to the degrees of divergence due to $\Phi\pi$ charge conservation.

Since the requirements (1) to (3) do not hide subtleties we dwell directly with the last three ones. We proceed by induction, assuming the existence of a unique solution to them up to and including order n-1 in $\hbar$. We assume in particular (requirements (4), (5)):

$$W_\alpha\Gamma = \hbar^n\Delta_\alpha(\Phi,c_+,A,\eta,\sigma,Y,p,z,u,v) + O(\hbar^{n+1}) , \tag{18.34}$$

$$\mathcal{S}(\Gamma) = \hbar^n\,\Delta(\Phi,c_+,A,\eta,\sigma,Y,p,z,u,v) + O(\hbar^{n+1}) \tag{18.35}$$

Table 18.1 UV-dimensions d and IR-dimensions r

	d	r		d	r
$\Phi - (\Phi)_{\Theta=0}$	0	0	ρ	2-m	2-m
$C \equiv (\Phi)_{\Theta=0}$	0	2	σ	3-2m	3-2m
A	1	1	Y	2-m	2-m
$Re(c_+)$	m	m	u	0	0
$Im(c_+) - Im(c_+)_{\Theta=0}$	m	m	v	m	m
$b+ \equiv Im(c_+)_{\Theta=0}$	m	m+2	$p \equiv \{a,e,\alpha,\mu^2\}$	0	0
c_-	1-m	1-m	$z \equiv \{x,y,\chi,\nu^2\}$	m	m
B	1	1	$\Theta,\bar{\Theta}$	-1/2	-1/2

where Δ_α, Δ are local functionals of their arguments: the breakings beginning at order n by assumption, we have retained the trivial graph contributions only. The UV- and IR-dimensions of Δ_α and Δ are bounded as follows (see Table 18.1)

$$\begin{aligned} \Delta_\alpha &: \quad d \leq 9/2 \quad , \quad r \geq 5/2 \\ \Delta &: \quad d \leq 4+m \quad , \quad r \geq 2+m \end{aligned} \tag{18.36}$$

These bounds are dictated by the dimensions of the functional operators W_α and $\mathcal{S}$ according to the action principle (11.5 - 11.15).

From the algebraic properties of the functional operators W and $\mathcal{S}$ (see (13.22) and (15.26)-(15.29)) follow the consistency conditions (see (13.27) and (15.33))

$$W_\alpha \bar{\Delta}_{\dot{\alpha}} + \bar{W}_{\dot{\alpha}} \Delta_\alpha = W_\alpha \Delta_\beta + W_\beta \Delta_\alpha = \bar{W}_{\dot{\alpha}} \bar{\Delta}_{\dot{\beta}} + \bar{W}_{\dot{\beta}} \bar{\Delta}_{\dot{\alpha}} = 0 \tag{18.37}$$

$$\mathcal{s}\Delta = 0 \tag{18.38}$$

$$W_\alpha \Delta - \mathcal{s}\Delta_\alpha = 0 \quad , \quad \bar{W}_{\dot{\alpha}} \Delta - \mathcal{s}\bar{\Delta}_{\dot{\alpha}} = 0 \tag{18.39}$$

with the nilpotent operator $\mathcal{s}$ defined in eqs. (15.30). The conditions (18.39) follow from the identity

$$W \mathcal{B}(\gamma) - \mathcal{B}_\gamma W\gamma = 0 \tag{18.40}$$

which is easily derived for any functional γ and for $W = W_\alpha$ or $\bar{W}_{\dot{\alpha}}$ by using the definitions (15.24, 15.25) with the operator $v\delta_u$ being added.

We are going to show that the general solution of the consistency conditions has the form

$$\begin{aligned} &\Delta_\alpha = W_\alpha \hat{\Delta} \quad , \quad \bar{\Delta}_{\dot{\alpha}} = \bar{W}_{\dot{\alpha}} \hat{\Delta} \\ &\Delta = \mathcal{s}\hat{\Delta} + r\mathcal{a} \quad ; \quad W_\alpha \mathcal{a} = \bar{W}_{\dot{\alpha}} \mathcal{a} = 0 \end{aligned} \tag{18.41}$$

where $\hat{\Delta}$ is a local functional of the fields and its UV- and IR-dimensions fulfill the bound $\underline{d \leq 4,\ r \geq 4}$. $\mathcal{a}$ is the supersymmetry-invariant chiral anomaly (15.62) and r its coefficient, computed in section 15.4 at the one-loop approximation (eq. (15.85)). If the anomaly is absent, $\hat{\Delta}$ can be absorbed as a counterterm in the Lagrangian (see eqs. (13.11 - 13.13) and (15.35 - 15.37)). This procedure leads to a corrected vertex functional fulfilling the Ward and Slavnov identities (18.33, 18.34) at the order n:

$$W_\alpha \Gamma = O(\hbar^{n+1}) \quad , \quad \mathcal{s}(\Gamma) = O(\hbar^{n+1}) \tag{18.42}$$

which is what we have to show.

We now turn to the proof of (18.41). The general solution of condition (18.37) is given by Theorem 13.3, eqs. (13.28 - 13.31):

$$\Delta_\alpha = W_\alpha \hat{\Delta}' \quad , \quad \bar{\Delta}_{\dot{\alpha}} = \bar{W}_{\dot{\alpha}} \hat{\Delta}' \tag{18.43}$$

where Δ' is a local functional of the fields, with dimensions bounded by

$$\hat{\Delta}' : d \leq 4 \quad , \quad r \geq 2 . \tag{18.44}$$

We note that $\hat{\Delta}'$ may have terms with IR-dimensions less than 4, which cannot be absorbed into the Lagrangian without spoiling IR-convergence. More precisely, $\hat{\Delta}'$ is rigid and R-invariant due to the requirements (1) to (3) and can have IR-dimensions $r = 2$ and $r \geq 4$. The dimension 2 contributions are of the form

$$\Delta_2 = \int dx (u_c(x))^k \, Q(x) \quad , \quad k \geq 0 \tag{18.45}$$

where u_c is the $\Theta = 0$ component of the superfield u and Q is a polynomial in the components of u, Φ, A and $\bar{A}$, except the $\Theta = 0$ components of u and Φ. Q has dimensions $d = r = 2$ and is rigid as well as R-invariant. But it is always possible to find an insertion Δ_4 ($d = r = 4$) and a supersymmetric insertion $\tilde{\Delta}$, both rigid and R-invariant, such that

$$\Delta_2 = \Delta_4 + \tilde{\Delta} \quad , \quad W_\alpha \tilde{\Delta} = \bar{W}_{\dot{\alpha}} \tilde{\Delta} = 0 \quad . \tag{18.46}$$

Let indeed $\tilde{Q}$ be the superfield polynomial having Q as its $\Theta = 0$ component. Recalling the definition of u' (18.7) we see that the solution to eq. (18.46) is given by

$$\begin{aligned} \Delta_4 &= - \frac{1}{8(k+1)\mu^2} \int dV u^{k+1} \tilde{Q} \quad , \\ \tilde{\Delta} &= \frac{1}{8(k+1)\mu^2} \int dV (u')^{k+1} \tilde{Q} \quad . \end{aligned} \tag{18.47}$$

As a consequence of (18.46) $\hat{\Delta}'$ in (18.43) can be chosen with IR dimension $r \geq 4$. We thus can absorb it as a counterterm, in such a way that the supersymmetry Ward identities are now fulfilled:

$$W_\alpha \Gamma = O(\hbar^{n+1}) \quad , \quad \bar{W}_{\dot{\alpha}} \Gamma = O(\hbar^{n+1}) \tag{18.48}$$

and the Slavnov identity reads

$$\mathcal{S}(\Gamma) = \hbar^n \Delta(\Phi, c_+, A, \eta, \sigma, Y, p_k, z_k, u', v') + O(\hbar^{n+1}) \ , \tag{18.49}$$

where the new Δ is rigid, R and supersymmetry invariant. It fulfills the consistency condition

$$\mathcal{b}\Delta = 0 \ , \tag{18.50}$$

and has dimensions bounded according to (18.36), i.e. $d \leq 4 + m$, $r \geq 2 + m$.

The cohomology problem (18.50) was solved in section 15 in the absence of the external superfields u' and v'. We are brought back to that case by Lemma 15.1 (page 202) and the following lemma:

Lemma 18.1 Let $\mathcal{b}_0$ be the linearization of $\mathcal{b}$ defined according to eq. (15.41) and Δ an integrated local functional which is polynomial in its arguments.

Then the general solution of

$$\mathcal{b}_0 \Delta(\Phi, c_+, A, \eta, \sigma, Y, p_k, z_k, u', v') = 0 \tag{18.51}$$

is of the form

$$\Delta = \mathcal{b}_0 \hat{\Delta}(\Phi, c_+, A, \eta, \sigma, Y, p_k, z_k, u', v') + \bar{\Delta}(\Phi, c_+, A, \eta, \sigma, Y, p_k, z_k) \tag{18.52}$$

with

$$\mathcal{b}_0 \bar{\Delta} = 0 \tag{18.53}$$

where $\hat{\Delta}$ and $\bar{\Delta}$ are integrated local functionals.

The proof is very similar to the one given at the beginning of appendix C for the elimination of the gauge parameters p_k, z_k.

The result of these considerations is that the general solution of (18.50) is

$$\Delta = \mathcal{b}\hat{\Delta}(\Phi,c_+,A,\eta,\sigma,Y,p_k,z_k,u',v') + r\,\mathcal{a}(\Phi,c_+) \tag{18.54}$$

where $\hat{\Delta}$ is a rigid, R and supersymmetry invariant insertion with dimensions

$$\hat{\Delta} : d \leq 4 \quad , \quad r \geq 2 \tag{18.55}$$

$\mathcal{a}$ is the supersymmetric chiral anomaly (15.62). We shall assume for the following that its coefficient r (15.85) is vanishing due to a suitable choice for the representation for the matter fields. Under this assumption the Slavnov identity (18.49) takes the form

$$\mathcal{s}(\Gamma) = \hbar^n \mathcal{b}\hat{\Delta} + O(\hbar^{n+1}) \ . \tag{18.56}$$

As noted above $\hat{\Delta}$ may still contain infrared anomalies, i.e. terms of infrared dimension $r < 4$ which therefore cannot be absorbed, and it remains to show that they are in fact absent. This will be done in the following subsection.

18.3 Higher orders: The absence of infrared anomalies and the Callan-Symanzik equation

In order to show the absence of IR-anomalies in the Slavnov identity (18.56) we first give the list of all possible terms of $\hat{\Delta}$ with IR dimension $r < 4$, rigid R and supersymmetry invariance being taken into account. They are of the form

$$\hat{\Delta}_i = \int dx\, u_c^k(x) Q_i(x) + \ldots \qquad k = 0, \cdot, \ldots \tag{18.57}$$

where the dots stand for contributions with $r \geq 4$ rendering the expression supersymmetric (see (18.44 - 18.46)). In the following we write only Q_i and drop always integral and trace symbol, we also do not write the complex conjugate terms. The component notation is self-explaining. The list contains 13 terms:

$$1)\ M\bar{M}\ ,\qquad 2)\ \text{ext. field } (u) \times \text{spinor} \times \text{spinor}\ ,$$
$$3)\ \chi\chi\bar{M}\ ,\qquad 4)\ u_\chi \chi \bar{M}\ , \tag{18.58A}$$

$$5)\ \eta_c a_+\ ,\qquad 6)\ v_\mu v^\mu\ ,\qquad 7)\ v_\mu \chi \sigma^\mu \bar{\chi}\ ,$$
$$8)\ v_\mu u_\chi \sigma^\mu \bar{\chi}\ ,\qquad 9)\ \chi\chi\bar{\chi}\bar{\chi}\ ,\qquad 10)\ u_\chi \chi\bar{\chi}\bar{\chi} \tag{18.58B}$$
$$11)\ u_\chi u_\chi \bar{\chi}\bar{\chi}\ ,\qquad 12)\ u_\chi \chi u_{\bar{\chi}} \bar{\chi}\ ,$$

$$13)\ A\bar{A}\ . \tag{18.58C}$$

We have to show that in

$$\mathfrak{s}(\Gamma) = \hbar^n\, \mathfrak{s}[\sum_{i=1}^{13} c_i \hat{\Delta}_i] \tag{18.59}$$

the coefficients c_i vanish. They are found by testing appropriately (18.59) at zero momentum. We divide the sum into three groups:

Group A: coefficients c_1 to c_4, which vanish due to the small momentum behaviour of vertex functions implied by IR power counting and/or due to R invariance.

By IR-power counting we mean the low momentum behaviour [III.18]:

$$\Gamma_{\varphi_1 \ldots \varphi_n}(p) = O(p^\omega) \quad \text{(modulo logarithms)} \tag{18.60}$$

with $\omega = 4 - \sum_1^n r(\varphi_i)$, if $\omega \geq 0$

and

$$N_\delta^\rho[O]\cdot\Gamma_{\varphi_1 \ldots \varphi_n}(p) = O(p^{\omega'}) \quad \text{(modulo logarithms)} \tag{18.61}$$

with $\omega' = \rho - \sum_1^n r(\varphi_i)$, if $\omega' \geq 0$.

This behaviour follows from the subtraction procedure described in Chapter III. The IR dimensions $r(\varphi_i)$ of the various fields φ_i are listed in Table 18.1 (see page 288).

Group B: coefficients c_5 to c_{12}, which vanish as a consequence of an improved small momentum behaviour of vertex functions deduced from a repeated use of the anomalous Slavnov identity (18.59) and IR power counting.

Group C: coefficient c_{13}, which requires an argument based on the Callan-Symanzik equation.

We shall restrict ourselves to the case $k = 0$ (see 18.57). The dimensions of u_c being zero as well as its R-weight the generalization to $k > 0$ is straightforward.

Group A

$$c_1: \quad \mathscr{b}(M\bar{M}) = F_+\bar{M} - \bar{F}_+M + \ldots \tag{18.62}$$

(A_+, ψ_+, F_+ are the components of c_+, we write only the bilinear terms). Then, up to a numerical factor

$$c_1 \sim \Gamma_{F_+\eta_c}\Gamma_{D\bar{M}} + \Gamma_{F_+\eta_M}\Gamma_{\bar{M}\bar{M}} + \Gamma_{F_+\eta_{\bar{M}}}\Gamma_{M\bar{M}} + \Gamma_{F_+\eta_\mu}\Gamma_{v_\mu\bar{M}}$$
$$+ \Gamma_{F_+\eta_D}\Gamma_{C\bar{M}} \quad . \tag{18.63}$$

All but the third term vanish due to R invariance which forbids $\Gamma_{D\bar{M}}$, $\Gamma_{\bar{M}\bar{M}}$, $\Gamma_{v_\mu\bar{M}}$ and $\Gamma_{C\bar{M}}$. On the other hand all terms but the last one vanish by IR power counting (18.60). Hence $c_1 = 0$.

We may already note at this stage that the only terms not vanishing by IR power counting are those involving vertex functions with fields of non-canonical IR dimensions, i.e. the fields C and b_+ according to Table 18.1. We shall make implicit use of this fact by not writing the contributions which vanish at $p = 0$.

c_2: The variation of Q_2 involves neither C nor b_+, furthermore no vertex function with C or b_+ occurs - due to the spin, hence $c_2 = 0$.

c_3, c_4: $\mathscr{s}(\chi\chi\bar{M}) = \bar{F}_+\chi\chi + \dots ,$

$$\mathscr{s}(u_\chi \chi \bar{M}) = v_\chi \chi \bar{M} + \dots , \tag{18.64}$$

$$c_3 \sim \Gamma_{\bar{F}_+\eta_D} \Gamma_{C\chi\chi}, \qquad c_4 \sim \Gamma_{v_\chi \chi \eta_D} \Gamma_{C\bar{M}} \tag{18.65}$$

Both vanish due to R invariance.

Group B

One finds in the same way

$$\begin{aligned} c_5 &\sim \Gamma_{a_+\eta_D} \Gamma_{CD} , & c_6 &\sim \frac{\partial}{\partial p_\mu} (\Gamma_{a_+\eta_D} \Gamma_{Cv_\mu}) , \\ c_7 &\sim \frac{\partial}{\partial p_\mu} (\Gamma_{a_+\eta_D} \Gamma_{C\chi\bar{\chi}}), & c_8 &\sim \Gamma_{v_\chi \bar{\chi}\eta_D} \Gamma_{Cv_\mu} , \\ c_9 &\sim \Gamma_{\psi_+\bar{\chi}\eta_D} \Gamma_{C\chi\bar{\chi}} , & c_{10} &\sim \Gamma_{v_\chi \bar{\chi}\eta_D} \Gamma_{C\chi\bar{\chi}} , \\ c_{11} &\sim \Gamma_{v_\chi \bar{\chi}\eta_D} \Gamma_{Cu_\chi \bar{\chi}} , & c_{12} &\sim \Gamma_{v_\chi \bar{\chi}\eta_D} \Gamma_{C\chi\bar{u}_\chi} . \end{aligned} \tag{18.66}$$

However neither R nor straightforward application of IR power counting make these expressions vanish. We therefore test the Slavnov identity (18.59) first with respect to C and a_+, second with respect to C, $\bar{\chi}$ and v_χ and third with respect to C, $\bar{\chi}$ and ψ_+. The result is

$$\begin{aligned} \Gamma_{a_+\eta_D} \Gamma_{CC} &= O(p^2) \\ \Gamma_{v_\chi \bar{\chi}\eta_D} \Gamma_{CC} &= O(p) \\ \Gamma_{\psi_+\bar{\chi}\eta_D} \Gamma_{CC} &= O(p) . \end{aligned} \tag{18.67}$$

Here we have not only used IR power counting, but also that at order $\hbar^n$ the r.h.s. of (18.59) is a $\mathscr{s}$-variation. This implies for some potentially dangerous terms that they contain derivatives. Those are

$$\mathscr{s}(C\partial v) = C\Box a_+ + \dots, \quad \mathscr{s}(Cu_\chi \gamma\bar{\chi}) = Cv_\chi \gamma\bar{\chi} + \dots , \qquad \mathscr{s}(C\chi\gamma\bar{\chi}) = C\psi_+\gamma\bar{\chi} + \dots , \tag{18.68}$$

respectively. Now from $\Gamma_{cc} \sim \mu^4 + O(\hbar)$ follows

$$\Gamma_{a_+\eta_D} = O(p^2) \ , \ \Gamma_{v_\chi\bar{\chi}\eta_D} = O(p) \ , \ \Gamma_{\psi_+\bar{\chi}\eta_D} = O(p) \ . \qquad (18.69)$$

This improved small momentum behaviour implies the vanishing of all coefficients (18.66).

Group C

The last surviving coefficient is c_{13}, denoted hereafter by c. Since neither R invariance nor small momentum behaviour imply that it vanishes, we shall investigate its dependence on the parameters of the theory. We shall show in particular that c does not depend on the parameters carrying dimension. The dimension of c being $[\text{mass}]^2$ it must then vanish. Parameters having dimension are the IR cut-off μ^2 - a gauge parameter - and the normalization parameters $\varkappa^2$ and μ_0^2. The dependence on the former is governed by the Slavnov identity, the dependence on the latter by renormalization group equations.

Let us first write the Slavnov identity proved thus far:

$$\mathcal{s}(\Gamma) = \hbar^n \mathcal{b}[c\int dx(A\bar{A} + \ldots) + O(\hbar^{n+1})]$$

$$= \hbar^n[\sum_k z_k \partial_{p_k} c\int dx\, A\bar{A} + c\mathcal{b}\int dx\, A\bar{A} + \ldots] + O(\hbar^{n+1}) \ . \qquad (18.70)$$

A test with respect to z_k, A, $\bar{A}$ yields

$$\partial_{z_k}\Gamma_{AY_F}\Gamma_{A\bar{A}} + \text{conj.} + \partial_{p_k}\Gamma_{A\bar{A}} = \hbar^n \partial_{p_k} c + O(\hbar^{n+1}) \qquad (18.71)$$

Due to IR power counting (18.60) the l.h.s. vanishes at zero momenta, hence

$$\partial_{p_k} c = 0 \quad , \quad p_k = \mu^2, \alpha, a_{\omega k}, e_{\omega k} \qquad (18.72)$$

i.e. c is gauge independent, in particular μ^2-independent.

To study the dependence on $\varkappa^2$ (normalization point) and μ_0^2 (value of μ^2 specified in the normalization conditions for the coupling constants) we observe that the differential operators $\lambda\partial_\lambda$ for $\lambda = \varkappa^2, \mu_0^2$ are symmetric, i.e. (see appendix D)

$$\lambda\partial_\lambda \mathcal{s}(\Gamma) = \mathcal{B}_{\bar{\Gamma}}\,[\lambda\partial_\lambda\Gamma] \tag{18.73}$$

and define rigid, R, supersymmetry and BRS invariant insertions of dimensions $d \leq 4$, $r \geq 4$. A basis of symmetric operators defining such insertions is given by (D.6). Therefore both $\lambda\partial_\lambda$ can be expanded in this basis, but only up to the order n-1 at which the Slavnov identity (18.70) holds. We then obtain the renormalization group equations - at vanishing Grassmann parameters z_k -

$$C_\lambda\Gamma \equiv [\lambda\partial_\lambda + \sum_i \beta_i{}^\lambda\partial_{g_i} - \gamma_\Phi{}^\lambda\mathcal{N}_\Phi - \gamma_A{}^\lambda\mathcal{N}_A - \gamma_+{}^\lambda\mathcal{N}_+$$
$$+ \sum_{\omega,k}(\gamma^\lambda_{a_{\omega k}}\partial_{a_{\omega k}} + \gamma^\lambda_{e_{\omega k}}\partial_{e_{\omega k}}) + \gamma_u{}^\lambda\mathcal{N}_u + O(u,v)] = O(\hbar^n)$$
$$(\lambda = \varkappa^2,\ \mu_o^2)\ . \tag{18.74}$$

The coefficients β and γ are of order $\hbar$ since the classical action does not depend on λ. The coupling constants are collectively denoted by g_i.

Before using (18.74) for controlling the λ-dependence of the coefficient c, we must deal with a problem concerning the very existence of its l.h.s. Its last term

$$\mathcal{N}_u\Gamma \equiv \int dV[u'\frac{\delta}{\delta u'} + v'\frac{\delta}{\delta v'}]$$
$$= \int dV[u\frac{\delta}{\delta u} + v\frac{\delta}{\delta v}] + 8\int dx\,[\mu^2\frac{\delta}{\delta u_D} + \nu^2\frac{\delta}{\delta v_D}] \tag{18.75}$$

contains pieces - the last two - which could have an IR dimension $r < 4$ and would thus not be defined. Let us attempt to get out of this difficulty by assuming for the moment that $\mathcal{N}_u\Gamma$ has in fact the right dimension $r \geq 4$ up to and including order n-1: (we use the N^ρ_δ notation of section 10)

$$\int dx\,\frac{\delta\Gamma}{\delta u_D} = N^4_4 - \text{insertion} + O(\hbar^n)$$
$$\int dx\,\frac{\delta\Gamma}{\delta v_D} = N^4_4 - \text{insertion} + O(\hbar^n)\ . \tag{18.76}$$

This assumption is to be added to the set of induction hypotheses (18.33, 18.34) and we shall have to chek it for the next order after we have proceeded with the Slavnov identity.

From (18.76) follows that the l.h.s. of (18.74) is an N_4^4-insertion at the order n since γ_u^λ is of order $\hbar$, and thus exists at this order (and is zero at lower orders):

$$C_\lambda \Gamma = \hbar^n N_4^4[Q]\cdot\Gamma + O(\hbar^{n+1})$$

$$= \hbar^n Q + O(\hbar^{n+1}) \tag{18.77}$$

where Q has dimensions $d \leq 4$, $r \geq 4$.

The action of C_λ on the Slavnov identity (18.70) and of the linearized Slavnov operator $\mathcal{B}_{\bar\Gamma}$ on the renormalization group equ. (18.77) yield the two equations - use is made of the already proven gauge independence of c which implies $\mathfrak{s}c = 0$ -

$$C_\lambda \mathcal{S}(\Gamma) = \hbar^n \lambda \partial_\lambda c\, \mathfrak{s}\int dx(A\bar A + \ldots) + O(\hbar^{n+1}) ,$$

$$\mathcal{B}_{\bar\Gamma} C_\lambda \Gamma = \hbar^n\, \mathfrak{s}Q + O(\hbar^{n+1}) . \tag{18.78}$$

Now, C_λ being a symmetric operator the two l.h.s. are equal. Thus, since both r.h.s. are $\mathfrak{s}$ variations and $\mathfrak{s}c = 0$, we can integrate:

$$\lambda\partial_\lambda c\int dx(A\bar A + \ldots) = Q - Q_{inv} . \tag{18.79}$$

Here Q_{inv} is invariant under all symmetries including BRS and thus has an IR dimension $r \geq 4$ (see Appendix D). Q has also $r \geq 4$ as we have shown whereas, for $A\ddot A$, $r = 2$. Hence both sides of (18.79) vanish. The conclusion is twofold. First

$$\partial_\lambda c = 0 \quad , \quad \lambda = \mu_0^2, \varkappa^2 \tag{18.80}$$

which ends the proof of $c = 0$, hence of the Slavhov identity at order n:

$$\mathcal{S}(\Gamma) = O(\hbar^{n+1}) . \tag{18.81}$$

The second consequence of (18.79) is that $Q = Q_{inv}$, with dimensions $d \leq 4$, $r \geq 4$. Q_{inv} can be expanded in the basis defined by the symmetric operators (D.6) applied to the classical action. We then insert this expansion in the r.h.s. of the renormalization group equ. (18.77) and, taking into account the factor $\hbar^n$, we absorb the r.h.s. in the l.h.s. by a redefinition of the coefficients β and γ at order $\hbar^n$. We finally obtain the renormalization group equation at the next order in $\hbar$:

$$C_\lambda \Gamma = O(\hbar^{n+1}). \tag{18.82}$$

We have still to verify the assumption (18.76) at the next order. Let us define two operators D and D'

$$\begin{aligned} D\,\Gamma &\equiv [\partial_{\mu^2} - 2\int dx \frac{\delta}{\delta u_D}]\Gamma = \Delta\cdot\Gamma + \hbar^n Q + Q(\hbar^{n+1}) , \\ D'\Gamma &\equiv [\partial_{\nu^2} - 2\int dx \frac{\delta}{\delta v_D}]\Gamma = \Delta'\cdot\Gamma + \hbar^n Q' + o(\hbar^{n+1}) , \end{aligned} \tag{18.83}$$

which commute with the infinitesimal rigid, R and supersymmetry transformations, in particular with the supersymmetry Ward identity operators W_α, $\bar{W}_\alpha$ (18.13). The insertion Δ and Δ' in the r.h.s. of (18.83) are N_4^4 from the assumption (18.76) whereas the correction terms Q, Q' of order n may have an IR dimension lower than four. $D\Gamma$ and $D'\Gamma$ being rigid, R and supersymmetry invariant, arguments completely analogous to those establishing the supersymmetry Ward identity (18.48) show that there exist at order n correction terms to Δ and Δ', of dimensions $d \leq 4$, $r \geq 4$, such that the corrected insertions are rigid, R and supersymmetry invariant. Hence Q and Q' possess these same symmetries.

It is easy to check by inspection that all possible contributions to Q' (Q' has $\Phi\pi$ charge -1) which have a UV dimension $d \leq 4$ have in fact an IR dimension $r \geq 4$. Q' is thus an N_4^4-insertion, which proves the second part of the assumption (18.76) at order n:

$$\int dx \frac{\delta}{\delta v_D}\,\Gamma = N_4^4\text{-insertion} + O(\hbar^{n+1}) . \tag{18.84}$$

Turning now to the first of equs. (18.83), we note that Q having $\Phi\pi$ charge zero can have terms with $r < 4$. These possible IR anomalies are in fact just the same as those encountered when dealing with the Slavnov identity: they are listed in (18.58). Calling them Q_i, we may write

$$D\Gamma = \Delta\cdot\Gamma + \hbar^n \sum_{i=1}^{13} x_i(Q_i + \ldots) + O(\hbar^{n+1}) \qquad (18.85)$$

where Δ is an N_4^4-insertion. We have to show that all x_i vanish. Now, from the validity of the Slavnov identity at order n we deduce

$$O(\hbar^{n+1}) = D'\mathcal{S}(\Gamma) = D\Gamma - \mathcal{B}_{\bar{\Gamma}}[D'\Gamma]. \qquad (18.86)$$

The coefficients x_i are evaluated by tests performed on this equality, in the same way as we did in the last subsection for the Slavnov identity (coefficients c_i). We find that x_1 and x_{13} vanish due to IR power counting (18.60, 18.61) and R invariance arguments. Note that x_{13}, unlike c_{13} falls in this class! For showing the vanishing of x_5 to x_{12} we rely as for the corresponding c's on the improved small momentum behaviour of certain vertex functions (c.f. (18.66 - 18.69)):

$$(1)\quad \Gamma_{a_+\eta_D} = O(p^2) \ ,$$

$$(2)\quad [D'\Gamma]_{\eta_D C} = O(p^0)\ , \qquad (3)\quad [D'\Gamma]_{\eta_D v_\mu} = O(p)\ , \qquad (18.87)$$

$$(4)\quad [D'\Gamma]_{\eta_D \chi\bar{\chi}} = O(p)\ , \qquad (5)\quad [D'\Gamma]_{\eta_D u_\chi \bar{\chi}} = O(p)\ .$$

The tests to be performed on (18.86) in order to show the last four are CC, Cv_μ, $C\chi\bar{\chi}$, $Cu_\chi\bar{\chi}$, respectively. The first was already shown (18.69). Without going into the details let us simply mention that (1) is used for the coefficient x_5, (2) and (3) for x_6, (4) for x_7 and x_9, (5) for x_8, x_{10}, x_{11} and x_{12}.

This ends the proof of the induction hypotheses (18.34, 18.35 and 18.76) at the next order, hence of the Ward and Slavnov identities at all orders.

It is clear at this point that we are still free to add Poincaré, rigid, R, supersymmetric and BRS invariant counterterms to the action without altering the Ward and Slavnov identities. The list of such symmetric counterterms Δ_i is given in Appendix D, eq. (D.8). Thus we will add the counterterms

$$\sum_i t_i \Delta_i \tag{18.88}$$

where the t_i are gauge independent. They are fixed by the normalization conditions (18.25, 18.33) in order to fulfill the last of our six requirements. We note that the Δ_i, except Δ_1 and Δ_8, involve arbitrary functions of the gauge parameters p_k: this is why the corresponding normalization conditions have to be imposed at all values of p_k - they determine these functions. In contrast to this, the coefficients of ∇_1 and ∇_8 are fixed by conditions at a normalization value p_{0k} of p_k.

In the course of the proof we were led to the renormalization group equ. (18.74) now valid at all orders, too. By summing the two equations (for $\lambda = \varkappa^2$ and μ_0^2) we get, setting the external fields u and v as well as the Grassmann parameters z_k equal to zero, the Callan-Symanzik equation

$$\begin{aligned}&[\varkappa^2\partial_{\varkappa^2} + \mu_0^2\partial_{\mu_0^2} + \sum_i \beta_i\partial_{g_i} - \gamma_\Phi \mathcal{N}_\Phi \\ &- \gamma_A\mathcal{N}_A - \gamma_+\mathcal{N}_+ + \gamma_u\mu^2\int dx\, \frac{\delta}{\delta u_D} \\ &+ \sum\gamma^a_{\omega k}\partial_{a_{\omega k}} + \sum\gamma^e_{\omega k}\partial_{e\omega k}]\Gamma = 0.\end{aligned} \tag{18.89}$$

One may note that the coefficients β_i are independent of the gauge parameters since no gauge parameter dependent factor appears in the symmetric operators ∂_{g_i}, i.e. the operators ∇_1 and ∇_8 of Appendix D, eq. (D.6). They may however depend on $\varkappa^2$ and μ_0^2.

18.4. Discussion of the result. Open questions

In the present section we have developed a scheme for dealing with the off-shell infrared problems in supersymmetric gauge theories. The main idea was to use an infrared cut-off μ^2 which is a gauge type parameter. Gauge independent quantities exist then infraredwise by construction. In the case of massless matter fields we have shown that the scheme yields IR-anomaly free Ward and Slavnov identities for Green's functions of elementary fields. These are a necessary prerequisite for a study of the only physical quantities in a massless theory: the Green's functions of gauge independent operators. Unfortunately the absence of IR-anomalies for the latter ones requires a proof case by case which we do not enter here. If IR-anomalies are absent our method shows that supersymmetry can be maintained since the explicit breaking by the IR-regulator occurs in the pure gauge sector. On the level of elementary fields the only physical trace left from supersymmetry is the fact that the massless matter fields remain massless i.e. the mass degeneracy in the multiplet is preserved.

What about generalizations to other models? The weak point of the above method is the use of an exact R-symmetry. It was needed for the elimination of some IR-anomaly candidates in the Slavnov identity. Theories of massive matter fields A without A^3 self-interaction are also covered by our treatment, since $n(A) = -1$ leads to an exact R-symmetry. There the mass degeneracy within the multiplets remains to be checked. The same comments apply for massive theories with an O'Raifeartaigh type R-symmetry (cf. section 3.3). The completely general case definitely requires a generalization of our method. This should then also permit to deal with other phases of the theory: when supersymmetry and/or gauge symmetry is spontaneously broken. Radiative mass generation (cf. section 17) and splitting is to be expected. One would have to show that all physical masses are independent of the gauge parameters including the IR-cut-off μ^2.

APPENDIX A NOTATIONS, CONVENTIONS AND USEFUL FORMULAE

A.1 σ-matrices, spinor calculus, covariant derivatives

The metric tensor $g_{\mu\nu}$ has diagonal elements (+1,-1,-1,-1). The σ-matrices are defined by

$$\sigma^{\mu} = (1,\underline{\sigma}), \ \bar{\sigma}^{\mu} = (1,-\underline{\sigma}), \ \bar{\sigma}^{\mu\dot{\beta}\alpha} = \sigma^{\mu\alpha\dot{\beta}} \tag{A.1}$$

where $\underline{\sigma} = (\sigma_1,\sigma_2,\sigma_3)$ are the Pauli matrices

$$\sigma_1 = \begin{pmatrix} 0 & 1 \\ 1 & 0 \end{pmatrix}, \quad \sigma_2 = \begin{pmatrix} 0 & -i \\ i & 0 \end{pmatrix}, \quad \sigma_3 = \begin{pmatrix} 1 & 0 \\ 0 & -1 \end{pmatrix}. \tag{A.2}$$

In addition the matrices $\sigma^{\mu\nu}$ and $\bar{\sigma}^{\mu\nu}$ are given by

$$\begin{aligned} \sigma^{\mu}_{\alpha\dot{\alpha}}\bar{\sigma}^{\nu\dot{\alpha}\beta} &= g^{\mu\nu}\delta_{\alpha}^{\ \beta} - i(\sigma^{\mu\nu})_{\alpha}^{\ \beta} \\ \bar{\sigma}^{\mu\dot{\alpha}\alpha}\sigma^{\nu}_{\alpha\dot{\beta}} &= g^{\mu\nu}\delta^{\dot{\alpha}}_{\ \dot{\beta}} - i(\bar{\sigma}^{\mu\nu})^{\dot{\alpha}}_{\ \dot{\beta}} . \end{aligned} \tag{A.3}$$

The summation convention is

$$\Theta\eta = \Theta^{\alpha}\eta_{\alpha} \qquad \bar{\Theta}\bar{\eta} = \bar{\Theta}_{\dot{\alpha}}\bar{\eta}^{\dot{\alpha}} \tag{A.4}$$

where lowering and raising of indices are effected through

$$\Theta^{\alpha} = \varepsilon^{\alpha\beta}\Theta_{\beta} \qquad \Theta_{\alpha} = \varepsilon_{\alpha\beta}\Theta^{\beta} \tag{A.5}$$

with $\varepsilon_{\alpha\beta} = -\varepsilon_{\beta\alpha}$, $\varepsilon^{12} = -\varepsilon_{12} = 1$ (the same for dotted indices).

Differentiation with respect to the anticommuting parameters $\Theta_{\alpha},\bar{\Theta}_{\dot{\alpha}}$ is defined by

$$\frac{\partial}{\partial\Theta^{\alpha}}\Theta^{\beta} = \delta_{\alpha}^{\ \beta} \qquad \frac{\partial}{\partial\bar{\Theta}^{\dot{\alpha}}}\bar{\Theta}^{\dot{\beta}} = \delta_{\dot{\alpha}}^{\ \dot{\beta}} . \tag{A.6}$$

They have the complex conjugation property

$$\begin{aligned} \left(\frac{\partial}{\partial\Theta^{\alpha}}\ \Phi\right)^{*} &= -\frac{\partial}{\partial\bar{\Theta}^{\dot{\alpha}}}\bar{\Phi} \quad \text{for } \Phi \text{ a Bose field} \\ \left(\frac{\partial}{\partial\Theta^{\alpha}}\ \psi\right)^{*} &= +\frac{\partial}{\partial\bar{\Theta}^{\dot{\alpha}}}\bar{\psi} \quad \text{for } \psi \text{ a Fermi field} . \end{aligned} \tag{A.7}$$

Covariant derivatives with respect to the supersymmetry transformations (1.20) are

$$D_\alpha \equiv \frac{\partial}{\partial\Theta^\alpha} - i\sigma^\mu_{\alpha\dot\alpha}\bar\Theta^{\dot\alpha}\partial_\mu$$

$$\bar D_{\dot\alpha} \equiv \frac{\partial}{\partial\bar\Theta^{\dot\alpha}} + i\Theta^\alpha\sigma^\mu_{\alpha\dot\alpha}\partial_\mu \quad . \tag{A.8}$$

They obey the anticommutation relations

$$\{D_\alpha,\bar D_{\dot\alpha}\} = 2i\sigma^\mu_{\alpha\dot\alpha}\partial_\mu \qquad \{D_\alpha,D_\beta\} = 0 = \{\bar D_{\dot\alpha},\bar D_{\dot\beta}\} \ . \tag{A.9}$$

As a consequence one can show

$$[D_\alpha,\bar D\bar D] = 4i\sigma^\mu_{\alpha\dot\alpha}\partial_\mu\bar D^{\dot\alpha}$$

$$[\bar D_{\dot\alpha},DD] = -4iD^\alpha\sigma^\mu_{\alpha\dot\alpha}\partial_\mu \tag{A.10}$$

$$D^\alpha\bar D_{\dot\alpha}D_\alpha = -\frac{1}{2}\bar D_{\dot\alpha}DD - \frac{1}{2}DD\bar D_{\dot\alpha}$$

$$\bar D_{\dot\alpha}D_\alpha\bar D^{\dot\alpha} = -\frac{1}{2}D_\alpha\bar D\bar D - \frac{1}{2}\bar D\bar D D_\alpha \tag{A.11}$$

$$[DD,\bar D\bar D] = -16\Box - 8i\bar D\bar\sigma^\mu D\partial_\mu$$

$$= +16\Box + 8iD\sigma^\mu\bar D\partial_\mu \tag{A.12}$$

$$D\bar D\bar D D = \bar D DD\bar D \tag{A.13}$$

$$D\bar D\bar D D - \frac{1}{2}\{DD,\bar D\bar D\} = 8\Box \tag{A.14}$$

Integration in superspace is defined by

$\int dV\Phi = \int dx DD\bar D\bar D\Phi$ for any superfield Φ

$\int dS = \int dx DD\Phi$ for a chiral superfield Φ ($\bar D\Phi = 0$)

$\int d\bar S\bar\Phi = \int dx\bar D\bar D\bar\Phi$ for an anti-chiral superfield ($D\bar\Phi = 0$). (A.15)

Functional differentiation in superspace is defined by

$$\frac{\delta\Phi(1)}{\delta\Phi(2)} = \delta_V(1,2) \quad \text{for a general superfield}$$

$$\frac{\delta\Phi(1)}{\delta\Phi(2)} = \delta_S(1,2) \quad \text{for a chiral superfield}$$

$$\frac{\delta\bar{\Phi}(1)}{\delta\bar{\Phi}(2)} = \bar{\delta}_S(1,2) \quad \text{for an anti-chiral superfield} \tag{A.16}$$

where

1) $\delta_V(1,2) = \frac{1}{16}\,\Theta_{12}^2\bar{\Theta}_{12}^2\delta^4(x_1 - x_2)$

and $\int dV\Phi(1)\delta_V(1,2) = \Phi(2)$

2) $\delta_S(1,2)\Big|_{\text{chiral representation}} = -\frac{1}{4}\,\Phi_{12}^2\delta^4(x_1 - x_2)$

and $\int dS(1)\Phi(1)\delta_S(1,2) = \Phi(2)$

3) $\bar{\delta}_S(1,2)\Big|_{\text{anti-chiral representation}} = -\frac{1}{4}\,\bar{\Theta}_{12}^2\delta^4(x_1 - x_2)$

and $\int d\bar{S}(1)\bar{\Phi}(1)\bar{\delta}_S(1,2) = \bar{\Phi}(2)$ (A.17)

with $\Theta_{ij} = \Theta_i - \Theta_j$, $\bar{\Theta}_{ij} = \bar{\Theta}_i - \bar{\Theta}_j$ and arguments (1), (2) refer to superspace points.

A.2 Field Transformations

Discrete Transformations

Parity transformations are given by

$$x^\mu = (x_0, \underline{x}) \longrightarrow x_P^\mu = (x_0, -\underline{x})$$

$$\Theta_\alpha \longrightarrow \Theta_\alpha^P = +\,i(\sigma^0\bar{\Theta})_\alpha$$

$$\bar{\Theta}_{\dot{\alpha}} \longrightarrow \bar{\Theta}_{\dot{\alpha}}^P = -\,i(\Theta\sigma^0)_{\dot{\alpha}} \tag{A.18}$$

on coordinates, and by

$$\partial_\mu \longrightarrow \partial^P_\mu = (-1)^{1+\delta_{\mu 0}} \partial_\mu$$

$$D_\alpha \longrightarrow D^P_\alpha = +\, i\sigma^0_{\alpha\dot\alpha}\bar D^{\dot\alpha}$$

$$\bar D_{\dot\alpha} \longrightarrow \bar D^P_{\dot\alpha} = -\, iD^\alpha\sigma^0_{\alpha\alpha} \tag{A.19}$$

on derivatives. If the Θ's and D's are contracted, one can simplify by the rule

$$\Theta_\alpha \leftrightarrow \bar\Theta^{\dot\alpha} \qquad D_\alpha \leftrightarrow \bar D^{\dot\alpha} \tag{A.20}$$

The σ-matrices obey

$$\begin{aligned} \sigma^o\bar\sigma^\mu\sigma^o &= (-1)^{1+\delta_{\mu 0}}\sigma^\mu \\ \bar\sigma^0\sigma^{\mu\nu}\sigma^0 &= (-1)^{\delta_{\mu o}+\delta_{\nu o}}\sigma^{\mu\nu} \end{aligned} \tag{A.21}$$

On fields parity acts by

$$P\Phi(x,\Theta,\bar\Theta)P^{-1} = \zeta\bar\Phi(x_p,\Theta^P,\bar\Theta^P) \tag{A.22}$$

with ζ being the intrinsic parity of the field Φ in question. The list of intrinsic parities reads as follows

field	Φ	A	c_+	c_-	ρ	σ
ζ	-1	+1	+1	-1	-1	+1

For charged chiral fields we have in addtion

$$A_\pm \leftrightarrow \bar A_\mp \qquad Y_\pm \leftrightarrow \bar Y_\mp \tag{A.23}$$

<u>Charge conjugation</u> is given by

$$\begin{aligned} &\Phi \to -\Phi^T \qquad A \to A^T \qquad \bar A \to \bar A^T \\ &c_\pm \to -c^T_\pm \qquad \bar c_\pm \to -\bar c_\pm{}^T \\ &\rho \to -\rho^T \qquad Y \to Y^T \qquad \sigma \to -\sigma^T \end{aligned} \tag{A.24}$$

If charged matter is present $\Phi \to -\Phi^T$ may enforce presence of a similarity transformation S in the representation space of the T's such that

$$\tilde{\Phi} \to -S^{-1}\tilde{\Phi}^T S \qquad (A.25)$$

Then

$$A_- \to A_+ S \qquad \bar{A}_- \to S^{-1}\bar{A}_+$$

$$Y_+ \to S^{-1}y_- \qquad \bar{Y}_+ \to \bar{Y}_- S \qquad Y_- \to Y_+ S \qquad \bar{Y}_- \to S^{-1}\bar{Y}_+ \qquad (A.26)$$

Continuous symmetry transformation

The generators of the superconformal algebra can be represented as linear superspace differential operators acting on a superfield Φ. That is the infinitesimal variation δ^X of the superfield Φ under generator X is defined by

$$[X,\Phi] = -i\delta^X\Phi \qquad (A.27)$$

For a general superfield Φ in the real representation the superconformal infinitesimal transformations are

$$\delta^P_\mu \Phi = \partial_\mu \Phi$$

$$\delta^M_{\mu\nu}\Phi = [x_\mu\partial_\nu - x_\nu\partial_\mu - \frac{i}{2}\Theta\sigma_{\mu\nu}\frac{\partial}{\partial\Theta} + \frac{i}{2}\bar{\Theta}\bar{\sigma}_{\mu\nu}\frac{\partial}{\partial\bar{\Theta}}]\Phi$$

$$\delta^D\Phi = [d + x^\lambda\partial_\lambda + \frac{1}{2}\Theta\frac{\partial}{\partial\Theta} - \frac{1}{2}\bar{\Theta}\frac{\partial}{\partial\bar{\Theta}}]\Phi$$

$$\delta^K_\mu\Phi = [2x_\mu\delta^D + 2x^\nu\delta^M_{\mu\nu} - 2x_\mu x^\lambda\partial_\lambda$$
$$+ x^2\partial_\mu + 2\Theta\sigma_\mu\bar{\Theta}\delta^R + in\Theta\sigma_\mu\bar{\Theta} + \Theta^2\bar{\Theta}^2\partial_\mu]\Phi$$

$$\delta^R\Phi = i[n + \Theta\frac{\partial}{\partial\Theta} + \bar{\Theta}\frac{\partial}{\partial\bar{\Theta}}]\Phi \qquad (A.28)$$

$$\delta^Q_\alpha \Phi = [\ \frac{\partial}{\partial\Theta} + i\sigma^\mu\bar{\Theta}\partial_\mu]_\alpha \Phi$$

$$\delta^{\bar{Q}}_{\dot{\alpha}} \Phi = [\ - \frac{\partial}{\partial\bar{\Theta}} - i\Theta\sigma^\mu\partial_\mu]_{\dot{\alpha}} \Phi$$

$$\delta^S_\alpha \Phi = [\ - x_\mu\sigma^\mu\delta^{\bar{Q}} + 2\Theta\delta^R - i\Theta^2 D$$

$$- \ 2i(d - \frac{n}{2})\Theta]_\alpha \Phi$$

$$\delta^{\bar{S}}_{\dot{\alpha}} \Phi = [\ - x_\mu\delta^Q\sigma^\mu + 2\bar{\Theta}\delta^R + i\bar{\Theta}^2\bar{D}$$

$$+ \ 2i(d + \frac{n}{2})\bar{\Theta}]_{\dot{\alpha}} \Phi \qquad \text{(A.28 cont.'d)}$$

where d is the scale dimension of the superfield (in the tree approximation this equals the canonical dimension) whereas n, defined in δ^R, is the γ_5-(chiral) or R-weight. The reality property implies that if n is the R-weight of Φ, the R-weight of the complex conjugate field $\bar{\Phi}$ is -n; hence n = 0 for a real superfield ("vector" superfield).

For a chiral superfield Φ in the chiral representation and for its complex conjugate anti-chiral superfield $\bar{\Phi}$ in the anti-chiral representation the superconformal infinitesimal transformations become

$$\delta^P_\mu \Phi = \partial_\mu \Phi \qquad \delta^P_\mu \bar{\Phi} = \partial_\mu \bar{\Phi}$$

$$\delta^M_{\mu\nu} \Phi = [x_\mu\partial_\nu - x_\nu\partial_\mu - \frac{i}{2}\Theta\sigma_{\mu\nu}\frac{\partial}{\partial\Theta}]\Phi \qquad \delta^M_{\mu\nu} \bar{\Phi} = [x_\mu\partial_\nu - x_\nu\partial_\mu + \frac{i}{2}\bar{\Theta}\bar{\sigma}_{\mu\nu}\frac{\partial}{\partial\bar{\Theta}}]\bar{\Phi}$$

$$\delta^D \Phi = [d + x^\lambda\partial_\lambda + \frac{1}{2}\Theta\frac{\partial}{\partial\Theta}]\Phi \qquad \delta^D \bar{\Phi} = [d + x^\lambda\partial_\lambda - \frac{1}{2}\bar{\Theta}\frac{\partial}{\partial\bar{\Theta}}]\bar{\Phi}$$

$$\delta^K_\mu \Phi = [2x_\mu(d + x^\lambda\partial_\lambda) - x^2\partial_\mu + x^\nu\Theta\sigma_\mu\bar{\sigma}_\nu\frac{\partial}{\partial\Theta}]\Phi \qquad \delta^K_\mu \bar{\Phi} = [2x_\mu(d + x^\lambda\partial_\lambda) - x^2\partial_\mu - x^\nu\bar{\Theta}\bar{\sigma}_\mu\sigma_\nu\frac{\partial}{\partial\bar{\Theta}}]\bar{\Phi}$$

$$\delta^R \Phi = i[-\frac{2}{3}d + \Theta\frac{\partial}{\partial\Theta}]\Phi \qquad \delta^R \bar{\Phi} = i[\frac{2}{3}d + \bar{\Theta}\frac{\partial}{\partial\bar{\Theta}}]\bar{\Phi} \qquad \text{(A.29)}$$

$$\delta^Q_\alpha \Phi = \frac{\partial}{\partial\Theta^\alpha}\,\Phi \qquad \delta^Q_\alpha \bar\Phi = 2i(\sigma^\mu\bar\Theta)_\alpha \partial_\mu \bar\Phi$$

$$\delta^{\bar Q}_{\dot\alpha} \Phi = -2i(\Theta\sigma^\mu)_{\dot\alpha}\partial_\mu \Phi \qquad \delta^{\bar Q}_{\dot\alpha}\bar\Phi = -\frac{\partial}{\partial\bar\Theta^{\dot\alpha}}\,\bar\Phi$$

$$\delta^S_\alpha \Phi = [-\,x_\mu \sigma^\mu \delta^{\bar Q} - 4id\Theta - 2i\Theta^2 \frac{\partial}{\partial\Theta}\,]_\alpha \Phi \qquad \delta^S_\alpha \bar\Phi = -x_\mu(\sigma^\mu\delta^{\bar Q})_\alpha \bar\Phi$$

$$\delta^{\bar S}_{\dot\alpha}\Phi = -x_\mu(\delta^Q\sigma^\mu)_{\dot\alpha}\Phi \qquad \delta^{\bar S}_{\dot\alpha}\bar\Phi = [-\,x_\mu \delta^Q \sigma^\mu + 4id\bar\Theta - 2i\bar\Theta^2 \frac{\partial}{\partial\bar\Theta}\,]_{\dot\alpha}\bar\Phi \qquad \text{(A.29 cont.'d)}$$

Note that due to the chirality of the fields the R-weight of Φ is $n = -2/3$ and the R-weight of $\bar\Phi$ is $n = +\,2/3$ d.

Supersymmetry transformations of component fields.

Chiral field

$$A = A_1(x,\Theta) = A + \Theta\psi + \Theta^2 F$$

$$\delta_\alpha A = \psi_\alpha$$

$$\delta_\alpha \psi_\beta = -2\varepsilon_{\alpha\beta} F$$

$$\delta_\alpha F = 0$$

$$\bar\delta_{\dot\alpha} A = 0$$

$$\bar\delta_{\dot\alpha}\psi_\alpha = 2i\sigma^\mu_{\alpha\dot\alpha}\partial_\mu A$$

$$\bar\delta_{\dot\alpha} F = i\partial_\mu \psi^\alpha \sigma^\mu_{\alpha\dot\alpha}$$

Anti-chiral field

$$\bar A = \bar A_2(x,\bar\Theta) = \bar A + \bar\Theta\bar\psi + \bar\Theta^2 \bar F$$

$$\delta_\alpha \bar A = 0$$

$$\delta_\alpha \bar\psi_{\dot\alpha} = 2i\sigma^\mu_{\alpha\dot\alpha}\partial_\mu \bar A$$

$$\delta_\alpha \bar F = -i\sigma^\mu_{\alpha\dot\alpha}\partial_\mu \bar\psi^{\dot\alpha}$$

$$\bar\delta_{\dot\alpha}\bar A = \bar\psi_{\dot\alpha}$$

$$\bar\delta_{\dot\alpha}\bar\psi_{\dot\beta} = 2\varepsilon_{\dot\alpha\dot\beta}\bar F$$

$$\bar\delta_{\dot\alpha}\bar F = 0 \qquad \text{(A.30)}$$

Real field

$$\Phi(x,\Theta,\bar\Theta) = C + \Theta\chi + \bar\Theta\bar\chi + \frac{1}{2}\Theta^2 M + \frac{1}{2}\bar\Theta^2 \bar M + \Theta\sigma^\mu\bar\Theta v_\mu + \frac{1}{2}\bar\Theta^2\Theta\lambda + \frac{1}{2}\Theta^2\bar\Theta\bar\lambda + \frac{1}{4}\Theta^2\bar\Theta^2 D$$

$$\delta_\alpha C = \chi_\alpha \qquad \bar{\delta}_{\dot\alpha} C = \bar{\chi}_{\dot\alpha}$$

$$\delta_\alpha \chi_\beta = -\varepsilon_{\alpha\beta} M \qquad \bar{\delta}_{\dot\alpha} \bar{\chi}_{\dot\beta} = \varepsilon_{\dot\alpha\dot\beta} \bar{M}$$

$$\delta_\alpha \bar{\chi}_{\dot\alpha} = \sigma^\mu_{\alpha\dot\alpha}(v_\mu + i\partial_\mu C) \qquad \bar{\delta}_{\dot\alpha} \chi_\alpha = -\sigma^\mu_{\alpha\dot\alpha}(v_\mu - i\partial_\mu C)$$

$$\delta_\alpha M = 0 \qquad \bar{\delta}_{\dot\alpha} \bar{M} = 0$$

$$\delta_\alpha \bar{M} = \lambda_\alpha - i(\sigma^\mu \partial_\mu \bar{\chi})_\alpha \qquad \bar{\delta}_{\dot\alpha} M = \bar{\lambda}_{\dot\alpha} + i(\partial_\mu \chi \sigma^\mu)_{\dot\alpha}$$

$$\delta_\alpha v_\mu = \frac{1}{2}(\sigma_\mu \bar{\lambda})_\alpha - \frac{i}{2}(\sigma^\nu \bar{\sigma}_\mu \partial_\nu \chi)_\alpha \qquad \bar{\delta}_{\dot\alpha} v_\mu = \frac{1}{2}(\lambda\sigma_\mu)_{\dot\alpha} + \frac{i}{2}\sigma^\nu_{\alpha\dot\alpha} \bar{\sigma}_\mu{}^{\dot\beta\alpha} \partial_\nu \bar{\chi}_{\dot\beta}$$

$$\delta_\alpha \lambda_\beta = -2\varepsilon_{\alpha\beta}(D + i\partial v) + \frac{1}{2}\sigma^{\mu\nu}_{\alpha\beta} v_{\mu\nu} \qquad \bar{\delta}_{\dot\alpha} \bar{\lambda}_{\dot\beta} = \varepsilon_{\dot\alpha\dot\beta}(D - i\partial v) - \frac{1}{2}\bar{\sigma}^{\mu\nu}_{\dot\alpha\dot\beta} v_{\mu\nu}$$

$$\delta_\alpha \bar{\lambda}_{\dot\alpha} = i\sigma^\mu_{\alpha\dot\alpha}\partial_\mu M \qquad \bar{\delta}_{\dot\alpha} \lambda_\alpha = i\sigma^\mu_{\alpha\dot\alpha}\partial_\mu \bar{M}$$

$$\delta_\alpha D = -i(\sigma^\mu \partial_\mu \bar{\lambda})_\alpha \qquad \bar{\delta}_{\dot\alpha} D = i(\partial_\mu \lambda \sigma^\mu)_{\dot\alpha} \tag{A.31}$$

APPENDIX B GENERATING FUNCTIONALS

The Green's functions

$$G(z_1\ldots,z_n) = \langle T(\Phi(z_1)\ \ldots.\Phi(z_n))\rangle \tag{B.1}$$

are defined via the Gell-Mann-Low formula (7.1). In order to deal with them collectively one introduces a generating functional:

$$\begin{aligned} Z \equiv Z(J) &= \langle T\, e^{\frac{i}{\hbar}\int dz\Phi(z)J(z)}\rangle \\ &= \sum_{n=0}^{\infty} \frac{1}{n!}\left(\frac{i}{\hbar}\right)^n \int dz_1\ldots dz_n J(z_1)\ldots J(z_n)\langle T(\Phi(z_1)\ldots\Phi(z_n))\rangle \end{aligned} \tag{B.2}$$

from which they are obtained by differentiation. The specific choice of (B.2) is guided by the fact that $T(\exp \frac{i}{\hbar}\int dz\Phi J)$ is the S-matrix if Φ is a free field interacting with the classical source J, hence satisfies the desired axioms. The source functions J are (commuting or anticommuting) test functions, whereas the G's are tempered distributions. The latter are calculated by use of Wick's theorem in the diagrammatic expansion including a subtraction procedure as explained in sections 7-9.

If there are local operators $Q_i(z_i)$ in the theory considered we know how to calculate diagrams containing them once their subtraction degrees are given (cf. section 10). The generating functional for their Green's functions is then in analogy to (B.1) (B.2) given by

$$Z = Z(J,\eta) = \langle T\, e^{\frac{i}{\hbar}\int dz(\Phi(z)J(z) + \eta(z)Q(z))}\rangle \tag{B.3}$$

and thus the Green's functions by

$$\begin{aligned} &G^{m,n}(z_1,\ldots,z_m;\ z_1',\ldots,z_n') \\ &\equiv \langle TQ(z_1)\ldots Q(z_m)\Phi(z_1')\ldots\Phi(z_n')\rangle \\ &= (-i\hbar)^{m+n} \left.\frac{\delta^{m+n}\ Z(J,\eta)}{\delta\eta(z_1)\ldots\delta\eta(z_m)\delta J(z_1')\ldots\delta J(z_n')}\right|_{J=\eta=0} \end{aligned} \tag{B.4}$$

(η denotes the sources for the operators Q).

Although the Green's functions are the basic objects of the theory since the physical axioms are formulated in terms of them, one needs for the purpose of renormalization (i.e. of making well-defined the theory) other quantities as well. These are the connected Green's functions and, ultimately, the one-particle-irreducible Green's functions.

The generating functional of connected Green's functions, $Z_c(J,\eta)$, is defined by

$$Z(J,\eta) = e^{\frac{i}{\hbar} Z_c(J,\eta)} \tag{B.5}$$

and one can show (cf. for instance [R.1]) that the contribution of connected diagrams to a Green's function is given by the derivatives of Z_c:

$$\langle TQ(z_1)\dots Q(z_m)\Phi(z_1')\dots\Phi(z_n')\rangle\Big|_{\text{conn.diagrams}} = (-i\hbar)^{m+n-1} \frac{\delta^{m+n} Z_c(J,\eta)}{\delta\eta(z_1)\dots\delta\eta(z_m)\delta J(z_1')\dots\delta J(z_n')} \tag{B.6}$$

The generating functional of vertex functions $\Gamma(\Phi_c,\eta)$ is obtained from $Z_c(J,\eta)$ by Legendre transformation. First of all one defines a field $\Phi_c(z,J,\eta)$ - "classical" field - by

$$\Phi_c(z,J,\eta) = \frac{\delta}{\delta J(z)} Z_c(J,\eta) \tag{B.7}$$

$$\Phi_c(z,0,0) = 0 \tag{B.8}$$

where (B.8) expresses the condition that the vacuum expectation value of the quantum field Φ should vanish. Assuming now that (B.7) can be solved with respect to J, for instance in the sense of a formal series, then Γ is given by the equation

$$\Gamma(\Phi_c,\eta) = Z_c(J,\eta) - \int dz\, J(z)\Phi_c(z) \tag{B.9}$$

with the understanding of J as solution of (B.7):

$$J = J(z,\Phi_c,\eta) \tag{B.10}$$

and

$$J(z,0,0) = 0 \tag{B.11}$$

Calculating $\delta\Gamma/\delta\Phi_c$ for $J = J(z,\Phi_c,\eta)$ one finds

$$\left.\frac{\delta\Gamma}{\delta\Phi_c}\right|_{J=J(z,\Phi_c,\eta)} = \frac{\delta J}{\delta\Phi_c}\cdot\frac{\delta Z_c}{\delta J} - \frac{\delta J}{\delta\Phi_c}\cdot\Phi_c - J(z,\Phi_c,\eta)$$

$$\left.\frac{\delta\Gamma}{\delta\Phi_c}\right|_{J=J(z,\Phi_c,\eta)} = -J(z,\Phi_c,\eta) \tag{B.12}$$

(by use of (B.7)). Analogously

$$\frac{\delta\Gamma}{\delta\eta} = \frac{\delta Z_c}{\delta\eta} \tag{B.13}$$

where the arguments Φ_c, J in the functionals have to be appropriately understood on both sides of the equation.

As far as the (diagrammatic) interpretation is concerned, one can show that Γ generates one-particle-irreducible (diagrams) Green's functions i.e. diagrams which remain connected when one line is cut, with external lines amputated [R.1].

$$\Gamma^{m,n}(z_1,\ldots,z_m,z_1'\ldots,z_n') \equiv \frac{\delta^{m+n}\Gamma(\Phi_c,\eta)}{\delta\eta(z_1)\ldots\delta\eta(z_m)\delta\Phi_c(z_1')\ldots\delta\Phi_c(z_n')} \tag{B.14}$$

$$= \left(\frac{i}{\hbar}\right)^{m-1} \left.\langle T\, Q(z_1)\ldots Q(z_m)\Phi(z_1')\ldots\Phi(z_n')\rangle\right|_{1\ \text{PI-diagrams}}$$

An instructive example is provided by the 2-point functions. Let us differentiate (B.7) with respect to $\Phi_c(z',J,\eta)$ for the solution $J = J(z,\Phi_c,\eta)$ (B.10):

$$\delta(z',z) = \int dz'' \frac{\delta J(z'')}{\delta\Phi_c(z')}\cdot\frac{\delta^2 Z_c}{\delta J(z'')\delta J(z)} = -\int dz'' \frac{\delta^2\Gamma}{\delta\Phi_c(z')\delta\Phi_c(z'')}\cdot\frac{\delta^2 Z_c}{\delta J(z'')\delta J(z)} \tag{B.15}$$

At $J = \eta = \Phi_c = 0$ this means that $\Gamma(z',z'')$ is the inverse (up to a factor) of the 2-point Green's function, since the latter is connected (for vanishing tadpoles (B.11)):

$$\int dz G^2(z_1,z)\Gamma^2(z,z_2) = i\hbar\delta(z_1,z_2) \tag{B.16}$$

In fact it is this relation which is exploited for calculating the propagators already for the free theory. In order to explain this, we first point out the connection between the functional Γ and the classical action of a theory. Let us consider a one-particle-irreducible diagram with m closed loops, I internal lines and V vertices and count the powers of $\hbar$ associated with it. We find

$$\hbar^{I} \cdot \hbar^{-V} \cdot \hbar \tag{B.17}$$

(the last factor arises from the relative $\hbar$ between Z and Z_c; cf. also (B.14)). But

$$I - V + 1 = m \tag{B.18}$$

is just the Euler formula i.e. the powers of $\hbar$ count the number of loops. Zero loop number means tree diagrams and thus we may identify $\Gamma^{(0)}$ with the classical action. This is, of course, the origin for the name "classical" field Φ_c. And this identification is also the basis for our recipe with which we have calculated propagators in chapter II. Solving the equation (B.12) with $\Gamma = \Gamma^{(0)}$ = classical action, for Φ_c in terms of J permits to calculate the 2-point function i.e. the propagator, as seen from (B.7), by differentiating with respect to J. In this way, one can effectively avoid to handle true quantum fields Φ directly, although this procedure is just equivalent to canonical quantization. Let us note that we usually omit the index c for the fields Φ_c.

Up to now we have never made explicit the different types of fields involved, but all collectively denoted by Φ, their sources by J and similarly for operators Q_i and their sources η. In table B.1 we display all superfields, insertions and sources occurring in SYM and indicate how they are Legendre transformed.

The Legendre transformation is given by

$$\begin{aligned} Z_c = \Gamma + \mathrm{Tr}\int dV J\Phi + \int dS(\mathrm{Tr}(\xi_+ c_- + \xi_- c_+) + J_A A + J_B B) \\ + \int d\bar{S}(\mathrm{Tr}(\bar{c}_- \bar{\xi}_+ + \bar{c}_+ \bar{\xi}_-) + \bar{J}_A \bar{A} + \bar{J}_B \bar{B}) \end{aligned} \tag{B.19}$$

and the relations (suitable arguments understood):

$$\frac{\delta Z_c}{\delta \xi_\pm} = c_\mp \qquad \frac{\delta \Gamma}{\delta c_\pm} = \xi_\mp$$
$$\frac{\delta Z_c}{\delta \bar{\xi}_\pm} = - \bar{c}_\mp \qquad \frac{\delta \Gamma}{\delta \bar{c}_\pm} = - \bar{\xi}_\mp \tag{B.20}$$

The Slavnov identity reads in terms of Γ (in the (α,B)-gauge formulation)

$$\mathcal{S}(\Gamma) \equiv \mathrm{Tr}\int dV \frac{\delta\Gamma}{\delta\rho}\frac{\delta\Gamma}{\delta\Phi} + \int dS\{\mathrm{Tr}\, B \frac{\delta\Gamma}{\delta c_-} + \mathrm{Tr}\frac{\delta\Gamma}{\delta\sigma}\frac{\delta\Gamma}{\delta c_+} + \frac{\delta\Gamma}{\delta Y}\frac{\delta\Gamma}{\delta A}$$
$$+ \int d\bar{S}\{\mathrm{Tr}\, \bar{B} \frac{\delta\Gamma}{\delta \bar{c}_-} + \mathrm{Tr}\frac{\delta\Gamma}{\delta\bar{\sigma}}\frac{\delta\Gamma}{\delta \bar{c}_+} + \frac{\delta\Gamma}{\delta \bar{Y}}\frac{\delta\Gamma}{\delta \bar{A}}\} = 0 \tag{B.21}$$

the ghost equation of motion

$$\frac{\delta\Gamma}{\delta c_-} = - \frac{1}{8}\bar{D}\bar{D}DD\frac{\delta\Gamma}{\delta\rho} \qquad \frac{\delta\Gamma}{\delta \bar{c}_-} = - \frac{1}{8}DD\bar{D}\bar{D}\frac{\delta\Gamma}{\delta\rho} \tag{B.22}$$

In terms of Z one finds

$$\mathcal{S}Z \equiv - \mathrm{Tr}\int dVJ\frac{\delta Z}{\delta\rho} + \int dS\{\mathrm{Tr}\xi_+\frac{\delta}{\delta J_B} + \mathrm{Tr}\xi_-\frac{\delta}{\delta\sigma} - J_A\frac{\delta}{\delta Y}\}Z$$
$$- \int d\bar{S}\{\mathrm{Tr}\bar{\xi}_+\frac{\delta}{\delta \bar{J}_B} + \mathrm{Tr}\bar{\xi}_-\frac{\delta}{\delta\bar{\sigma}} + \bar{J}_A\frac{\delta}{\delta\bar{Y}}\}Z = 0 \tag{B.23}$$

and for the ghost equation of motion

$$\xi_+ Z = - \frac{1}{8}\bar{D}\bar{D}DD\frac{\delta Z}{\delta\rho}$$
$$\bar{\xi}_+ Z = \frac{1}{8}DD\bar{D}\bar{D}\frac{\delta Z}{\delta\rho} \tag{B.24}$$

TABLE B.1

field	Φ	A	c_+	$\bar{c}_+$	c_-	$\bar{c}_-$	B
source	J	J_A	ξ_-	$\bar{\xi}_-$	ξ_+	$\bar{\xi}_+$	J_B

composite operator	Q	c_+c_+	$\tilde{c}_+A$	$\bar{A}\tilde{\bar{c}}_+$
external field	ρ	σ	Y	$\bar{Y}$

APPENDIC C $\mathscr{b}_0$-COHOMOLOGY

As an intermediate step in the course of solving the BRS consistency condition (section 15.2) we have to solve the $\mathscr{b}_0$-cohomology equation

$$\mathscr{b}_0\Delta(\Phi,c_+,A,\eta,\sigma,Y,p_k,z_k) = 0 \tag{C.1}$$

for a classical insertion Δ which has dimension 4 and $\Phi\Pi$-charge 1, is supersymmetric, rigid- and R-invariant. $\mathscr{b}_0$ is the nilpotent variational operator defined in (15.41, 15.42). Any insertion of the form $\Delta = \mathscr{b}_0\hat{\Delta}$ solves (C.1) since $\mathscr{b}_0$ is nilpotent, but solutions which are not $\mathscr{b}_0$-variations also exist, as we shall see. We shall use as in section 15 the notation $\overset{\mathscr{b}}{\approx}$ for equality up to $\mathscr{b}_0$variations.

We begin with the gauge parameter (p_k, z_k)-dependence of Δ. Let us decompose $\mathscr{b}_0$ in

$$\begin{aligned} \mathscr{b}_0 &= \bar{\mathscr{b}}_0 + d \\ d &= z_k\partial_k \quad , \quad \partial_k = \partial/\partial p_k \\ d^2 &= 0 \quad , \quad \bar{\mathscr{b}}_0^2 = 0 \quad , \quad dp_k = z_k \ . \end{aligned} \tag{C.2}$$

The most general form of Δ is, its field dependence being dropped:

$$\Delta = \sum_{n=1}^{N} z_{k_1}\dots z_{k_n}\,\Delta_{[k_1\dots k_n]}(p) + \Delta_0(p) \tag{C.3}$$

Δ is thus a sum of n-forms ω_n in p-space. The number of parameters p being finite for any order in the fields and in $\hbar$, the relevant p-space has finite dimension, the summation in (C.3) is finite and the Δ's are polynomials in p. As a consequence the cohomology for such n-forms is trivial:

$$d\omega_n = 0 \Rightarrow \exists\,\hat{\omega}_{n-1} : \omega_n = d\hat{\omega}_{n-1} \tag{C.4}$$

Writing (C.3) as

$$\Delta = \sum_{n=0}^{N} \omega_n \tag{C.5}$$

we conclude from (C.1):

$$\sum_{n=0}^{N} (d\omega_n + \bar{\delta}_o\omega_n) = 0 \ . \qquad (C.6)$$

At the highest degree in z this gives $d\omega_N = o$ and, using (C.4),

$$\omega_N = d\hat{\omega}_{N-1} = \delta_o\hat{\omega}_{N-1} - \bar{\delta}_o\hat{\omega}_{N-1} \qquad (C.7)$$

The last term being a (N-1)-form, we have

$$\Delta = \delta_o\hat{\omega}_{N-1} + \sum_{n=0}^{N-1} \omega_n' \ \hat{\approx} \ \sum_{n=0}^{N-1} \omega_n' \ . \qquad (C.8)$$

Pursuing the argument down to the lowest order in z we get

$$\Delta \ \hat{\approx} \ \Delta_o \quad , \quad \partial_k\Delta_o = 0 \qquad (C.9)$$

We have thus eliminated the gauge parameters:

$$\Delta \ \hat{\approx} \ \Delta(\Phi, c_+, A, \eta, \sigma, Y) \quad , \quad \delta_o\Delta = 0 \ . \qquad (C.10)$$

N.B. This argument would not work for the δ-cohomology. The reason is that the operator δ-d, contrary to $\bar{\delta}_o = \delta_o$-d, may increase the order in z since the classical action (15.12) depends on p_k and z_k, whereas its bilinear terms do not.

We are now left with field dependence only. The following 2 lemmas will be useful:

Lemma C.1

Let X^q be a homogeneous polynomial of degree n ($n \geq q$) in Φ, c_+, $\bar{c}_+$, of dimension o and ΦΠ-charge q. Then

$$\delta_o X^q = 0 \qquad (C.11)$$

implies the existence of a function X^{q-1} and of a number x such that

$$X^q = \delta_o X^{q-1} + x(c_+)^q \, \delta_{nq} \quad , \qquad (C.12)$$

where $(c_+)^q$ denotes some polynomial of order q in the $\Phi\Pi$-fields c_{+i}.

For the proof, we first note that

$$\{\delta_0, \frac{\partial}{\partial \bar{c}_{+i}}\}X = - \frac{\partial}{\partial \Phi_i} X \tag{C.13}$$

for any function $X(\Phi, c_+)$ of dimension 0. In particular, due to the hypotheses (C.11),

$$\delta_0 \frac{\partial}{\partial \bar{c}_{+i}} X^q = - \frac{\partial}{\partial \Phi_i} X^q \tag{C.14}$$

Therefore

$$\begin{aligned} \Phi_i \frac{\partial}{\partial \Phi_i} X^q &= -\Phi_i \delta_0 \frac{\partial}{\partial \bar{c}_{+i}} X^q \\ &= - \delta_0 (\Phi_i \frac{\partial}{\partial \bar{c}_{+i}} X^q) + (c_+ - \bar{c}_+)_i \frac{\partial}{\partial \bar{c}_{+i}} X^q \end{aligned} \tag{C.15}$$

Writing

$$X^q = X_{(o)} + O(c_+) \tag{C.16}$$

where $X_{(o)}$ denotes a polynomial of degree q in $\bar{c}_+$, i.e. of degree o in c_+, we find by inserting (C.16) into (C.15)

$$\Phi_i \frac{\partial}{\partial \Phi_i} X_{(o)} + \bar{c}_{+i} \frac{\partial}{\partial \bar{c}_{+i}} X_{(o)} = - \delta_0(\Phi_i \frac{\partial}{\partial \bar{c}_{+i}} X^q) + O(c_+) \tag{C.17}$$

i.e.

$$(n+q)X_{(o)} = \delta_0(-\Phi_i \frac{\partial}{\partial \bar{c}_{+i}} X^q) + O(c_+) \tag{C.18}$$

Thus

$$X^q = \delta_0 \hat{X}_{(o)} + X_{(1)} + O(c_+^2) \tag{C.19}$$

where $X_{(1)}$ is of degree 1 in c_+, hence degree q-1 in $\bar{c}_+$. Reasoning as before, one finds

$$X_{(1)} = \delta_0 \hat{X}_{(1)} + X_{(2)} + O(c_+^3) \ . \tag{C.20}$$

This process continues until the last step and finally

$$X^q = \mathcal{b}_0\hat{X} + X_{(q)} \tag{C.21}$$

where $X_{(q)}$ is of degree q in c_+. Now

$$\mathcal{b}_0 \frac{\partial}{\partial \bar{c}_{+i}} X_{(q)} = 0 \tag{C.22}$$

due to the absence of $\bar{c}_+$'s in $X_{(q)}$. Hence (C.14) applied to $X_{(q)}$ - which is $\mathcal{b}_0$-invariant - implies

$$\frac{\partial}{\partial \Phi_i} X_{(q)} = 0 \tag{C.23}$$

i.e.

$$X_{(q)} = x(c_+)^q \tag{C.24}$$

(C.21) and (C.24) are the result (C.12) we wanted to prove.

Lemma C.2

Let $A^q_\alpha(z)$, $F^q_\alpha(z)$ and $H^q(z)$ be local, rigidly invariant functionals of degree n in Φ, c_+, $\bar{c}_+$ with dimensions 1/2, 3/2 and 2, R-weights 1, -1 and o, and arbitrary $\Phi\Pi$-charge q. Then the general solution of the cohomologies

$$\mathcal{b}_0 A^q_\alpha = 0 \quad , \quad \mathcal{b}_0 F^q_\alpha = 0 \quad , \quad \mathcal{b}_0 H^q = 0 \tag{C.25}$$

is

$$A^q_\alpha = \mathcal{b}_0 A^{q-1}_\alpha \ , \ F^q_\alpha = \mathcal{b}_0 F^{q-1}_\alpha + x\mathrm{Tr}[\bar{D}\bar{D}D_\alpha\Phi(c_+)^q]\delta_{n-1,q}$$

$$H^q = \mathcal{b}_0 H^{q-1} + x\mathrm{Tr}[D\bar{D}\bar{D}D\Phi(c_+)^q]\delta_{n-1,q} \ . \tag{C.26}$$

Proof

a) The general form of A^q_α is dictated by its dimension and R-weight requirements:

$$A^q_\alpha = \mathrm{Tr}[D_\alpha\Phi X^q + D_\alpha c_+ Y^{q-1}] \tag{C.27}$$

X^q and Y^{q-1} are homogeneous polynomials of degree n-1 in Φ, c_+, $\bar{c}_+$. Applying $\mathcal{b}_o$ we find

$$0 = \mathrm{Tr}[D_\alpha c_+(X^q - \mathcal{b}_o Y^{q-1}) + D_\alpha\Phi\ \mathcal{b}_o X^q] \qquad (C.28)$$

Since the two terms are independent, there follows

$$X^q = \mathcal{b}_o Y^{q-1} \qquad (C.29)$$

from the first term - a solution which makes disappear the second term. With this

$$A^q_\alpha = \mathcal{b}_o A^{q-1}_\alpha \quad , \quad A^{q-1}_\alpha = \mathrm{Tr}[D_\alpha\Phi\ Y^{q-1}] \qquad (C.30)$$

b) For F^q_α (dimension 3/2 and R-weight -1) we write its general form as

$$F^q_\alpha = \mathrm{Tr}[D_\alpha\bar{D}\bar{D}\Phi\ X^q_1 + D_\alpha\bar{D}\bar{D}\bar{c}_+X^{q-1}_2 + \bar{D}\bar{D}D_\alpha\Phi X^q_3$$
$$+ D_\alpha\Phi X^q_4 + D_\alpha c_+X^{q-1}_5 + \bar{D}D_\alpha\Phi X^q_6 + \bar{D}D_\alpha c_+X^{q-1}_7$$
$$+ D_\alpha\bar{D}\Phi X^q_8 + D_\alpha\bar{D}\bar{c}_+X^{q-1}_9] \qquad (C.31)$$

The condition (C.25) implies

$$X^q_1 = -\mathcal{b}_o X^{q-1}_2 \quad , \quad X^q_4 = \mathcal{b}_o X^{q-1}_5$$
$$X^q_6 = \mathcal{b}_o X^{q-1}_7 \quad , \quad X^q_8 = -\mathcal{b}_o X^{q-1}_9 \qquad (C.32)$$
$$\mathcal{b}_o X^q_3 = 0$$

The last equation is solved according to Lemma C.1 by (the dimension of X^q being o and its degree n-1)

$$X^q_3 = \mathcal{b}_o X^{q-1} + x(c_+)^q\ \delta_{n-1,q} \qquad (C.33)$$

Then from (C.32, C.33) follows

$$F_\alpha^q = \mathfrak{b}_0 F_\alpha^{q-1} + x\mathrm{Tr}[\bar{D}\bar{D}D_\alpha \Phi (c_+)^q]\, \delta_{n-1,q}$$

with

$$\begin{aligned} F_\alpha^{q-1} &= \mathrm{Tr}[-D_\alpha \bar{D}\bar{D}\Phi X_2^{q-1} + \bar{D}\bar{D}D_\alpha \Phi X^{q-1} \\ &+ D_\alpha \Phi X_5^{q-1} + \bar{D}D_\alpha \Phi X_7^{q-1} - D_\alpha \bar{D}\Phi X_9^{q-1}] \end{aligned} \qquad (C.34)$$

c) The case of H^q (dimension 2) is treated in the same manner. Since the detailed proof is rather lengthy due to the high dimension involved and the vanishing R weight, we shall not present it explicitly.

Let us now return to the cohomology problem (C.1), with Δ depending only on the fields due to the result (C.10). We begin by writing the most general insertion $\Delta(\Phi,c_+,A,\eta,\sigma,Y)$ of dimension 4, $\Phi\Pi$-charge 1, R-weight o, supersymmetric and rigidly invariant (see Table 1 on page 62 for the dimensions, $\Phi\Pi$-charges and R-weights):

$$\begin{aligned} \Delta(\Phi,c_+,A,\eta,\sigma,Y) &= x\mathrm{Tr}\int dS\sigma c_+{}^3 + x'\mathrm{Tr}\int d\bar{S}\bar{\sigma}\bar{c}_+{}^3 \\ &+ \mathrm{Tr}\int dV\eta F(\Phi,c_+) + \int dS[yY\tilde{c}_+{}^2 A + l_{abci} A_a A_b A_c\, c_{+i}] \\ &+ \int d\bar{S}[y'\bar{A}\tilde{\bar{c}}_+{}^2 \bar{Y} + l'_{abci} \bar{A}_a \bar{A}_b \bar{A}_c \bar{c}_{+i}] \\ &+ \int dV\bar{A}\tilde{G}(\Phi,c_+)A \\ &+ \Delta'(\Phi,c_+) \quad . \end{aligned} \qquad (C.35)$$

The matrices $F = F_i\tau^i$, $\tilde{G} = G_i T^i$ have dimension o and $\Phi\Pi$-charge 2 and 1, respectively, and transform appropriately under the rigid group. x, x', y and y' are arbitrary coefficients; the coefficients l_{abci} and l'_{abci} are subject to the conditions (15.55) due to rigid invariance (15.20). Let us compute the variation of Δ under $\mathfrak{b}_0$ with the help of (15.42) and impose the $\mathfrak{b}_0$-invariance condition (C.10). Considering first the dependence on η and σ we obtain

$$0 = \mathrm{Tr}\int dV\eta[-xc_+^{\ 3} + x'\bar{c}_+^{\ 3} - \mathscr{b}_0 F] + \eta,\sigma\text{-indep. terms} \tag{C.36}$$

hence

$$\mathscr{b}_0 F(\Phi,c_+) = -xc_+^{\ 3} + x'\bar{c}_+^{\ 3} \tag{C.37}$$

The condition for the r.h.s. to be a $\mathscr{b}_0$-variation as indicated by the l.h.s. is readily seen to be $x' = x$. Indeed, a necessary condition for a function of Φ, c_+, $\bar{c}_+$ to be the $\mathscr{b}_0$-variation of a function with the same arguments is that it vanishes for $c_+ = \bar{c}_+ =$ constant matrix since $\mathscr{b}_0\Phi = c_+ - \bar{c}_+$ and $\mathscr{b}_0\, c_+ = \mathscr{b}_0\, \bar{c}_+ = 0$. Then (C.37) is solved by

$$F = -x(c_+^{\ 2}\Phi - c_+\Phi c_+ + \Phi\bar{c}_+^{\ 2}) + F' \tag{C.38}$$

with

$$\mathscr{b}_0 F'(\Phi,c_+) = 0 \tag{C.39}$$

The general solution of the last equation is given by the Lemma C.1:

$$F' = \mathscr{b}_0\hat{F}(\Phi,c_+) + tc_+^{\ 2} \tag{C.40}$$

where $\hat{F}$ is some function of $\Phi\Pi$-charge 1 and t a constant. Inserting these results in (C.35) and noting that

$$\begin{aligned} \mathrm{Tr}\int dV\eta\mathscr{b}_0\hat{F}(\Phi,c_+) &= -\mathscr{b}_0\mathrm{Tr}\int dV\eta\hat{F} + \eta,\sigma\text{-indep. terms} \\ \mathrm{Tr}\int dV\eta c_+^{\ 2} &= -\mathscr{b}_0\mathrm{Tr}\int dS\sigma c_+^{\ 2} + \eta,\sigma\text{-indep. terms} \end{aligned} \tag{C.41}$$

we get

$$\Delta(\Phi,c_+,A,\eta,\sigma,Y) \approx xK(\Phi,c_+,\eta,\sigma) + \Delta''(\Phi,c_+,A,Y) \tag{C.42}$$

with

$$
\begin{aligned}
K &= Tr\int dS\sigma\bar{c}_+{}^3 + c.c. - Tr\int dV\eta(c_+{}^2\Phi - c_+\Phi\bar{c}_+ + \Phi\bar{c}_+{}^2) \\
s_0 K &= -\frac{1}{64g^2}\, Tr\int dV\; D\bar{D}\bar{D}D\Phi(c_+\phi^2 - c_+\phi\bar{c}_+{}^2 + \phi\bar{c}_+{}^2) \\
&\equiv L(\Phi,c_+)
\end{aligned}
\tag{C.43}
$$

s_0-invariance of Δ yields now

$$
xL(\Phi,c_+) + s_0\Delta''(\Phi,c_+,A,Y) = 0 \, . \tag{C.44}
$$

However L cannot be the variation of a function of Φ, c_+, A and Y since it does not vanish for $c_+ = \bar{c}_+ =$ constant. Hence $x = o$ and

$$
\Delta \circeq \Delta(\Phi,c_+,A,Y) \ , \tag{C.45}
$$

where Δ has the form (C.35) with the σ and η terms dropped.

Let us go on with the terms depending on A and Y. We first observe that the cubic terms are identically s_0-invariant, hence we get no further restriction on their coefficients l_{abci} and l'_{abci} beyond those implied by rigid invariance (15.55). For the bilinear terms we obtain from s_0-invariance the condition

$$
\begin{aligned}
&\int dV\bar{A}\ [\frac{y}{16}\, c_+{}^2 + \frac{y'}{16}\,\tilde{\bar{c}}_+{}^2 + s_0\tilde{G}(\Phi,c_+)]A \\
&\qquad + Y,A\text{ - indep. terms} \ = \ 0
\end{aligned}
\tag{C.46}
$$

hence

$$
s_0\tilde{G} = -\frac{y}{16}\,\tilde{c}_+{}^2 - \frac{y'}{16}\,\tilde{\bar{c}}_+{}^2 \ . \tag{C.47}
$$

Again, as for (C.37), the condition that the r.h.s. be a s_0-variation leads to $y' = -y$. The equation can then be integrated:

$$
\begin{aligned}
\tilde{G} &= -\frac{y}{16}\,(\tilde{\Phi}\tilde{c}_+ - \tilde{\bar{c}}_+\tilde{\Phi}) + \tilde{G}' \\
s_0\tilde{G}'(\Phi,c_+) &= 0 \ .
\end{aligned}
\tag{C.48}
$$

Using again the Lemma C.1 for G' we get

$$
\tilde{G}' = s_0\hat{G}(\Phi) + z\tilde{c}_+ \ , \tag{C.49}
$$

where $\hat{G}(\Phi)$ has ΦΠ-charge o and z is a constant. Thus the terms of (C.35) bilinear in $\bar{A}$, A read

$$\int dV\bar{A}\,[\tfrac{y}{16}(\tilde{\tilde{\Phi c}}_+ - \tilde{\tilde{\bar{c}}}_+\tilde{\Phi}) + \mathscr{s}_o\hat{G}(\Phi) + z\tilde{c}_+]A$$

$$= -\tfrac{y}{16}\int dV\bar{A}(\tilde{\tilde{\Phi c}}_+ - \tilde{\tilde{\bar{c}}}_+\tilde{\Phi})A + \mathscr{s}_o[\int dV\bar{A}\hat{G}(\Phi)A + 16z\int dSY\tilde{c}_+A] \qquad (C.50)$$

Collecting the results (C.45, C.50) we obtain

$$\Delta(\Phi,c_+,A,\eta,\sigma,Y) \approx$$

$$y\{\int dSY\tilde{c}_+{}^2A - \int d\bar{S}\bar{A}\tilde{\tilde{\bar{c}}}_+{}^2\bar{Y} - \tfrac{1}{16}\int dV\bar{A}(\tilde{\tilde{\Phi c}}_+ - \tilde{\tilde{\bar{c}}}_+\tilde{\Phi})A\}$$

$$+ l_{abci}\int dSA_aA_bA_cc_{+i} + l'_{abci}\int d\bar{S}\bar{A}_a\bar{A}_b\bar{A}_c\bar{c}_{+i}$$

$$+ \Delta'''(\Phi,c_+) \quad , \quad \mathscr{s}_o\Delta''' = 0 \ . \qquad (C.51)$$

We are finally left with terms depending only on Φ, c_+, $\bar{c}_+$:

$$\Delta'''(\Phi,c_+) = \int dVH(\Phi,c_+),\ \mathscr{s}_o\Delta''' = 0\ , \qquad (C.52)$$

where H has dimension 2, R-weight o, ΦΠ-charge 1 and is rigid-invariant. $\mathscr{s}_o$-invariance of Δ''' implies that the $\mathscr{s}_o$-variation of H is a total derivative

$$\mathscr{s}_oH = D^\alpha F^2_\alpha - \bar{D}_{\dot{\alpha}}G^{2\dot{\alpha}} \qquad (C.53)$$

where F^2_α and $\bar{G}^2_{\dot{\alpha}}$ are local functionals of Φ, c_+ of dimension 3/2 and ΦΠ-charge 2 (here and in the following the superscript indicates the ΦΠ-charge). The solution of eq. (C.53) proceeds by repeated application of $\mathscr{s}_o$, projection with derivatives D_α ($\bar{D}_{\dot{\alpha}}$) and "integration" - with respect to $\mathscr{s}_o$ and D_α ($\bar{D}_{\dot{\alpha}}$). Application of $\mathscr{s}_o$ to (C.53) yields first

$$D^\alpha\mathscr{s}_oF^2_\alpha - \bar{D}_{\dot{\alpha}}\mathscr{s}_oG^{2\dot{\alpha}} = 0\ . \qquad (C.54)$$

The general solution of (C.54) is given by

$$\mathcal{b}_0 F^2_\alpha = \bar{D}\bar{D}A^3_\alpha - (D\bar{D} + 2\bar{D}D)_{\alpha\dot{\alpha}} B^{3\dot{\alpha}} + xX^3_\alpha$$

$$\mathcal{b}_0 G^{2\dot{\alpha}} = DDB^{3\dot{\alpha}} - (\bar{D}D + 2D\bar{D})^{\dot{\alpha}\alpha} A^3_\alpha + yY^{3\dot{\alpha}} \tag{C.55}$$

where A and B are dimension 1/2 local functionals of Φ, c_+; x, y are arbitrary coefficients and

$$X^3_\alpha = \mathrm{Tr}[D_\alpha \bar{D}\bar{c}_+ (\bar{c}_+ \bar{D}\bar{c}_+ - \bar{D}\bar{c}_+ \bar{c}_+)]$$

$$Y^3_{\dot{\alpha}} = \mathrm{Tr}[\bar{D}_{\dot{\alpha}} Dc_+ (c_+ Dc_+ - Dc_+ c_+)] \tag{C.56}$$

are the general solutions of the equations

$$D^\alpha X^3_\alpha = 0 \quad , \quad \bar{D}_{\dot{\alpha}} Y^{3\dot{\alpha}} = 0 \tag{C.57}$$

for X (resp. Y) with dimension 3/2 and R-weights -1 (resp. +1) (Δ and H have R-weight o).

$\mathcal{b}_0$-variation of (C.55) yields the equations

$$\bar{D}\bar{D}\mathcal{b}_0 A^3_\alpha - (D\bar{D} + 2\bar{D}D)_{\alpha\dot{\alpha}} \mathcal{b}_0 B^{3\dot{\alpha}} = 0$$

$$DD\mathcal{b}_0 B^{3\dot{\alpha}} - (\bar{D}D + 2D\bar{D})^{\dot{\alpha}\alpha} \mathcal{b}_0 A^3_\alpha = 0 \tag{C.58}$$

which are solved by

$$\mathcal{b}_0 A^3_\alpha = D_\alpha E^4$$

$$\mathcal{b}_0 B^3_{\dot{\alpha}} = -\bar{D}_{\dot{\alpha}} E^4 \tag{C.59}$$

the local functional E^4 having dimension 0.

Applying now $\mathcal{b}_0$ to (C.59) we get

$$D_\alpha \mathcal{b}_0 E^4 = 0 \quad , \quad \bar{D}_{\dot{\alpha}} \mathcal{b}_0 E^4 = 0 \tag{C.60}$$

hence

$$\mathcal{b}_o E^4 = 0 \tag{C.61}$$

The cohomology (C.61) is solved by Lemma C.1, eqs. (C.11, C.12): there exists a local functional E^3 such that (one has $Tr(c_+)^4 = 0$)

$$E^4 = \mathcal{b}_o E^3 \tag{C.62}$$

Eqs. (C.59) become

$$\begin{aligned} &\mathcal{b}_o(A^3_\alpha - D_\alpha E^3) = 0 \\ &\mathcal{b}_o(B^3_{\dot\alpha} + \bar{D}_{\dot\alpha} E^3) = 0 \end{aligned} \tag{C.63}$$

This cohomology is solved by Lemma C.2, eqs. (C.25, C.26): there exist local functionals A^2, B^2 such that

$$\begin{aligned} &A^3_\alpha = D_\alpha E^3 + \mathcal{b}_o A^2_\alpha \\ &B^3_{\dot\alpha} = -\bar{D}_{\dot\alpha} E^3 + \mathcal{b}_o B^2_{\dot\alpha} \end{aligned} \tag{C.64}$$

Inserting this into eqs. (C.55) yields

$$\begin{aligned} &\mathcal{b}_o F^2_\alpha = \mathcal{b}_o[\bar{D}\bar{D}A^2_\alpha - (D\bar{D} + 2\bar{D}D)_{\alpha\dot\alpha} B^{2\dot\alpha} + xX^2_\alpha] \\ &\mathcal{b}_o G^{2\dot\alpha} = \mathcal{b}_o[DDB^{2\dot\alpha} - (\bar{D}D + 2D\bar{D})^{\dot\alpha\alpha} A^2_\alpha + yY^{2\dot\alpha}] \end{aligned} \tag{C.65}$$

We have used here the fact that X^3 and Y^3 as defined by (C.56) can be expressed as

$$\begin{aligned} &X^3_\alpha = \mathcal{b}_o X^2_\alpha \quad ; \quad X^2_\alpha = -Tr[D_\alpha \bar{D}\Phi(\bar{c}_+ \bar{D}\bar{c}_+ - \bar{D}\bar{c}_+ \bar{c}_+)] \\ &Y^3_{\dot\alpha} = \mathcal{b}_o Y^2_{\dot\alpha} \quad ; \quad Y^2_{\dot\alpha} = Tr[\bar{D}_{\dot\alpha} D\Phi(c_+ Dc_+ - Dc_+ c_+)]\,. \end{aligned} \tag{C.66}$$

The cohomology (C.65) is solved by Lemma C.2, too:

$$\begin{aligned} F^2_\alpha = {} &\bar{D}\bar{D}\, A^2_\alpha - (D\bar{D} + 2\bar{D}D)_{\alpha\dot\alpha} B^{2\dot\alpha} + xX^2_\alpha \\ &+ \mathcal{b}_o F^1_\alpha + \alpha Tr(\bar{D}\bar{D}D\Phi\bar{c}^2_+)_\alpha \end{aligned} \tag{C.67}$$

$$G^{2\dot{\alpha}} = DDB^{2\dot{\alpha}} - (\bar{D}D + 2D\bar{D})^{\dot{\alpha}\alpha}A^2_\alpha + yY^{2\dot{\alpha}}$$

$$+ \delta_0 G^{1\dot{\alpha}} + \beta Tr(DD\bar{D}\Phi)c_+{}^2)^{\dot{\alpha}} \qquad \text{(C.67 cont.'d)}$$

where F^1 and G^1 are local functionals of dimension 3/2; α and β arbitrary coefficients.

We now insert this result into eq. (C.55). Observing that

$$D^\alpha X^2_\alpha = \delta_0 \bar{U} \quad , \quad \bar{D}_{\dot{\alpha}} Y^{2\dot{\alpha}} = \delta_0 U$$

$$U = Tr\left[-(\bar{D}\bar{D}D\Phi D\Phi + D\Phi\bar{D}\bar{D}D\Phi)c_+ + \bar{D}D\Phi\bar{D}D\Phi c_+\right] , \qquad \text{(C.68)}$$

we thus obtain

$$\delta_0 H = \delta_0[D^\alpha F^1_\alpha - \bar{D}_{\dot{\alpha}} G^{1\dot{\alpha}} + x\bar{U} - yU]$$

$$+ Tr[D\bar{D}\bar{D}D\Phi(\alpha c_+{}^2 - \beta\bar{c}_+{}^2)] \qquad \text{(C.69)}$$

The last line being a δ_0-variation as indicated by the remainder of the equation, this implies $\beta = \alpha$. Then

$$Tr[D\bar{D}\bar{D}D\Phi(c_+{}^2 - \bar{c}_+{}^2)] = \delta_0 Tr[D\bar{D}\bar{D}D\Phi(\Phi c_+ - \bar{c}_+\Phi)]$$

(C.69) becomes a cohomology problem, solved by

$$H = D^\alpha F^1_\alpha - \bar{D}_{\dot{\alpha}} G^{1\dot{\alpha}} + x\bar{U} - yU + \alpha Tr[D\bar{D}\bar{D}D\Phi(\Phi c_+ - \bar{c}_+\Phi)]$$

$$+ \delta_0\hat{H} + \gamma Tr(D\bar{D}\bar{D}D\Phi c_+) \qquad \text{(C.70)}$$

due to Lemma C.2, $\hat{H}$ being a local functional of $\Phi, c_+, \bar{c}_+$ and γ an arbitrary constant.

The superspace integral (C.52) finally reads

$$\Delta'''(\Phi, c_+) \,\hat{\approx}\, x\bar{X}_3 - yX_3 \qquad \text{(C.71)}$$

with

$$X_3 = \int dVU = Tr\int dS\bar{D}\bar{D}D\Phi\bar{D}\bar{D}D\Phi c_+ \qquad \text{(C.72)}$$

To obtain (C.71) we have used the identity

$$Tr\int dV\, D\bar{D}\bar{D}D\Phi(\Phi c_+ - \bar{c}_+\Phi) = -\delta_0 Tr\int dV\, DD\bar{D}\Phi\bar{D}\Phi\Phi \qquad \text{(C.73)}$$

The results (C.10, C.51, C.71) put together yield the expression (15.54) of section 15: this is the general solution of the cohomology (C.1).

APPENDIX D SYMMETRIC INSERTIONS AND DIFFERENTIAL OPERATORS

In this appendix we extend the discussion of section 12 incorporating all gauge parameters. Like in section 12 a BRS invariant differential operator ∇ is defined as an operator "commuting" with $\mathcal{S}$:

$$\nabla\mathcal{S}(\Gamma) = \mathcal{B}_{\bar{\Gamma}}\nabla\Gamma \tag{D.1}$$

where $\mathcal{S}$ and $\mathcal{B}_{\bar{\Gamma}}$ are the Slavnov operators (18.19, 18.20). $\nabla\Gamma$ is further assumed to obey the homogeneous gauge condition and ghost equation:

$$\frac{\delta}{\delta B}\nabla\Gamma = 0 \quad , \quad \mathcal{G}\nabla\Gamma = 0 \ , \tag{D.2}$$

where $\mathcal{G}$ is the operator (18.17). If Γ fulfills the Slavnov identity, then

$$\mathcal{B}_{\bar{\Gamma}}\nabla\Gamma = 0 \ . \tag{D.3}$$

The insertion Δ defined by

$$\nabla\Gamma = \Delta\cdot\Gamma \tag{D.4}$$

via the action principle is thus a BRS invariant insertion:

$$\begin{aligned} &\mathcal{B}_{\bar{\Gamma}}[\Delta\cdot\Gamma] = 0 \\ &\frac{\delta}{\delta B}[\Delta\cdot\Gamma] = 0 \ , \quad \mathcal{G}[\Delta\cdot\Gamma] = 0 \ . \end{aligned} \tag{D.5}$$

The following differential operators ∇_i define a basis for the rigid, R, supersymmetry and BRS invariant insertions of dimension $d \leq 4$ and $\Phi\Pi$-charge 0:

$$\begin{aligned} \nabla_1 &= \partial_g \\ \nabla_2 &= c(p)\mathcal{N}_+ + \mathcal{b}c(p)\mathcal{N}_+^{(-)} = \{\mathcal{B}_{\bar{\Gamma}}, c(p)\mathcal{N}_+^{(-)}\} \\ \nabla_3 &= d_1(p)\mathcal{N}_\Phi + \mathcal{b}d_1(p)\mathcal{N}_\Phi^{(-)} = \{\mathcal{B}_{\bar{\Gamma}}, d_1(p)\mathcal{N}_\Phi^{(-)}\} \\ \nabla_4 &= e(p)\mathcal{N}_u + \mathcal{b}e(p)\mathcal{N}_u^{(-)} = \{\mathcal{B}_{\bar{\Gamma}}, e(p)\mathcal{N}_u^{(-)}\} \end{aligned} \tag{D.6}$$

$$\nabla_5 = d^a_{\omega k}(p)\partial_{a_{\omega k}} + \mathscr{b} d^a_{\omega k}(p)\partial_{x_{\omega k}} = \{\mathscr{B}_{\bar\Gamma}, d^a_{\omega k}\partial_{x_{\omega k}}\}\ , \quad k \geq 2$$

$$\nabla_6 = d^e_{\omega k}(p)\partial_{e_{\omega k}} + \mathscr{b} d^e_{\omega k}(p)\partial_{y_{\omega k}} = \{\mathscr{B}_{\bar\Gamma}, d^e_{\omega k}\partial_{y_{\omega k}}\}\ , \quad k \geq 2$$

$$\nabla_8 = \partial_{h_{abc}} \qquad \text{(D.6 cont.'d)}$$

$$\nabla_9 = f(p)\mathscr{N}_A + \mathscr{b} f(p)\mathscr{N}^{(-)}_A = \{\mathscr{B}_{\bar\Gamma}, f(p)\mathscr{N}^{(-1)}_A\}$$

where

$$\begin{aligned}
\mathscr{N}_\Phi\Gamma &= [N_\Phi - N_\rho - N_{c_-} - N_{\bar c_-} - N_B - N_{\bar B} + 2\alpha\partial_\alpha + 2\chi\partial_\chi]\Gamma \\
\mathscr{N}^{(-)}_\Phi\Gamma &= \text{Tr}\int dV\rho\Phi - [\text{Tr}\int dSc_- \frac{\delta}{\delta B} - \text{Tr}\int d\bar S\bar c_- \frac{\delta}{-\delta\bar B} - 2\alpha\partial_\chi]\Gamma \\
\mathscr{N}_+\Gamma &= [N_{c_+} + N_{\bar c_+} - N_\sigma - N_{\bar\sigma}]\Gamma \\
\mathscr{N}^{(-)}_+\Gamma &= -\text{Tr}\int dS\sigma c_+ - \text{Tr}\int d\bar S\bar\sigma\bar c_+ \qquad \text{(D.7)} \\
\mathscr{N}_A\Gamma &= [N_A + N_{\bar A} - N_Y - N_{\bar Y}]\Gamma \\
\mathscr{N}^{(-)}_A\Gamma &= \int dSYA + \int d\bar S\bar Y\bar A \\
\mathscr{N}_u\Gamma &= N_{u'} + N_{v'} \\
\mathscr{N}^{(-)}_u\Gamma &= \int dVu' \frac{\delta}{\delta v} \\
N_\psi &= \int \psi \frac{\delta}{\delta\psi}\ , \quad \psi = \Phi, c_+, c_-, B, A, \rho, \sigma, Y, u', v'\ .
\end{aligned}$$

The coefficients c(p), d(p), e(p) in (D.6) are arbitrary functions of the gauge parameters p_k, and $\mathscr{b}c(p) = \Sigma z_k \partial_{p_k} c(p)$, etc. The ∇_i defined in (D.6) will be called symmetric operators and the insertions Δ_i defined from them by (D.4) symmetric insertions. The latter are given below in the tree approximation, together with some of their lowest order terms:

$$\begin{aligned}
\Delta_1 &= -\frac{2}{g}\Gamma_{YM}(\mathscr{F}(\Phi)) = \frac{-1}{64g^3}\text{Tr}\int dV\Phi D\bar D\bar D D\Phi + \ldots \\
\Delta_2 &= -\mathscr{b}\,[c(p)\text{Tr}\int dS\sigma c_+ + \text{c.c.}] = c(p)\text{Tr}\int dS\sigma c_+ c_+ + \ldots \\
\Delta_3 &= \mathscr{b}[d_1(p)\text{Tr}\int dV\eta\Phi] = -d_1(p)\text{Tr}\int dV\eta(c_+ - \bar c_+) + \ldots \qquad \text{(D.8)}
\end{aligned}$$

$$\Delta_4 = \mathfrak{s}[e(p)\mathrm{Tr}\int dVu' \frac{\delta\Gamma^{(o)}}{\delta v}] = e(p)\mathrm{Tr}\int dVu\eta(c_+ - \bar{c}_+) + \dots$$

$$\Delta_5 = \mathfrak{s}[d^a_{\omega k}\partial_{x_{\omega k}}\Gamma^{(o)}] = -d^a_{\omega k}(p)\,\mathfrak{s}\mathrm{Tr}\int dVs^{\omega}_{ii_1\dots i_k}\eta_i\Phi_{i_1}\dots\Phi_{i_k} + \dots$$

$$= -kd^a_{\omega k}(p)\mathrm{Tr}\int dVs^{\omega}_{ii_1}\dots{}_{i_k}\eta_i c_{+i_1}\Phi_{i_2}\dots\Phi_{i_k} + \dots$$

$$k \geq 2\ . \qquad \text{(D.8 cont.'d)}$$

$$\Delta_6 = \mathfrak{s}[d^e_{\omega k}(p)\partial_{y_{\omega k}}\Gamma^{(o)}]$$

$$= -kd^a_{\omega k}(p)\mathrm{Tr}\int dVus^{\omega}_{ii_1\dots i_k}\eta_i c_{+i_1}\Phi_{i_2}\dots\Phi_{i_k} + \dots\ , \qquad k \geq 2$$

$$\Delta_7 = \mathfrak{s}\mathrm{Tr}\int dV(u')^m s^{\omega}_{ii_1\dots i_k}\eta_i\Phi_{i_1}\dots\Phi_{i_k}$$

$$= O(u^{m-1}) \quad , \quad m \geq 2,\ k \geq 1$$

$$\Delta_8 = \int dSA_aA_bA_c + \text{c.c.}$$

$$\Delta_8 = \mathfrak{s}[f(p)\int dSYA] + \text{c.c.} = f(p)T^i_{ab}\int dSY_a c_{+i}A_b + \dots\ .$$

The terms of lowest order in the fields displayed in the expressions above show their one-to-one correspondence with the normalization conditions (18.25, 18.33).

These insertions form a basis for the symmetric insertions of UV-dimension up to four. One can check that their IR-dimension is in fact not less than four. They generate thus all possible symmetric counterterms left free at each order in $\hbar$ once the Ward- and Slavnov-identities are fulfilled.

The careful reader may have noticed that no differential operator related to the 7th insertion (D.8) has been given in (D.5). The reason is that this insertion generates couplings of order greater than one in u' and v', i.e. vanishing with u and v: such uninteresting couplings were fixed arbitrarily by the 7th normalization condition (18.25).

APPENDIX E SOLUTION OF SOME SUPERFIELD CONSTRAINTS

In this appendix we prove, firstly, that the constraints (16.3.41, 16.3.42)

$$\bar{D}\bar{D}\Delta_\alpha = 0 \tag{E.1}$$

$$\int d^4x(D^\alpha\Delta_\alpha - \bar{D}_{\dot{\alpha}}\bar{\Delta}^{\dot{\alpha}}) = 0 \tag{E.2}$$

have the general solution (16.3.43)

$$\Delta_\alpha = \bar{D}^{\dot{\alpha}}V_{\alpha\dot{\alpha}} + 2D^\beta S_{\alpha\beta} \tag{E.3}$$

with $V^\mu \equiv \sigma^\mu_{\alpha\dot{\alpha}}V^{\alpha\dot{\alpha}}$ axial and $S_{\alpha\beta}$ chiral and, secondly, that the general solution of the constraints (16.3.87)

$$\bar{D}\bar{D}B_\alpha = 0 \tag{E.4}$$

$$D^\alpha B_\alpha - \bar{D}_{\dot{\alpha}}\bar{B}^{\dot{\alpha}} = 0 \tag{E.5}$$

is given by

$$B_\alpha = \bar{D}\bar{D}D_\alpha B + D^\beta S_{(\alpha\beta)} \tag{E.6}$$

with B scalar and $S_{(\alpha\beta)}$ chiral, symmetric in α,β.

I.e. for the superfields Δ_α, B_α which are local functionals of the elementary superfields we want to show the existence of solutions V,S,B belonging to the same class.

Let us begin by reducing the first problem (E.1, E.4) to the second one (E.4, E.5). The vanishing of the space-time integral (E.2) implies that its integrand is a total derivative:

$$D^\alpha\Delta_\alpha - \bar{D}_{\dot{\alpha}}\bar{\Delta}^{\dot{\alpha}} = i\partial_\mu U^\mu \tag{E.7}$$

$$= \frac{1}{4}\{D^\alpha,\bar{D}^{\dot{\alpha}}\}\sigma_{\mu\alpha\dot{\alpha}}U^\mu = \frac{1}{2}\{D^\alpha,\bar{D}^{\dot{\alpha}}\}U_{\alpha\dot{\alpha}}$$

with U^{μ} being a real, local superfield functional.
That U^{μ} is indeed a superfield is a consequence of that $\partial_{\mu}U^{\mu}$ is a superfield and that any superfield is uniquely determined by its $\Theta = 0$ component. Defining now

$$B_{\alpha} = \Delta_{\alpha} - \frac{1}{2}\bar{D}^{\dot{\alpha}}U_{\alpha\dot{\alpha}} \tag{E.8}$$

we see that we are led to solve for a local superfield functional B_{α} obeying the constraints (E.4, E.5).

Solving (E.4, E.5) is the same type of problem as solving the supersymmetry consistency condition (13.27) and we shall therefore briefly sketch the method [IV.4, III.25].

If B_{α} is made up from purely chiral fields only, one lists all possible terms (dim $B_{\alpha} = 7/2$, $n(B_{\alpha}) = -1$, $Q_{\Phi\Pi}(B_{\alpha}) = 0$) and checks that the desired decomposition (E.6) holds. If B_{α} contains with chiral fields also at least one antichiral or vector superfield then we write B_{α} in the following way:

$$\begin{aligned} B_{\alpha}(z) &= \sum_{n} B_{\alpha}^{n}(z) \\ B_{\alpha}^{n}(z) &= \int dz_1 \ldots dz_n \ldots \varphi_1 \ldots \varphi_n \cdot \\ &\cdot \sum_{d\geq 0} \sum_{|M|=d} b_{\alpha M}^{n}(\tau,\tau_i) \prod_{i=1}^{n} \partial^{M_i}\delta^4(y_i) \ . \end{aligned} \tag{E.9}$$

The notation is as follows:

$\varphi_i = \varphi_{k_i}(z_i)$ is a superfield with internal index k_i, at superspace point z_i;
dz_i is the appropriate integration measure;
$M = (M_1,\ldots,M_n)$, M_i = multiindex $(\mu_1 \ldots \mu_p)$
$|M| = \Sigma|M_i|$, $|M_i| = p$.

We have also chosen special variables

$$\begin{aligned} z &= (x,\Theta,\bar{\Theta}) \ , \quad z_i = (x_i,\Theta_i,\bar{\Theta}_i) \ , \\ \tau &= \Theta-\Theta_1 \ , \quad \bar{\tau} = \bar{\Theta}-\bar{\Theta}_1 \ , \quad \tau_i = \Theta_i-\Theta_1 \ , \quad \bar{\tau}_i = \bar{\Theta}_i-\bar{\Theta}_1 , \\ y_i &= x_i - x + i\Theta\sigma\bar{\Theta}_i - i\Theta_i\sigma\bar{\Theta} \ . \end{aligned} \tag{E.10}$$

Here z_1 is the argument of a vector superfield - if any is present - chosen to be φ_1. Otherwise we may choose φ_1 chiral and φ_2 anti-chiral and (E.10) is modified accordingly

$$\tau = \Theta-\Theta_1, \quad \bar{\tau} = \Theta-\bar{\Theta}_2, \quad \tau_i = \Theta_i-\Theta_1, \quad \bar{\tau}_i = \bar{\Theta}_i- \bar{\Theta}_2 . \tag{E.10'}$$

The Θ-structure of the representation (E.9) is necessary and sufficient for B_α to be a superfield, and its locality is expressed by the kernel being a polynomial of derivatives of space-time δ-functions.

Since the differential constraints (E.4, E.5) do not mix the B^n with different degrees n, we shall work at fixed degree and omit the superscript n. Having in mind a proof by induction in the number of space-time derivatives $|M| = d$, we start with $d = 0$ and expand the corresponding kernel $b_\alpha(\tau,\tau_i)$ in its variable τ:

$$b_\alpha(\tau,\tau_i) = C_\alpha + \tau^\beta \chi_{\alpha\beta} + \bar{\tau}_{\dot{\beta}}\psi_\alpha{}^{\dot{\beta}} + \ldots \tag{E.11}$$

where the coefficients C_α, $\chi_{\alpha\beta}$, ... are functions of the τ_i, $\bar{\tau}_i$. The constraints (E.4, E.5) yield at this order the equations

$$\bar{\partial}\bar{\partial}b_\alpha = 0 \quad , \quad \partial^\alpha b_\alpha + \bar{\partial}_{\dot{\alpha}}\bar{b}^{\dot{\alpha}} = 0 \tag{E.12}$$

($D_\alpha, \bar{D}_{\dot{\alpha}}$ reduce to $\partial = \partial/\tau$, $-\bar{\partial} = -\bar{\partial}/\bar{\tau}$ up to space-time derivatives). A little algebra shows that they are solved by

$$b_\alpha(\tau,\tau_i) = \bar{\partial}\bar{\partial}\partial_\alpha b(\tau,\tau_i) + \partial^\beta\bar{\partial}\bar{\partial}c_{(\alpha\beta)}(\tau,\tau_i)$$

$$+ \bar{\partial}_{\dot{\alpha}} e_\alpha{}^{\dot{\alpha}}(\tau,\tau_i) \tag{E.13}$$

with $b = \bar{b}$ and $e_\alpha{}^{\dot{\alpha}} = -\bar{e}^{\dot{\alpha}}{}_\alpha$. Moreover the τ-expansion (see (E.11)) of $e_\alpha{}^{\dot{\alpha}}$ has terms in τ^2, $\bar{\tau}^2$, $\bar{\tau}\tau^2$, $\tau\bar{\tau}^2$ and $\tau^2\bar{\tau}^2$ only.

Let us define the <u>local superfield functionals</u> $B^{(o)}(z)$, $C^{(o)}_{(\alpha\beta)}(z)$ and $E^{(o)}_{\alpha\dot{\alpha}}(z)$ using the representation (E.9) with the kernel $b_{\alpha M}$ replaced by b, $c_{(\alpha\beta)}$ and $e_{\alpha\dot{\alpha}}$ respectively for $|M| = 0$ and by zero for $|M| \geq 1$. Then the substitution of the result (E.13) into (E.9) yields

$$B_\alpha(z) = \bar{D}\bar{D}D_\alpha B^{(0)}(z) + D_\alpha \bar{D}\bar{D}C^{(0)}_{(\alpha\beta)}(z) - \bar{D}_{\dot\alpha} E^{(0)\dot\alpha}_\alpha(z) + B'_\alpha(z) \tag{E.14}$$

where B'_α contains at least one space-time derivative in its kernel and fulfills (E.4). Although the first two terms have just the form (E.6) - hence are solutions of the constraints - the E-term does not: it violates the constraints (E.5). In fact it vanishes, as we shall see now by an excursion in the next order in the number of derivatives, i.e. at the order d = 1.

Let us consider B'_α in the representation (E.9), the lowest order term having now one derivative:

$$B'_\alpha = \int dz_1 \ldots dz_n \varphi_1 \ldots \varphi_n \sum_{k=1}^{n} b'_{\alpha\mu,k}(\tau,\tau_i) \cdot \cdot \partial^\mu \delta(y_k) \prod_{\substack{i=1 \\ i\neq k}}^{n} \delta(y_i) + \ldots \tag{E.15}$$

where the dots here and in the following represent terms with more than one derivative. The constraint (E.5) applied to (E.14) reads

$$2i\sigma_\mu{}^\alpha{}_{\dot\alpha} \partial^\mu E^{(0)\dot\alpha}_\alpha = D^\alpha B'_\alpha - \bar{D}_{\dot\alpha} \bar{B}'^{\dot\alpha} + \ldots \tag{E.16}$$

and we used the fact that $E^{(0)\dot\alpha}_\alpha$ is an axial vector: $E^{(0)\dot\alpha}_\alpha = -\bar{E}^{(0)\dot\alpha}{}_\alpha$. At order 1 it yields the equation

$$2i\sigma_\mu{}^\alpha{}_{\dot\alpha} e_\alpha{}^{\dot\alpha} = \partial^\alpha b'_{\alpha\,\mu,k} + \bar\partial_{\dot\alpha} \bar{b}'^{\dot\alpha}{}_{\mu,k}, \quad k = 1,\ldots n \tag{E.17}$$

$$\bar\partial\bar\partial b'_{\alpha\mu,k} = 0 \tag{E.18}$$

the last one following from (E.4), still valid for B'_α. The τ-expansion of the r.h.s. of (E.17) contains only terms in $\tau^0, \tau, \bar\tau$ and $\tau\bar\tau$ due to (E.18). Recalling that no such terms are present in the expansion of the l.h.s. (see remark following (E.13)), we conclude that $e_{\alpha\dot\alpha} = 0$, hence

$$E^{(0)}_{\alpha\dot\alpha} = 0 \, . \tag{E.19}$$

We have thus shown that (E.6) holds at order d = 0:

$$B_\alpha = \bar{D}\bar{D}D_\alpha B^{(0)} + D^\beta \bar{D}\bar{D}C^{(0)}_{(\alpha\beta)} + (d\geq 1) \text{ terms} \tag{E.20}$$

The proof can be iterated in exactly the same way, leading to the result (E.6) with $S_{(\alpha\beta)} = \bar{D}\bar{D}C_{(\alpha\beta)}$. Recall that this holds for the generic case of functionals not exclusively built up with chiral fields. In the purely chiral case $S_{(\alpha\beta)}$ can be any product of chiral fields and space-time derivatives of them, e.g.

$$S_{(\alpha\beta)} = \sigma^{\mu\nu}_{\alpha\beta}\partial_\mu A_1 \partial_\nu A_2 \tag{E.21}$$

But usually severe restrictions follow from the assignments of quantum numbers and dimensions in the specific situations and they enforce the solution in the above sense.

Having now solved the problem (E.4, E.5) we can go back to (E.1, E.2). Inserting the solution (E.6) into (E.8) yields

$$\Delta_\alpha = \bar{D}\bar{D}D_\alpha B + D^\beta S_{(\alpha\beta)} + \frac{1}{2}\bar{D}^{\dot\alpha}U_{\alpha\dot\alpha} \tag{E.22}$$

which can be brought to the form (E.3) by using the identity

$$\bar{D}_{\dot\alpha}[D^\alpha, \bar{D}^{\dot\alpha}]B = -\frac{1}{2}D^\alpha \bar{D}\bar{D}B - \frac{3}{2}\bar{D}\bar{D}D^\alpha B \tag{E.23}$$

and the definitions.

REFERENCES

REVIEWS, BOOKS ON SUPERSYMMETRY AND/OR RENORMALIZATION

R.1 C. Becchi, The renormalization of gauge theories, Les Houches 1983, eds. B.S. DeWitt, R. Stora (Elsevier 1984)

R.2 P. Fayet, S. Ferrara, Supersymmetry, Phys. Reps. 32C (1977) 250

R.3 S.J. Gates, M.T. Grisaru, M. Roček, W. Siegel, Superspace: or one thousand and one lessons in supersymmetry, London (Benjamin/Cummings 1983)

R.4 K. Hepp, Renormalization theory in statistical mechanics and quantum field theory, Les Houches 1970, eds. C. DeWitt, R. Stora (Gordon and Breach, New York, 1971)

R.5 P.V. Nieuwenhuizen, Supergravity, Phys. Reps. 68C (1981) 189

R.6 H.P. Nilles, Supersymmetry and phenomenology, Phys. Reps. 110C (1984) 1

R.7 O. Piguet, A. Rouet, Symmetries in perturbative quantum field theory, Phys. Reps. 76C (1981) 1

R.8 O. Piguet, Renormalisation en théorie quantique des champs I II, Cours du 3ème Cycle de la Suisse Romande 1983

R.9 O. Piguet, K. Sibold, On the renormalization of N=1 rigid supersymmetric theories, Karpacz 1983, ed. B. Milewski, World Scientific Publ., Singapore 1983

R.10 A. Salam. J. Strathdee, Supersymmetry and superfields, Fortschr. d. Physik 26 (1978) 57

R.11 M.F. Sohnius, Introducing supersymmetry, Phys. Reps. 128C (1985) 39

R.12 O. Steinmann, Perturbation expansions in axiomatic field theory, Lecture Notes in Physics II, Springer, Berlin 1971

R.13 R. Stora, Algebraic structure and topological origin of anomalies, Cargèse Institute of Theoretical Physics "Progress in Gauge Field Theory" 1983

R.14 G. Velo, A.S. Wightman (eds.), Renormalization theory, D. Reidel Publ. Co., Dordrecht Holland 1976

R.15 J. Wess, J. Bagger, Supersymmetry and supergravity, Princeton University Press 1983

CHAPTER I

I.1 Yu.A. Gol'fand, E.P. Likhtman, Extension of the algebra of Poincaré group generators and violation of P-invariance, JETP Letts. 13 (1971) 323 (engl.ed.)

I.2 D.V. Volkov, Y.P. Akulov, Is the neutrino a Goldstone particle?, Phys. Lett. 46B (1973) 109

I.3 J. Wess, B. Zumino, Supergauge transformations in 4 dimensions, Nucl. Phys. B70 (1974) 39

I.4 A. Salam, J. Strathdee, Supergauge transformations, Nucl. Phys. B76 (1974) 477

I.5 S. Ferrara, J. Wess, B. Zumino, Supergauge multiplets and superfields, Phys. Lett. 51B (1974) 239

I.6 T.E. Clark, O. Piguet, K. Sibold, Supercurrents, renormalization and anomalies, Nucl. Phys. B143 (1978) 445

CHAPTER II

II.1 F.A. Berezin, The method of second quantization, Acad. Press, New York (1966)

II.2 A. Salam, J. Strathdee, Superfields and Fermi-Bose-symmetry, Phys. Rev. D11 (1975) 1521

II.3 L. O'Raifeartaigh, Spontaneous symmetry breaking for chiral superfields, Nucl. Phys. B96 (1975) 331

II.4 J. Wess, B. Zumino, A Lagrangian model invariant under supergauge transformations, Phys. Lett. 49B (1974) 52

II.5 A. Salam, J. Strathdee, On Goldstone fermions, Phys. Lett. 49B (1974) 465

II.6 J. Wess, B. Zumino, Supergauge invariant extension of quantum electrodynamics, Nucl. Phys. B78 (1974) 1

II.7 S. Ferrara, O. Piguet, Perturbation theory and renormalization of supersymmetric Yang-Mills theories, Nucl. Phys. B93 (1975) 261

II.8 B. de Wit, D.Z. Freedman, Combined supersymmetric and gauge-invariant field theories, Phys. Rev. D12 (1975) 2286

II.9 P. Fayet, J. Iliopoulos, Spontaneously broken supergauge symmetries and Goldstone spinors, Phys. Lett. 51B (1974) 961

II.10 A. Salam, J. Strathdee, Supersymmetric and non-Abelian gauges, Phys. Lett. 51B (1974) 353

II.11 S. Ferrara, B. Zumino, Supergauge invariant Yang-Mills theories, Nucl. Phys. B79 (1974) 413

II.12 C. Becchi, A. Rouet, R. Stora, Renormalization of gauge theories, Ann. of Phys. 98 (1976) 287

II.13 O. Piguet, K. Sibold, Renormalization of N=1 supersymmetric Yang-Mills theories. I. The classical theory, Nucl. Phys. B197 (1982) 257

II.14 O. Piguet, K. Sibold, Gauge independence in N=1 supersymmetric Yang-Mills theories, Nucl. Phys. B248 (1984) 301

II.15 S. Ferrara, B. Zumino, Transformation properties of the supercurrent, Nucl. Phys. B87 (1975) 207

II.16 T.E. Clark, O. Piguet, K. Sibold, The renormalized supercurrents in supersymmetric QED, Nucl. Phys. B172 (1980) 201

II.17 O. Piguet, K. Sibold, The supercurrent in SYM. I. The classical approximation, Nucl. Phys. B196 (1982) 428

CHAPTER III

III.1 W. Zimmermann, The power counting theorem for Minkowski metric, Comm. Math. Phys. 11 (1968) 1

III.2 N.N. Bogoliubov, D.V. Shirkov, Introduction to the theory of quantized fields, J. Wiley, New York (1959) 10

III.3 J.H. Lowenstein, E. Speer, Distributional limits of renormalized Feynman integrals with zero-mass denominators, Comm.Math. Phys. 47 (1976) 43

III.4 W. Zimmermann, Convergence of Bogoliubov's method of renormalization in momentum space, Comm. Math. Phys. 15 (1969) 208

III.5 J. Honerkamp, F. Krause, M. Scheunert, M. Schlindwein, Perturbation theory in superfields, Phys. Lett. 53B (1974) 60

III.6 D.M. Capper, G. Leibbrandt, On the degree of divergence of Feynman diagrams in superfield theories, Nucl. Phys. B85 (1975) 492

III.7 A. Salam, J. Strathdee, Feynman rules for superfields, Nucl. Phys. B86 (1975) 142

III.8 K. Fujikawa, W. Lang, Perturbation calculations for the scalar multiplet in a superfield formulation, Nucl. Phys. B88 (1975) 61

III.9 A.A. Slavnov, Renormalization of supersymmetric gauge theories, Nucl. Phys. B97 (1975) 155

III.10 M.T. Grisaru, M. Roček, W. Siegel, Improved methods for supergraphs, Nucl. Phys. B159 (1979) 429

III.11 T.E. Clark, O. Piguet, K. Sibold, Renormalization theory in superspace, Ann. of Phys. 109 (1977) 418

III.12 J. Lowenstein, W. Zimmermann, On the formulation of theories with zero-mass propagators, Nucl. Phys. B86 (1975) 77

III.13 J. Lowenstein, Convergence theorems for renormalized Feynman integrals with zero-mass propagators, Comm. Math. Phys. 47 (1976) 53

III.14 J.H. Lowenstein, W. Zimmermann, The power counting theorem for Feynman integrals with massless propagators, Comm. Math. Phys. 44 (1975) 73

III.15 O. Piguet, A. Rouet, Supersymmetric BPHZ renormalization. I. SQED, Nucl. Phys. B99 (1975) 458

III.16 W. Zimmermann, Composite operators in the perturbation theory of renormalizable interactions, Ann. of Phys. 77 (1973) 536

III.17 J.H. Lowenstein, Differential vertex operations in Lagrangian field theory, Comm. Math. Phys. 24 (1971) 1

III.18 T.E. Clark, J.H. Lowenstein, Generalization of Zimmermann's normal-product identity, Nucl. Phys. B113 (1976) 109

III.19 Y.M.P. Lam, Perturbation Lagrangian theory for scalar fields - Ward-Takahashi identity and current algebra, Phys. Rev. D6 (1972) 2145

III.20 Y.M.P. Lam, Perturbation Lagrangian theory for Dirac fields - Ward-Takahashi identity and current algebra, Phys. Rev. D6 (1972) 2161

III.21 J.H. Lowenstein, Normal product methods in renormalized perturbation theory, Maryland, Technical Rep. 73-068, December 1972, Seminars on Renormalization Theory, Vol. II

III.22 T.E. Clark, O. Piguet, K. Sibold, The gauge invariance of the supercurrent in SQED, Nucl. Phys. B169 (1980) 77

III.23 J. Iliopoulos, B. Zumino, Broken supergauge symmetry and renormalization, Nucl. Phys. B76 (1974) 310

III.24 L. Bonora, P. Pasti, M. Tonin, Cohomologies and anomalies in supersymmetric theories, Padua University preprint DFPD 5/84

III.25 O. Piguet, Algebraic Poincaré Lemmata, (in preparation)

CHAPTER IV

IV.1 C. Becchi, A. Rouet, R. Stora, Renormalizable theories with symmetry breaking, in "Field Theory, Quantization and Statistical Physics", ed. E. Tirapegui, D. Reidel Publ. Co., 1981

IV.2 J. Wess, B. Zumino, Consequences of anomalous Ward-identities, Phys. Lett. 49B (1974) 52

IV.3 K. Symanzik, Renormalization of theories with broken symmetries, in: "Cargèse Lectures in Physics 1970", ed. D. Bessis (Gordon and Breach, New York, 1972), Vol. 5, 179

IV.4 O. Piguet, M. Schweda, K. Sibold, General solutions of the supersymmetry consistency conditions, Nucl. Phys. B174 (1980) 183

IV.5 T.E. Clark, O. Piguet, K. Sibold, Infrared anomaly induced by spontaneous breakdown of supersymmetry, Nucl. Phys. B119 (1977) 292

IV.6 T.E. Clark, O. Piguet, K. Sibold, The renormalized supercurrents in SQED, Nucl. Phys. B172 (1980) 201

IV.7 K. Sibold, Renormalization of extended QED with spontaneously broken supersymmetry, Nucl. Phys. B174 (1980) 491

IV.8 G. Bandelloni, C. Becchi, A. Blasi, R. Collina, Renormalization of models with radiative mass generation, Comm. Math. Phys. 67 (1979) 147

IV.9 O. Piguet, M. Schweda, A complete BPHZ-version of the supersymmetry and broken chiral symmetry model, Nucl. Phys. B92 (1975) 334

IV.10 O. Piguet, K. Sibold, The anomaly in the Slavnov identity, Nucl. Phys. B247 (1984) 484

IV.11 C. Becchi, A. Rouet, R. Stora, Broken symmetries: Perturbation theory of renormalizable models, Lectures given at Baško Polje, Yugoslavia 1974

IV.12 T.E. Clark, O. Piguet, K. Sibold, The gauge invariance of the supercurrent in SQED, Nucl. Phys. B169 (1980) 77

IV.13 O. Piguet, K. Sibold, Conditions for the Adler-Bardeen theorem, Max Planck-Inst. preprint MPI-PAE/PTh 53/84, München

IV.14 O. Piguet, K. Sibold, The supercurrent in N=1 SYM theories II, Nucl. Phys. B196 (1982) 447

IV.15 T.E. Clark, O. Piguet, K. Sibold, Absence of radiative corrections to the axial current anomaly, Nucl. Phys. B159 (1979) 1

IV.16 K.G. Wilson, W. Zimmermann, Operator product expansions and composite field operators in the general framework of quantum field theory, Comm. Math. Phys. 24 (1972) 87

IV.17 W. Zimmermann, Normal products and the short distance expansion in the perturbation theory of renormalizable interactions, Ann. of Phys. 77 (1973) 570

CHAPTER V

V.1 W.A. Bardeen, O. Piguet, K. Sibold, Mass generation in a normal product formulation, Phys. Lett. 72B (1977) 231

V.2 O. Piguet, M. Schweda, K. Sibold, Radiative mass generation in perturbation theory and the renormalization group, Nucl. Phys. B168 (1980) 337

V.3 K. Pohlmeyer, Large momentum behaviour of Feynman amplitudes, Lecture Notes in Physics No. 39, ed. H. Araki, Springer 1975

V.4 K. Symanzik, Massless ϕ^4 theory in 4-ε dimensions, Cargèse Lectures 1973, DESY preprint

V.5 O. Piguet, K. Sibold, The off-shell infrared problem in SYM. I. Pure super Yang-Mills, Nucl. Phys. B248 (1984) 336

V.6 O. Piguet, K. Sibold, The off-shell infrared problem in SYM. II. All massless models, Nucl. Phys. B249 (1984) 396

SUBJECT INDEX

abelian gauge symmetry
  23, 25, 29, 31, **34**, 35, 36, 40, **165**, **187**, 192

action principle
  152, 161, 169, 172, 181, 182, 183, 186, 190, 199, 212, 213, 214, 216, 221, 224, 232, 236, 245, 253, 288, 298, 329

anomalous dimensions
  217, 227, 228, 229, 262

anomalous R-weights
  217, 227, 228

anomaly 167
  axial current- 230, 243, 247
  chiral- 207
  infrared- 170, 174, 176, **267**, 271, **293**
  - in Slavnov identity 207
  ultraviolet- 170, **173**, 207, 230, 243, 247

anticommutator
  1, 4, 9, 10, 168, 169, 173

asymptotic behaviour
  124, 128, 143, 174, 294

Becchi-Rouet-Stora
  - doublets 73, 281, 285
  - transformation **46**, **53**, 56, 60, 199, 202, 281, 285

β-function
  217, 218, 219, 220, 222, 228, 229, 232, 239, 240, 241, 242, 261, 262, 263, 277, 278, 298, 302

Bogoliubov-Parasiuk-Hepp-Zimmermann
  128

Callan-Symanzik equation
  - SQED 229, 239
  - SYM 262, 302
  - Wess-Zumino model 220

charge conjugation
  307, 308

chiral superfield
  6, 8, 14, 15, 16, 310

chiral symmetry
  92, 230, 243, 247

classical field
  313, 315

cohomology
  201, 202, 204, 286, 292, 317

conformal invariance
  9, 10, 19, 78, 82, 83, 84, 85, 91, 92, 94, 103, 308
  (s.a. Callan-Symanzik equ.)

connected Green's functions
  313, 315

convergence theorem
  (for diagrams) 136, 137

counterterms
  222, 223, 270, 285, 329

covariant derivative
  gauge covariant- 43
  susy covariant- 5

CP-invariance
  192

current
  79, 104, 109

degree of divergence
  124

dilatation current
  82, 83, 85, 108, 216, 218, 229

dimension
  canonical (naive)- 62, 89, 94, 215, 235,264, 308, 309
  infrared- 124, 126, 144, 175, 176, 270, 289
  ultraviolet- 124, 126, 144, 175, 176, 270, 289

Dirac spinor
  33, 34

effective action
  152, 153, 154, 155, 158, 190, 192
  - O'Raifeartaigh model 270
  - SQED 223
  - SYM 96, 286
  - Wess-Zumino model 213

effective ptotential
  20, 21, 37, 39

electromagnetic field
  24

electron
  34

energy-momentum tensor
  79, 82, 83, 85, 92, 103, 106, 107, 108, 215, 217, 219

equation of motion
  renormalized- 153, 154
  - free chiral field 17

-free vector field 28
- SQED 89
- SYM 96, 97
- Wess-Zumino model 81, 84, 213

equivalence theorem
71

Euclidean region
167, 174, 179, 182, 197, 250, 262, 263

exceptional momenta
134

Faddeev-Popov ($\Phi\pi$) charge
54, 62, 289

Faddeev-Popov ($\Phi\pi$) ghosts
47, 53, 54, 55, 289, 315, 316

Feynman diagrams, rules
119

Feynman gauge
26, 119

Fock space
165

forest formula
129, 132

functionals
312

gauge fixing
α-gauge 26, 54, 55
(α,B) gauge 29, 54, 55
B-gauge 30, 54, 55

gauge independence
72, 166, 190, 191, 207, 237, 239, 247

gauge parameters
26, 29, 68, 73, 196, 281, 285

Gell-Mann Low formula
113

generating functional
312

Goldstone fermion
22, 38

Grassmann variables
3, 4, 304
(s.a. BRS-doublets)

Green's functions
312

hard symmetry breaking
167, 170, 174, 207

Hilbert space
16, 165

infrared
- convergence 136, 177, 179, 225
- cut-off 279
- divergences 141, 177, 187, 188, 278

insertion
integrated- 122, 145
local- 121, 145

Landau gauge
26

Legendre transformation
17, 27, 313, **315**

loop-wise expansion
315

Lorentz transformations
308, 309

mass
- insertion 147, 187
- generation 271

massive vector field
26, 27, 39, 223, 231, 234

minimal coupling
35, 41

Noether's theorem
79, 111

non-abelian gauge transformations
43, 45

non-renormalization theorem
- for chiral vertices 118, **126**, 219, 263
- for the axial anomaly 243

normalization conditions
- Wess-Zumino model 20, 213
- O'Raifeartaigh model 23, 271
- SQED 223
- S'QED 40
- SYM 75, 284, 287

normal product
144, 145

one-particle-irreducible diagrams (Green's functions)
314

operator
gauge independent- 166
gauge invariant- 165, 166

renormalization of a composite-
55, 72, 282

symmetric- 162, 163, 164
symmetric differential- 162, 275, 299, 330

overlapping divergences
129

parity
18, 35, 208, **307**

Pauli matrices
304

power counting
124

propagator (free)
- chiral field 119
- component fields 280
- vector field 119

quantization
canonical- 315

renormalization group equation
- O'Raifeartaigh model 277, 278
- SYM 298
- Wess-Zumino model 220

rigid gauge symmetry
19, 34, 35, 41, 52, 65, 180, 181

R-invariance
18, 21, 22, 36, 77, 78, 91, **182**, 192, 197, 218, 250, 251, 262, 270, 287

R-weight (naive)
19, 21, 36, 62, 81, 86, 90, 94, 215, 229, 235, 264, 308, 309
(s.a. anomalous R-weight)

short-distance expansion
253

Slavnov identity
56, 57, 59, 61, 69, 70, 73, 77, 194, 198, 199, 207, 250, 281, 283, 288, 292, 299, 316

soft insertion
174

soft symmetry breaking
174

spontaneous symmetry breaking
- gauge symmetry 39, 45, 170
- rigid symmetry 170, 181
- supersymmetry 22, 39, 45, 170, 172, 176, 178, 193, 267

subtraction operator
132, 143, 147, 178

superconformal
- algebra 9
- transformations 308, 309, 310

supercurrent
- chiral models 86
- SQED 90, 226, 228, 233, 235, 241
- SYM 102, 259, 266
- Wess-Zumino model 82, 215, 217

superficial degree of divergence
124, 127

superfield
chiral- 6
linear- 6
vector- 6
quasi- 11

superspace
4

supersymmetry
1, 5, 14, 19, 22, 39, 45, 171, 174, 175, 176, 179, 193, 197, 271, 290, 310

symmetry transformations
- for fields s. transformation laws
- for functionals s. Ward-identity

transformation laws
BRS- **46**, **53**, 56, 60, 199, 202, 281, 285
gauge- 25, 34, 41, 42
R- 21, 36, 37, 308, 309
superconformal 308, 309, 310
supersymmetry 308, 309, 310

ultraviolet
- convergence 136, 177, 179
- divergence 116

unitarity
27, 112

Ward-identity for
- abelian gauge invariance 27, 29, 31, 36, 40, 165, 187
- rigid gauge symmetry 19, 180, 181
- R-invariance 19, 22, 36, 77, 78, 84, 91, **182**, 192, 197, 218, 250, 251, 262, 270, 287
- for supersymmetry 19, 22, 39 174, 175, 176, 179, 193, 197, 271, 290
(s.a. Slavnov identiy)

wave function renormalization
222, 279

Wess-Zumino gauge
33

Weyl spinor
1

Wick's theorem
113, 114

Wilson-Zimmermann expansion
253

Yang-Mills field
44

zero mass limit
220, 241, 242, 272

Zimmermann identities
147, **150**, **151**, 189, 190, 214, 225, 231, 234, 236, 237, 239

Progress in Physics

1 Collet/Eckmann. Iterated Maps on the Interval as Dynamical Systems
ISBN 3-7643-3026-0
2 Jaffe/Taubes. Vortices and Monopoles, Structure of Static Gauge Theory
ISBN 3-7643-3025-2
3 Manin. Mathematics and Physics
ISBN 3-7643-3027-9
4 Atwood/Bjorken/Brodsky/Stroynowski. Lectures on Lepton Nucleon Scattering and Quantum Chromodynamics
ISBN 3-7643-3079-1
5 Dita/Georgescu/Purice. Gauge Theories: Fundamental Interactions and Rigorous Results
ISBN 3-7643-3095-3
6 Frampton/Glashow/van Dam. Third Workshop on Grand Unification, 1982
ISBN 3-7643-3105-4
7 Frohlich Scaling and Self-Similarity in Physics (Renormalization in Statistical Mechanics and Dynamics)
ISBN 0-8176-3168-2
8 Milton/Samuel. Workshop on Non-Perturbative Quantum Chromodynamics
ISBN 0-8176-3127-5
9 Weldon/Langacker/Steinhardt. Fourth Workshop on Grand Unification
ISBN 0-8176-3169-0
10 Fritz/Jaffe/Szasz. Statistical Physics and Dynamical Systems: Rigorous Results
ISBN 0-8176-3300-6
11 Ceausescu/Costache//Georgescu. Critical Phenomena: 1983 Brasov School Conference
ISBN 0-8176-3289-1
12 Piguet/Sibold. Renormalized Supersymetry: The Perturbation Theory of N = 1 Supersymmetric Theories in Flat Space-Time
ISBN 0-8176-3346-4